G IS FOR GEOGRAPHY:

CHILDREN'S LITERATURE AND THE FIVE THEMES

by Laurel R. Singleton

Social Science Education Consortium
Boulder, Colorado
1993

ORDERING INFORMATION

This publication is available from:

Social Science Education Consortium
3300 Mitchell Lane, Suite 240
Boulder, CO 80301-2272

ISBN 0-89994-370-5

ACKNOWLEDGMENTS

Thanks are due the companies that provided review copies of children's books for use in writing this book:

> Atheneum
> Bradbury Press
> Carolrhoda Books
> Four Winds Press
> Greenwillow Books
> Gulliver Books
> Harcourt Brace Jovanovich
> Henry Holt and Co.
> Holiday House
> Little, Brown
> Lothrop, Lee and Shepard Books
> Macmillan Publishing
> Putnam Publishing Group
> Wm. Morrow and Co.

The comments of reviewers Germaine Taggart and Paula Nelson were helpful in improving the book. Appreciation is also extended to Cindy A.E. Cook for her efforts in designing and producing the final product and to John Martin for his cover design.

CONTENTS

INTRODUCTION

> *When I found myself on my feet, I looked about me, and must confess I never beheld a more entertaining prospect. The country round appeared like a continued garden, and the inclosed fields, which were generally forty foot square, resembled so many beds of flowers. These fields were intermingled with woods of half a stang, and the tallest trees, as I could judge, appeared to be seven foot high. I viewed the town on my left hand, which looked like the painted scene of a city in a theatre.*
>
> Jonathan Swift, "A Voyage to Lilliput: Chapter II," *Gulliver's Travels* (1726).

Children's Literature and Geography

A good book, story, or poem can create an entire world, often complete with well-defined physical and cultural characteristics. The reader may be able to describe in detail where the action took place, the landscape in that place, the homes of characters inhabiting this fictional (or real) world, the means of transportation the characters used, and the links between this place and others. While we seldom think of reading trade books (books that are not textbooks, including works of fiction, poetry, drama, and nonfiction) as a way of learning about geography, immersion in a well-described literary world, whether real or imaginary, can develop significant geographic understandings.

With the growth of the whole language movement, social studies educators have begun reconsidering the role literature might play in developing young people's understanding of history and geography, as well as other aspects of the social world. McGowan and Guzzetti (1991) suggest several reasons why trade books can enhance social studies instruction:

1. Trade books are more readily comprehensible than textbooks, and they are available to suit a wide range of ability levels. Thus, trade books can increase the likelihood that all students will experience success in social studies.

2. Trade books are high interest. Their engaging writing style and "sense of direction" make them more enjoyable than textbooks and therefore more likely to be read by students.

3. Trade books help students relate their own experiences to classroom content, a key factor in learning. Textbooks are rarely able to make these kinds of links between concepts and everyday experience.

These factors are especially relevant to geography. Many geography textbooks are compendia of factual information, with little to engage student interest or relate student experiences to the information provided. Yet, as Dowd (1990) points out, "quality literature, whether fiction or nonfiction, has the power to capture students' interest in the places associated with the characters." Because literary works virtually always have a setting, they virtually all include geographic information with which student readers can enhance their geographic understanding.

This resource book is based on the belief that children's literature and geography should be linked in elementary classrooms. The children's books mentioned are only a few of the many that could be used to develop geographic understandings; our suggestions for their use are intended to serve as models. While children's literature *can* enhance geography instruction, some cautions are in order. As Hepler (1988) has pointed out, activities developed around works of children's literature "should serve the reader and the book first, not some particular curriculum area. When the latter happens, *Charlotte's Web* becomes the basis for a science unit on spiders or a social studies theme of 'farming,' although to a child, the story may be about the power of friendship, or learning to stand up on your own piggy trotters."

A second caution is the need to insure quality in the works selected. Several sources are available to help teachers in this task. Annually, a National Council for the Social Studies committee, with assistance from the Children's Book Council, prepares a list of recommended children's trade books in the social studies; the list appears in the April/May issue of *Social Education*. Many of the books mentioned in this publication were taken from the past three years' versions of this list. Another journal of NCSS, *Social Studies and the Young Learner*, carries a regular column on children's literature and social studies. *Booklist*, *The School Library Journal*, *The Horn Book*, and the October issue of *The Reading Teacher* carry reviews of children's literature. Recently, *The Journal of Geography* has had fairly frequent articles on geography and children's literature that have included rather extensive annotated bibliographies (see, for example, Oden 1992). These sources and other critical bibliographies of children's literature can make teachers aware of high-quality literature useful in developing geographic understanding.

Organization of This Book

This book has three major sections. The first provides a brief overview of each of the five themes of geography. These themes were developed by a joint committee of the American Association of Geographers and the National Council on Geographic Education and first appeared in the two organizations' *Guidelines for Geographic Education*. Since their publication in 1984, the themes have served as the centerpiece of a resurgence in geographic education, assisted with generous support from the National Geographic Society. It is important to point out that the themes are not intended to be taught for themselves; rather, they provide guides for identifying important understandings to be developed.

The overview of each theme in the first section of this book is accompanied by suggestions for using works of children's literature to develop understanding of important ideas related to that theme. Some specific titles are annotated in conjunction with each theme. While the annotations are organized by grade level, we recognize that every class includes students with widely varying reading skills. Thus, teachers may want to read annotations for other grade levels, may choose to read books aloud to the class, or may decide to make a range of books available for students to select themselves and then complete projects on. The suggestions for using these specific titles are generally brief. Keeping our first caution above in mind, the suggested activities are not intended to replace enjoyment of the books for their literary merit or their exploration of other important themes, such as family, friendship, and diversity. Rather, they are designed to show how geographic understandings can be developed through engaging activities that do *not* distract from the books' other merits.

While we may include a title under **place**, teachers will likely note that it could also be used in teaching about **movement**; our inclusion of a book under one theme is not intended to be exclusive. In reality, of course, all the themes are related, and thus many books will provide insights into several or all of the themes.

The book's second major section provides guides for using specific works of children's literature to teach geographic understandings. These guides are based on a format developed by Hepler (1988). Each guide includes a summary of the book, initiating activities, discussion questions intended to stimulate critical thinking, and follow-up activities. The books included represent a variety of genres and are suitable for a range of grade levels.

The final section of the book presents directions for three thematic geography units based on children's literature. The themes of these units are environmental protection, farming, and the Great Lakes.

An index of the books mentioned in the text concludes the book.

Throughout the book, we have tried to provide examples suitable for all grade levels, K-6. We have also drawn on a variety of literary genres, including picture books, realistic fiction, historical fiction, biographies, and informational books. Most of the books included were published since 1988, but a few are older. We make no claim that the books we have included are the *best* books for use in developing geography understandings. Many other works may be equally or better suited for these purposes. We hope that our suggestions will spark your own ideas for using other works.

References

Dowd, Frances, "Geography Is Children's Literature, Math, Science, Art and a Whole World of Activities," *Journal of Geography* (March-April 1990).

Guidelines for Geographic Education: Elementary and Secondary Schools (Washington, DC: Association of American Geographers, and Macomb, IL: National Council for Geographic Education, 1984).

Hepler, Susan, "A Guide for the Teacher Guides: Doing It Yourself," *The New Advocate* (Summer 1988).

McGowan, Tom, and Barbara Guzzetti, "Promoting Social Studies Understanding Through Literature-Based Instruction," *The Social Studies* (January-February 1991).

Oden, Pat, "Geography Is Everywhere in Children's Literature," *Journal of Geography* 91, no. 4 (1992), pp. 151-158.

USING LITERATURE TO TEACH
THE FIVE THEMES OF GEOGRAPHY

Location

According to Hill and McCormick (1989), location is "the most basic of geographical themes. Our ability to locate ourselves in geographic space satisfies a deep human need. Without this spatial ability, we would be literally lost." Being able to locate places and "behave as if we had maps in our heads" is a fundamental human skill. Certainly, the notion of being lost is one that both children and adults can easily relate to; it is also one that is reflected in literature.

Usually, however, educators deal with what Hill and McCormick deem a "more intellectual concept," called locational knowledge. Judging from media coverage of the "crisis" in geographic education, one might assume that many Americans believe that locational knowledge *is* geography. While geography is much more than place location, having some basic locational knowledge and the ability to locate places that are unfamiliar to us provide a foundation for further geographic learning.

According to the National Council for Geographic Education and the Association of American Geographers, the theme of location, or position on the earth's surface, involves understanding that "absolute and relative location are two ways of describing the positions of people and places on the earth's surface" (*Guidelines for Geographic Education* 1984). In expanding on the theme's curricular implications for elementary students, geographers (*K-6 Geography* 1987) identified three key ideas related to location:

- Location of places can be described using reference systems. The use of reference systems is related to the concept of absolute location, which involves pinpointing a location using a reference system such as latitude and longitude.

- Location of places can be described using relative terms. This kind of description involves specifying the relationship of one place to some other place or places. For example, a place might be described as "in the panhandle of Oklahoma," "70 miles south of Dayton," "in the Black Hills," or "on Florida's Gulf Coast." Relative location is the type of location that people deal with on a day-to-day basis.

- Reasons can be identified for the location of places. This key idea involves answering the basic geographic question: "Why there?"

One of the most worthwhile and simplest things teachers can do with respect to the theme of location and their use of literature is to develop a norm in their classrooms that students locate places they read about on a map. Display a large map in the classroom specifically for the purpose of showing where books students have read were set. These locations could be marked with pushpins or stickers on which the names of the books are written. If space is available around the margins of the map, students could clip pictures of the setting from old textbooks, travel or photography magazines, or photo service catalogs; mount the pictures around the edges of the map; and attach the picture to the appropriate location on the map using string or yarn. As students mark locations on the map, they might describe each location, in relative or absolute terms.

Very young students can describe the relative location of items in a book's illustrations. For example, "the book is on the table," or "the shoes are under the bed."

Another general strategy that can be useful in developing locational understanding is having students map houses, neighborhoods, or towns described in books. They might then be given the task of deciding where something new would be added to the setting. For example, if a book is set in a well-described neighborhood, have students consider where they would build a swimming pool, movie theater, or park in that neighborhood.

Some suggestions for using specific titles in developing key ideas about location follow.

Grades K-2

As the Crow Flies: A First Book of Maps, by Gail Hartman, illustrated by Harvey Stevenson (New York: Bradbury Press, 1991), is a very simple book that can be used to help students develop locational understanding. The book shows the routes taken by five animals—an eagle, a rabbit, a crow, a horse, and a seagull—first in pictures and then in picture maps. A larger map at the end of the book shows how the locations of the five places in the smaller maps are related. Students could use any of the maps to describe the relative location of various features. They could also explain why each of the animals lives where it does. They could draw similar picture maps to show the relative locations of some of their own favorite places.

Bread, Bread, Bread, by Ann Morris (New York: Lothrop, Lee and Shepard, 1989), is a charming book with full-color photographs by Ken Heyman. The focus is making, eating, and selling bread in diverse cultures around the world. A key at the back of the book tells in what country each photograph was taken. After students have enjoyed the text and photos, they could use the key and locate each country listed on a wall map; symbols used to identify each country could be shaped like the bread made there. Students could then describe the locations by naming the continent on which each country is found.

Robin and Sally Hirst take a whimsical but informative look at the question "Where do you live?" in *My Place in Space*, illustrated by Roland Harvey with Joe Levin (New York: Orchard Books, 1990). When a bus driver asks Henry and Rosie where they live, Henry provides a detailed answer pinpointing their Australian hometown's place in the universe. The illustrations are an unexpected but effective combination of cartoon-style drawings of the town and air-brushed astronomical paintings. Before reading the story, you might ask students to answer the question, "Where is our classroom?" How many different answers are given? Are they all correct? Draw a set of concentric circles on the chalkboard showing the classroom in the school in the city in the state and so on, using as many of the answers students gave as possible. After reading the story, have students draw a similar set of concentric circles showing the various answers Henry gave to the question of where he lived. Which answer was most exact or specific? Which answers would also include your classroom's location? The story could also be used as the basis for a science lesson on the solar system and universe.

Grades 3-4

Let's Go Traveling, written and illustrated by Robin Rector Krupp (New York: Morrow Junior Books, 1992), tells the story of a youngster traveling in search of mysteries of the ancient world, including Stonehenge, the Caves of France, the Great Wall of China, the pyramids, and Mayan temples. The book is illustrated with maps, photographs, drawings, and collages of such items as ticket stubs, newspaper clippings, bottle caps, and postcards. Students could find the locations on a map, calculate distances between them, and describe each site's location relative to the previous site. They could also create their own "fantasy trips," describing the locations they would visit and creating collages representing the sites.

A book with a similar travel theme is *Kate Heads West* (New York: Bradbury, 1990). One of a series of books about the well-traveled Kate written by Pat Brisson and illustrated by Rick Brown, this book is presented as a series of letters written by Kate on a vacation in the American Southwest. Addressed to family members (including the family fish), friends, and such diverse others as Kate's teacher and dentist, the letters describe each stop Kate has made on her trip with her best friend Lucy's family. As with the previous book, students could locate the sites visited on a map, calculate distances between them, and describe locations in relation to each other. They could also deal with the question "Why there?" by writing questions on note cards that they could attach to the map for other students to answer. For example, students locating Oklahoma City might ask "Why is the Cowboy Hall of Fame in Oklahoma City?" or "Why was Oklahoma chosen as the place the Cherokee were forced to move to?" Other students can then attempt to answer the questions, perhaps writing their answers on notecards to be attached near the map. Class discussion of the students' responses will help them understand factors related to location.

Hector Lives in the United States Now: The Story of a Mexican-American Child, by Joan Hewett, photographs by Richard Hewett (New York: Lippincott, 1990), tells the story of a Los Angeles family that applied for permanent residency under the amnesty program offered to illegal immigrants in the 1980s. In the book, Hector's fifth-grade class is studying immigration, which gives students in your class an opportunity to locate continents, countries, regions, and states of origin for the students described in the text. This information could be mapped and compared with information on the countries of origin of your students' families. Students could also look at the locations in Los Angeles important to the Almaraz family—their church, schools, library, laundromat, park, and so on. Students could then identify and map the places in the community most important to their own families. Are these places conveniently located? Which places require more travel? Would a closer location meet the same need for the family? Why or why not?

Grades 5-6

In *The American Promise: Voices of a Changing Nation, 1945-Present* (New York: Bantam Books, 1990), Milton Meltzer presents source documents from recent U.S. history, organized in a combined chronological/thematic format. Although most students are unlikely to read the book from cover to cover, it can be used with small groups to develop locational knowledge, as well as historical understanding and research skills. Each of the book's ten sections could be assigned to a small group. Each small group would read the material in its section; for example, the group assigned the section entitled "Only in America: Immigrants and Refugees" would read testimony by a Japanese-American interned during World War II, an oral history excerpt from a Mexican-American community organizer, the words of a minister whose church provides sanctuary to illegal aliens from Central America, and an interview with a Native American. Each group would then create a symbol representing their topic or issue and use the symbol to show areas affected by the issue on a map. The groups would then do some additional research on the topic or theme, focusing particularly on identifying locations affected; the immigration group, for instance, might use the almanac to identify the parts of the country having the highest percentage of immigrants in the school population. Additional locations discovered in the research would also be mapped. Finally, the groups would make presentations to the class, explaining the importance of their topic or issue and why it relates to the areas they have shown on the map.

Since Europeans came to the Americas, the history of Panama has been a case study in the importance of location. This point is well made in Judith St. George's *Panama Canal: Gateway to the World* (New York: Putnam's, 1989), which provides a detailed account of the building of the canal, including the hardships posed by the natural environment. While reading the book, students could construct a timeline of Panamanian history, marking events affected by location with a special symbol. What does their timeline indicate about the importance of Panama's location? Has the impact been mostly positive or mostly negative? After students have read the book, they might look for other locations in the world that would have similar strategic importance and then discuss them to see if their hypotheses were correct.

Place

Place is a theme that Hill and McCormick (1989) have described as "deceptively simple." This theme is defined in the *Guidelines for Geographic Education* as follows: "All places on the earth have distinctive tangible and intangible characteristics that give them meaning and character and distinguish them from other places." Key ideas related to this theme include the following (*K-6 Geography*):

- Places have physical characteristics. These characteristics, which are products of natural processes, include such things as landforms, bodies of water, natural resources, natural plant and animal life, and climate.

- Places have human characteristics. These characteristics include the aspects of the environment created by humans, such as buildings, roads, pollution, and so on.

- Places may be described or represented in different ways. This apparently simple statement embodies the idea that people invest places with meanings, the "human intellectual and emotional responses to the place" (Hill and McCormick 1989). These responses represent the intangible aspects of the theme of place; although often neglected in geography classes, they can be addressed well using literature, which often reflects human love of place. This statement also includes a notion key to understanding maps—the idea that places can be represented through symbols.

To draw students' attention to the three different aspects of place, teachers might display a book with an evocative cover picture of the book's setting (*Antler, Bear, Canoe,* annotated below, is an example). Ask students to spend a few moments imagining themselves in the scene shown in the picture. Then ask students to free write for two to three minutes, describing the setting and their responses to it. They can use any format they want but should try to write as steadily as possible. When the time is up, ask students to share some of the words or phrases they used to describe the setting. Post these words in three columns—one for the physical environment (e.g., cold, snowy), one for the built environment (e.g., cabin, swing), and one for human emotional and intellectual responses to the environment (e.g., cozy, lonely). After posting a selection of words and phrases, discuss with students the importance of all three categories of information in understanding what a place is like. Once students have read the book you displayed, encourage them to reflect on whether the book changed their emotional response to the cover picture and, if so, why.

Students' observation of characteristics of place could be reinforced by having them create new dust jackets for books (including those whose jackets have long since been discarded and those that students think have ugly or undescriptive jackets). Encourage students to show as much about the setting as possible in their designs. As another option, students might create coat-hanger mobiles representing book settings; the mobiles could include drawings, as well as such real objects as leaves, flowers, pinecones, feathers, or small rocks.

One way to develop students' understanding of the intangible aspects of place is by encouraging them to imagine themselves in the settings of books they read. What would they see, smell, hear, etc.? After students have read several books, ask them to consider in which setting they would rather live and why. They might write real estate ads for homes in particular literary settings. What characteristics of the place do students think would be most attractive? What characteristics would they not emphasize? Why?

Having students compare various representations of the same place—ground-level, aerial, and satellite photographs, drawings, picture maps, and maps using symbols—is a good first step in developing the idea that places can be represented in many different ways. Books that show fairly detailed and large-scale photographs or drawings of places can be useful in having students draw picture maps of the places illustrated; they may then move on to drawing maps that use symbols. *As the Crow Flies* (annotated above) is a good resource for starting this process with very young children, as it includes drawings and picture maps, which students could convert to maps using symbols. Another example would be *The Secret of Nimh*, by Robert C. O'Brien (New York: Scholastic, 1982); students could map Mrs. Frisby's view from atop Jeremy's back.

Some suggestions for using specific books to develop key ideas about place follow.

Grades K-2

Mr. Jordan in the Park, written and illustrated by Laura Jane Coats (New York: Macmillan, 1989), is the literary embodiment of the theme of place. Mr. Jordan is an older man who has enjoyed activities in the park since he was very young. The book shows the evolution of those activities—from feeding bread crumbs to birds, flying a kite on the open lawn, boating on the lake with a sweetheart, pushing his son in a swing, and playing chess with friends—as Mr. Jordan and the park both change with time. Illustrations show the physical and human characteristics of the park, including changing technologies. Students could draw a mural showing change in the park over time, ending with pictures of what they think the park will be like in the future. They could write or illustrate a story about what they do in their favorite places and why they like those places. They might interview people about their favorite places; interviews with older people could focus on how a favored place has changed over time.

Antler, Bear, Canoe: A Northwoods Alphabet Year, by Betsy Bowen (Boston: Little, Brown, 1991), is actually a rather complex alphabet book. Each letter of the alphabet reflects some aspect of life in the northwoods of Minnesota; however, the information is also arranged to show how the northwoods change as the seasons pass from winter to spring to summer to fall and back to winter again. Beautifully illustrated with painted woodblock prints, the book provides insight into all three aspects of this place; students should have plenty of information to discuss their responses to the northwoods as well as to compare the northwoods with their own neighborhood or community. In small groups or as a class, students might create similar alphabet books to reflect the physical and human characteristics of their community.

Eve Bunting's *Fly Away Home*, illustrated by Ronald Himler (New York: Clarion, 1991), provides an unusual perspective on a place many students have likely visited—an airport. The book's young hero and his father *live* in an airport because they cannot afford an apartment. The text and watercolor illustrations poignantly convey how Andrew copes while hoping that he and his dad will soon have a place of their own. The story is likely to evoke many feelings among students, who will need time to talk about their responses. Focus some attention on the characteristics of the airport that caused Andrew's dad to decide they should live there. What other options do people who cannot afford housing have? Let students write a happy ending for the story illustrated with the floor plan for a small apartment Andrew's dad might be able to rent.

A much more positive view of a family's special place is provided in *We Keep a Store*, by Ann Shelby, illustrated by John Ward (New York: Orchard Books, 1990). The storekeepers are an African-American family whose members work together to operate the business. Their store also serves as a gathering place for people in the rural community; men sit on the porch, whittling and telling stories, women chat as they prepare fruits and vegetables for canning, and children play in the yard. Encourage students to compare the store in the book with stores where they shop: how are the natural settings in which the stores are located alike and different? How are the human-made features of the settings similar and different? Do people do the same things in the stores? Why or why not? Do students know any stores where people gather to talk or work together? Encourage students to design a store where people in their community could gather. What would be sold there? How would the store be laid out?

Grades 3-4

Brooklyn Dodger Days, written and illustrated by Richard Rosenblum (New York: Atheneum, 1991), is a tribute to a place obviously well-loved by the author: Ebbets Field. Through the story of a boy's afternoon at the ballpark, Rosenblum conveys a detailed sense of what this special human-made place was like. For many students, the book will evoke memories of their own special experiences at ball fields or stadiums, whether at the professional or Little League level; these experiences can provide the basis for a discussion of how all the aspects of place add to our experiences. Because the book focuses on baseball, students might create a baseball card for the field, showing a picture on one side and on the reverse providing such information as the location, how the field got its name, the number of people it would hold, what team played there, and so on. Students particularly interested in baseball might create cards for other historic and current baseball fields. Through the drawings and information provided, these cards could provide a basis for discussing how places created for the same purpose can still be very different.

The young African-American hero of *Evan's Corner*, by Elizabeth Starr Hill, illustrated by Sandra Speidel (New York: Viking, 1991), longs for a place of his own. His mother gives him his choice of one corner of their two-room apartment, which he proceeds to decorate with much ingenuity. Simply having a special place of his own is not as satisfying as he had anticipated, however; he is happier when his mother suggests helping younger brother Adam fix up *his* corner. The book should stimulate young readers' thought on the importance of having personal space (as well as family and helping). For some students, having a corner would be much less space than they currently have to themselves, while for others, having such a corner would be a luxury. Students could design their own corners, deciding what they would need to make that place uniquely their own.

Many of Constance Levy's poems in *I'm Going to Pet a Worm Today*, illustrated by Ronald Himler (New York: McElderry Books, 1991), provide a close-up look of a small portion of a child's environment; for example, the very first poem in the collection, "Tree Coming Up," describes the place where the shoot of an oak tree breaks through the earth. Other poems focus on a child's reaction to nature—the questions a child might ask a butterfly, sneezing and scratching around weeds, feeling lonely outside in the fall when all the insects have disappeared. Following a reading of some of the poems, you might take the class into the schoolyard and assign pairs of students to focus on small segments of the yard. Students should observe the area as closely as possible, as Levy must have done before writing her poems. They could share what they observe in their assigned area through drawings or poems. Some students may discover that their assigned areas could be the places described in such poems as "Connection" or "Old Leaves." Student responses to many of the poems could serve as the basis for bulletin boards or posters. For example, "Out" tells why a young person prefers being outdoors, even when the weather is bad. Students could create drawings or poems that explain their own preferences and display them on a bulletin board with two headings, "Out" and "In."

Also in poetic form is *The Cataract of Lodore*, by Robert Southey, illustrated by David Catrow (New York: Henry Holt, 1992). The poem, which describes a waterfall in the English Lake District, was written in the 1800s by Southey, then Poet Laureate of England, in response to a request by his children. Southey's delightful, rhythmic words inspired Catrow "to celebrate in color and form the beauty of nature and the imagination of all children." The illustrations achieve those goals by depicting Southey and his children in an overstuffed armchair, riding the river through the English countryside. After students have had a chance to enjoy the text and illustrations, engage them in analyzing more closely how the words convey the feeling of the river as it journeys from mountain tarn, through meadows, to the waterfall. What words don't they know? What do they mean? How do they contribute to understanding what the cataract is like? Students might imagine that a member of their family is a poet. What place would they like to have him/her describe in a poem? Encourage students to write and illustrate poems about favorite places.

Grades 5-6

In *Children of the Fire* (New York: Atheneum, 1991), Harriette Gillem Robinet presents an incredible, detailed description of one place, Chicago, during a night when it changed drastically—October 8, 1871, the first night of the great Chicago fire. The story is told from the perspective of an 11-year-old African-American girl named Hallelujah. By tricking her foster parents into letting her go out with their son and then escaping from his care, Hallelujah is able to spend the entire night away from home. Her perceptions of the fire and Chicago itself change as she watches the destruction of the city and interacts with rich and poor from other neighborhoods. Students who have experienced an event that changed their own city or neighborhood, such as an earthquake, fire, tornado, or flood, may compare their own perceptions with those of Hallelujah. Did their perceptions of the place change? Did their views of themselves or of other people change? Students might also imagine themselves to be Hallelujah ten years later, writing a letter to a friend who has moved away. How has Chicago changed? How has it stayed the same? Could another fire affect Chicago as drastically as this one did? Why or why not?

The biography *Julia Morgan: Architect of Dreams*, by Ginger Wadsworth (New York: Lerner, 1990), can be used to illustrate how human-made features of the environment can be designed to complement natural features. Achieving such design, as Morgan—who helped to open the field of architecture to women—did, requires sensitivity to the natural environment, as well as artistic vision. Among the more than 700 buildings designed by Morgan are the Asilomar Conference Center, the Greek theater at the University of California-Berkeley, San Francisco's Fairmont Hotel, and the Hearst Castle. The biography's many photographs, some in full color, will provide students with opportunities to identify aspects of the natural environment reflected in the buildings. As a follow-up to reading the book, students may design buildings that would blend well with the natural environment in your area. What building materials would they use? What colors? How tall would their buildings be? Would they use sharp angles or rounded edges? Display the students' completed drawings for group enjoyment and discussion.

Relationships Within Places

The theme of relationships within places looks at the interactions between humans and the environment. These interactions have three significant aspects (Hill and McCormick 1989):

- How people depend on the environment. People obviously depend on the environment in many ways—for food, building materials, water, air, and on and on. While the highly technological nature of contemporary society may suggest that people are less dependent on the environment than they once were, this is an illusion. As Hill and McCormick (1989) point out, "each intervention we make with nature comes with a cost and, rather than eliminating environmental limits, imposes a new set of conditions."

- How people adapt to the environment. The world's environments are widely different, some being more easily settled by humans than others. The solving of environmental barriers to settlement is called *adaptation.*

- How people change the environment. Through the process of adaptation, humans change the physical environment. Technology is important to this aspect of human/environment relationships, since increasing levels of technological development tend to result in larger changes to the environment.

This geographic theme is obviously linked to many of the environmental problems that we face today and thus is critical to our shared future. It is also well-suited not only to treatment through literature but also to interdisciplinary science/social studies lessons.

To develop awareness of how people depend on the environment, students could build a pyramid showing the environmental base on which human activities rest. To start this project, cover a bulletin board with paper and trace the outline of a pyramid on the paper. Lightly trace the outline of blocks or bricks over the surface of the pyramid; in the triangle formed at the peak of the pyramid, place a picture of several fictional characters who are favorites of the students or a picture of the students themselves. Tell students that they are going to build a pyramid showing how people depend on or need the environment; give or have students give some examples from their own lives. Explain that the pyramid will be built using examples from books that students read. Each time they find an example of how characters in a book depend on the environment, they can list the example and the title of the book in one of the squares on the pyramid. A student should only use one example per book, although another student may use an example from the same book. When students complete the pyramid, conduct a class discussion of the various examples they found.

Through discussion of works of historical fiction that involve settlement of new lands (a rather common genre), students can be helped to understand both how humans adapt to different environments and how they change the environment. Following such a discussion, students might draw a picture of a place from the book, accompanied by a picture of what they think the place looks like today. In the caption, they could tell why the place changed.

The dust jackets from books that focus on environmental problems could be removed and displayed on a bulletin board. Students who have read the books could then look for and clip newspaper or magazine articles about the same problems. The articles could be displayed on the bulletin board with the dust jackets. Students might follow up on a particular problem through such activities as surveying other students and adults to determine their views on the issue; writing letters to the editor about the problem and its relevance to their community; creating posters about the problem to be displayed around the school; or writing "grant proposals" describing the problem and suggesting solutions.

Some suggestions for using specific books to develop key ideas about relationships within places follow.

Grades K-2

A Country Far Away, by Nigel Gray, illustrated by Philippe Dupasquier (New York: Orchard Books, 1989), presents a single narrative of events in a young boy's life. The dual illustrations—one set of a boy in an African village, the other of a boy in a British or American city—show that people in diverse settings have different ways of living but also share many common aspects of culture, from celebrating the birth of a family member, to playing soccer, to reading books and wishing to make friends with someone in another country. Students could draw a third set of illustrations to show their own lives. They might also want to undertake a pen pal project in which they could share information about similarities and differences in lifestyles with children from other nations. Organizations that coordinate pen pal programs include the following:

People-to-People International
501 East Armour Boulevard
Kansas City, MO 64109

Student Letter Exchange
630 Third Avenue
New York, NY 10017

Many students will be able to identify with the troubles of the Sheldon family described in *Aurora Means Dawn*, by Scott Russell Sanders, illustrated by Jill Kastner (New York: Bradbury Press, 1989). Moving from Connecticut to Ohio in 1800, the family faces a terrifying thunderstorm on the first night in their new homeland. The effects of the storm and how the Sheldons worked to overcome the difficulties it posed illustrate well the theme of relationships and should provide ample material for discussing how storms affect people, both now and in the past. Students might create a bulletin board display showing the effects of storms today and illustrating how technology helps people deal with those effects. The bulletin board should also illustrate how people work cooperatively to solve problems caused by storms, as shown in the book.

The Stream, written and illustrated by Naomi Russell (New York: Dutton, 1990), provides a simple introduction to the water cycle and thus could be tied to science, as well as geography. Some of the ways in which humans use water are mentioned. Selected pages fold out to show the underwater perspective. Before reading the book with students, have students brainstorm all the ways they use water. Ask them where the water they use comes from. List students' answers to both questions on the chalkboard. Read the book with students; discuss the water cycle with the class, drawing the circular movement of water on the chalkboard. Have students add water uses from the book to their list. Follow up the class activities by having students do a simple inventory of water uses at home, showing the results in a drawing.

How humans change the environment is beautifully illustrated in *Heron Street*, by Ann Turner, illustrated by Lisa Desimini (New York: Harper and Row, 1989). The first few pages of the book describe a marsh and its inhabitants, undisturbed by humans. As settlers arrive, however, the marsh changes until, by the present, it has lost virtually all of its original character. Only one small patch of grass remains to whisper to the single heron flying overhead. The simple yet vivid illustrations underline the changes described in the text. The book could lead to a discussion on how changing a natural environment means that wildlife loses habitat. The result has sometimes been the extinction or endangering of a species. Although the heron is not endangered, several other North American bird species are. Interested students might look at wildlife magazines for children for evidence that people are now attempting to change the environment positively, by protecting or restoring habitats.

Grades 3-4

Although it contains virtually no text and thus could be used with even the youngest students, *Window*, by Jeannie Baker (New York: Greenwillow Books, 1991), should also be considered by teachers of older students. Through her marvelous collage constructions, Baker shows the view through a boy's window, from the time he is a baby until young adulthood. The scene changes from one that is rural to one that is modestly developed to a highly urbanized vista. At the end of the book, the young man and *his* new baby look out the window of a new home into another rural scene with the city visible in the distance. The illustrations should provide ample stimulus for class discussion of the changes humans made in the scene. Students could be encouraged to make their own collages showing how the scene through the window at the new house might change over the course of the new baby's childhood.

As its subtitle suggests, *Sugar Season: Making Maple Syrup*, by Diane Burns, photographs by Cheryl Walsh Bellville (Minneapolis: Carolrhoda Books, 1990), provides a detailed but understandable description of making maple syrup. The book shows clearly human dependence on nature, as well as human adaptations and changing technologies. Students could make a chart showing ways that people who make maple syrup depend on nature. They could also look for evidence that these people try not to make unnecessary changes in the environment. Students might create their own books providing the same kind of detail about an agricultural activity in your community. A work of fiction that could be used as a companion to *Sugar Season* is the Newbery-winning *Miracles on Maple Hill*, by Virginia Sorensen, illustrated by Beth and Joe Krush (New York: Harcourt Brace, 1956).

Grades 5-6

Students can see both positive and negative human effects on the environment in *Riverkeeper*, by George Ancona (New York: Macmillan, 1990). The author's text and photographs focus on John Cronin, whose job is to protect the Hudson River ecosystem, a job that involves dealing with many kinds of pollution. Readers learn not only how pollution occurs, but strategies for fighting pollution. Because the book connects geography and government, students could be encouraged to show what they learned from the book through posters, bumper stickers, or other methods used to convey political messages. You might invite a local resource person whose job is to protect the environment to visit the class and discuss his/her work with the students. The book might also serve as a starting point for a class study of a local body of water.

In Two Worlds: A Yup'ik Eskimo Family, by Aylette Jenness and Alice Rivers, with photographs by Aylette Jenness (Boston: Houghton Mifflin, 1989), provides an in-depth look at the life of a family living on Scammon Bay in western Alaska and comparing their lives with those of previous generations. Several geographic concepts are well presented in the book: humans' dependence on and adaptation to the environment, as well as the effects of changing technologies on human/environment relations. Students could construct retrieval charts showing the ways in which the Yup'ik people depend on the environment and the technologies they have used in adapting to it. Near the end of the book, the authors point out how a collection of items on the family's television symbolizes the modern and traditional aspects of their lives; students might create a display of items that they think symbolizes the relationship between the Rivers family and the environment around Scammon Bay.

Earth to Matthew, by Paula Danzinger (New York: Delacorte Press, 1991), is a light-hearted look at how Matthew and his friends in the sixth-grade deal with such problems as a sister who shaves her head and pierces her nose, divorcing parents, and budding boy-girl relationships. At the same time, they are engaged in a class study of ecosystems that culminates in an overnight trip to the Franklin Institute. A great deal of information about ecosystems is embedded in the story, which also makes interesting comparisons between ecosystems and family relationships. After reading the book, students might make posters illustrating how families and ecosystems are alike, in that the actions of one person affect many other parts of the system. As an accompaniment or introduction to a unit on ecosystems, the book could provide students with ideas for research and action.

Movement

Movement refers to all kinds of human interactions between and among places. As described in the *Guidelines for Geographic Education*, "they [people in various places] travel from one place to another, they communicate with each other or they rely upon products, information, and ideas that come from beyond their immediate environment." Key ideas related to this theme (*K-6 Geography*) include the following:

- Movement demonstrates interdependence. People seek contact with other places because they cannot meet all their needs in the immediate environment; at the same time, they can help people in other places meet their needs.

- Movement involves linkages between places. Transportation and communication systems are obvious examples, and ones that demonstrate again the importance of technology.

- Patterns of movement involve people, ideas, and products.

Primary students who are learning about transportation could create a bulletin board bar graph showing types of transportation used in the books they read. Give the bulletin board a title, such as "How Book Characters Move From Place to Place"; with students' help, decide what kinds of transportation should be shown on the graph (e.g., foot, bicycle, wagon, car, truck, bus, train, airplane, boat). Put a label for each type of transportation at the bottom of the bulletin board. Demonstrate for students how to affix notecards to the graph to show how characters in a book got around; for example, a student who read *Aurora Means Dawn* (annotated above) would place a card in the "foot" column, as well as in the "animal-drawn wagon" column. The name of the book should be written on each card placed on the graph. When the graph has had time to grow, give students practice reading bar graphs by discussing the data they have generated.

Vacations are a reason for movement that is interesting to students, as well as a subject for many books. Students could create a chart showing vacation destinations and the attractions that draw book characters to those destinations. For example, a student who had read *Fudge-a-Mania*, by Judy Blume (New York: Dell, 1990), would enter "Maine" on the chart, along with "to get away from the city" as the reason for choosing Maine for a vacation. Students could compare charted information with their own vacation preferences and could then design their own "ideal" vacation, plotting their route on a map, calculating distances, deciding what types of transportation they would use, and so on.

Some suggestions for using specific books to develop key ideas related to movement follow.

Grades K-2

The title character in *Chita's Christmas Tree*, by Elizabeth Fitzgerald Howard, illustrated by Floyd Cooper (New York: Bradbury Press, 1989), is a young African-American girl in turn-of-the-century Baltimore. The focal point of the story is a trip to get a Christmas tree. Traveling by horse and buggy, Chita and her father leave the city and go to a woods where they pick out a tree but do not cut it down; on Christmas morning, however, that very tree appears in their living room. This example of movement to obtain something not available in one place will have relevance to many students, who may enjoy drawing pictures that illustrate how selecting a Christmas tree at the beginning of the century was different from selecting a tree at the end of the century. Students who do not celebrate Christmas could identify examples of items they use to celebrate particular holidays that are not available without traveling (note that they may travel to find the item or the item may travel before it is available to them).

Araminta's Paint Box, by Karen Ackerman, illustrated by Betsy Lewin (New York: Atheneum, 1990), is set in 1847. Araminta Darling's family is migrating from Boston to California, where her father will become one of the first doctors. Along the way, Araminta loses her special paint box. The story traces the route of the paint box as it becomes a tool box, a baby bassinet, and a prospector's mining box before being given to Araminta's father in payment for medical treatment. A map compares Araminta's route with that of the box. Each person or family that encountered the box on its journey had a different reason for moving. Small groups of students could each be assigned one of these people/families; each group could write a letter to Araminta, explaining the reasons for their journey. What did they hope to find in their new home that they did not have before? These letters could be displayed in a cardboard box cut and decorated to resemble the paint box.

I Go with My Family to Grandma's, by Riki Levinson, illustrated by Diane Goode (New York: Dutton, 1986), is another historical story, this time set in New York City around the turn of the century. Five young cousins living in the five boroughs travel with their families to grandmother's house. Each family uses a different mode of transportation on the trip. Before reading the story, you may need to explain that New York City has five sections; this story is about five children who lived in the different sections of New York City about 100 years ago. Let students speculate on why the children are traveling to grandma's and how they might get there. Students should consider the information that the children live in New York City in giving their answers. Read the story and discuss the forms of transportation used by the five families. Which of these forms of transportation would probably still be used today? Which would not? How might the story have been different if the grandmother had lived in a small town or on a farm? Encourage students to draw a contemporary version of the story, showing how members of their extended family might travel to a family reunion.

While a young man named Zack builds a sailboat, his dog Salty rides a ferry across the bay to visit him. That is the story told in *Salty Dog*, by Gloria Rand, illustrated by Ted Rand (New York: Henry Holt, 1989). Readers learn about a special form of transportation, the ferry, but also see the preparation for another kind of sea voyage—a sail around the world. Locate Mercer Island, Washington, where the Rands live, on a map, asking students why people who live there would need the services of a ferry. Have students search the map for other places where a ferry would be helpful. Students could also try drawing a floor plan or blueprint for a ferry based on the information in the book. You might also focus class discussion on planning a long journey, such as a sail around the world. What route would students take? In what season would they start the trip? What supplies would they need? Where would they stop along the way? What kinds of things would be easiest to get along the way? What would be hardest?

Grades 3-4

The desire for freedom is a strong motivation for movement. Students can explore the dangerous journeys people are willing to take in *Escape from Slavery: Five Journeys to Freedom*, by Doreen Rappaport, illustrated by Charles Lilly (New York: HarperCollins, 1991). Based on newspaper stories of the time and accounts written by people involved with the Underground Railroad, the book dramatically tells of such escapes as a woman and her baby crossing the Ohio River as the ice breaks up, a man traveling for three days in a box too small to allow movement and with limited air holes, and two girls masquerading as boys to elude slave hunters. The courage of those escaping slavery and those helping them should inspire students. Divide the class into five groups, assigning each group one of the stories to present to the class through a simulated press conference with the people involved. Make a map available so that the escaped slaves can show where they started their journey, how they traveled, and where they finally settled down.

The City Girl Who Went to Sea, by Rosmarie Hausherr (New York: Four Winds Press, 1990), is the story of Alicia Belford's summer in the small fishing village of Salvage. Alicia, who lives in a one-room apartment in New York, learns about the very different life in Newfoundland, a geography lesson in itself. Her story also illustrates the theme of movement, in that even this rather remote community is connected to other parts of the world in a variety of ways. On a world map, students could plot the movement of people, ideas, and products to and from Salvage. They could create similar maps showing movement of people, ideas, and products that impact their own lives.

Grades 5-6

Patricia and Fredrick McKissack's *W.E.B. DuBois* (New York: Franklin Watts, 1990) illustrates well how biographies can be used to develop understanding of movement without interfering with another reason for reading. Before students begin reading the life story of this civil rights leader, give them blank world maps, with instructions to use different colored lines to show movement. For example, they might use blue lines to show movement of DuBois' ancestors; these lines would encompass not only various areas in the United States, but Africa and Latin America as well. Green lines could be used to connect DuBois' various homes throughout his life; these would include locations in the Northern and Southern United States, as well as Europe and Africa. Along these lines, students might want to note DuBois' reasons for moving. Orange lines could be used to show the movement of DuBois' ideas; these lines might include his travels to important conferences and to meet world leaders, as well as arrows emanating from Atlanta and New York to show the dissemination of his ideas through his writings. Upon completing their reading, students would have a graphic representation of the notion that people and ideas move from place to place. A similar strategy could be used with other biographies of noted thinkers, activists, or inventors, such as Jean Fritz's *The Great Little Madison* (New York: Putnam's, 1989).

A sequel to the popular *Dear Mr. Henshaw*, *Strider*, by Beverly Cleary, illustrated by Paul O. Zelinsky (New York: Morrow, 1990), continues the story of Leigh Botts into high school. In the course of a year, Leigh endures more crises with his divorced parents, gains a pet, takes up track, and has his first date (to pull weeds!). The book's themes of friendship and growing up should not be subverted in order to develop geographic understanding, but a brief small-group webbing activity can serve both purposes. Have students write *Leigh Botts, garden cottage, Pacific Grove* in the center of a sheet of posting paper. Around the center, each student should write the names of several people (or animals) important to Leigh, connecting those people to Leigh with lines. Each student should then pick a name written by another student and write one or more places that person connected Leigh to; for example, Leigh's friend Barry connected him to Los Angeles, where Barry visited his mother; Leigh's dad connected him to Bakersfield, Gilroy, Cholame, and other places he drove his truck; Leigh's dog Strider connected him to the beach where Leigh and Barry found him and golf courses where they looked for lost balls to clean and sell. The completed webs will show the network of friends and family members surrounding Leigh, as well as the less obvious geographic connections.

Regions

Because environments are so varied and the number of possible places of study so huge, geographers must find some way to organize their data to make it manageable. The region is a device they have created to do that. A region is an "area that displays unity in terms of selected criteria." For example, a mountain region is defined by its landforms, while a state is a region defined by a shared government.

Hill and McCormick (1989) differentiate regions from places by explaining that while a place has a "material existence," a region is a "mental construct," an intellectual tool. The concept of region is useful to us in comparing places and identifying reasons for differences and similarities. Like places, however, regions also evoke emotional responses. As Joel Garreau says in describing the regions he calls the "nine nations of North America," "When you're from one, and you're in it, you know you're home" (Garreau 1981, p. 13).

Key ideas related to the theme of region (*K-6 Geography*) are:

- Regions are a way to organize information.
- A region has common characteristics.
- Regions change.

As with location, it may first be useful to help students become familiar with what is meant by commonly used regional terms, such as Latin America, the Middle East, the former Soviet Union, the Rocky Mountain region, and the like. One way to develop this type of familiarity might be to develop a bingo-style board that students would fill in as they read books that took place in certain regions. For example, such a board might be structured as shown below.

Mountain region of the U.S.	Latin America	An urban area
Grasslands region of the U.S.	Eastern Europe	A rural area
Desert region of the U.S.	Middle East	A region with a hot, humid climate
Forest region of the U.S.	Southern Africa	A region with a cold, dry climate

Older students might make an effort to read folktales, myths, legends, or fables from as many different regions as possible. These could be recorded in a "Regions Log" in which each student noted the region, what distinguished the region, the story read, and any perceived relationship between the story and the special characteristics of the region; for example, an animal story might reflect the special wildlife of a region. Students could also, of course, write about their own responses to the stories. The list of recently published folktales, myths, legends, fables, and stories provided in the appendix may be a useful starting point for such an activity.

When students have read a book that has provided considerable information about a region, they could compare that region with their own, identifying common features and those that distinguish the two. Students could create a scavenger hunt, listing wildlife they have learned are common to a particular region; they could then look for those items in their own community, identifying which are common to both regions and which appear to be unique to the region under study. Students might also plan a trip to the region about which they have read, listing the things they would need to take and what they might bring back with them that would represent the region especially well.

Some suggestions for using specific books to develop key ideas about regions follow.

Grades K-2

Surrounded by Sea: Life on a New England Fishing Island, by Gail Gibbons (Boston: Little, Brown, 1991), takes the reader through the seasons on an island where fishing and tourism are the major sources of income. An island is a clearly defined region, set apart from other areas because it is surrounded by water. Through the details provided in the book, students will be able to note many similarities and differences between their own lives and those of the island children. Students might make drawings showing the similarities and differences; they might speculate on similarities between the lives of children on this New England island and children on a small island in a different setting, such as Hawaii. Students could create a table display representing jobs on the fishing island and jobs in their own communities—which are the same? which are different?

My Prairie Christmas, by Brett Harvey, illustrated by Deborah Kogan Ray (New York: Holiday House, 1990), recounts a family's first Christmas after leaving Maine for the prairie. Central to the story, which is one of several prairie stories by this author, is the father's failure to come home after going out to find a Christmas tree on Christmas Eve. The comparisons between this holiday and previous ones make clear some of the differences between the two regions, as well as adaptations the family made to the new environment. Before students read the story, it would be helpful to locate Maine on a U.S. map and to point out those areas of the United States considered prairies. Showing students pictures of Maine and of the grasslands would also be helpful. After reading the story, students can decide whether they would rather celebrate a holiday important to their family in Maine or on the prairie; they might also speculate on whether the holiday celebration would be different in subsequent years, when the family had been on the prairie longer. Students could also consider what kinds of decorations, foods, or homemade gifts could be made in their own region if they had access to only those things that occurred in nature or could be grown and preserved.

Somewhere in Africa, by Ingried Menned and Niki Daly, illustrated by Nicolaas Maritz (New York: Dutton, 1990), directly confronts stereotypes about Africa. The book is the story of a young boy who lives in urban South Africa but enjoys reading about the Africa of lions, crocodiles, and zebras. Before reading the story, ask students to draw pictures showing what it would be like to live in Africa. After reading the book, talk with students about what in the story surprised them. Did any of their drawings of life in Africa show city life? Why or why not? How is city life in Africa like city life in the United States? How are they different? Do students like to read about places in the United States that are very different from where they live, like Ashraff liked to read the library book about Africa? Another view of South Africa can be found in *At the Crossroads*, written and illustrated by Rachel Isadora (New

York: Greenwillow Books, 1991). This book is the story of children in a shanty-town in a segregated township, waiting for their fathers to come home after months of working in the mines. Students can compare and contrast the lives of the young people in the two books.

Grades 3-4

Writer Cynthia Rylant and artist Barry Moser, both natives of Appalachia, have created a tribute to that region in *Appalachia: The Voices of Sleeping Birds* (New York: Harcourt Brace Jovanovich, 1991). After students have enjoyed the drawings and the language, let them identify the region talked about in the book. The author refers to the Appalachian Mountains but does not otherwise define the region, so students will need to make some minor inferences to identify the region on a map. When they have done so, encourage them to look for more information about the region from the map and try to make additional inferences from that information. For example, they might note that most of the larger towns in the region are located in river valleys and infer that life there would be different from the book's description or that the region stretches many miles from north to south and thus probably varies in climate from place to place within the region. Groups of students might create two murals showing similarities and differences between your region and Appalachia. If your school is located in Appalachia, students could compare their own experience of the region with that of the author and illustrator. Why might students see some things differently?

A great contrast with Appalachia can be seen in *Arctic Memories*, written and illustrated by Normee Ekoomiak (New York: Henry Holt, 1990). The author describes, in both English and Inuktitut, the Arctic lifestyle of his childhood on James Bay in Canada. The book is illustrated with Ekoomiak's innovative paintings and fabric artworks. Before students read the book, read the material on the back cover with them and point out where Ekoomiak lived as a child on the map provided or on a wall map. Ask students to list everything they know about this region, as well as the people who lived there, called Inuits. Then ask students to list what they would like to know about this region and its people. Finally, create a column for students to enter what they learn as they read the book. When they have finished, discuss the charted information with them. What items that they "knew" were incorrect, according to the book? Which of their questions about the region and its people were they able to answer using the book? What new questions did the book raise? Encourage interested students to pursue those questions using other sources. Students could also create dioramas or murals showing this lifestyle and comparing it with ways of life in their own region.

An even greater contrast can be seen in *The Day of Ahmed's Secret*, by Florence Parry Heide and Judith Heide Gilliland, illustrated by Ted Lewin (New York: Lothrop, Lee, and Shephard, 1990), which is set in Cairo. The book follows Ahmed through his day delivering gas canisters around the city, all the while reveling in his secret—he can write his name. The detailed paintings bring the colors of the city to life, as the text does for its sounds. Help students locate Cairo and the Nile River on a map and identify the desert referred to in the text. Encourage them to speculate on the climate in Cairo. Why do people dress as shown in the pictures? What are some of the ways people transport themselves and goods from place to place? Have students make a daily schedule for Ahmed. When do they think he had time to learn to write? How is Ahmed's schedule different from their own? Is Ahmed's life similar to theirs in any way?

Grades 5-6

Mildred D. Taylor's *Mississippi Bridge* (New York: Dial Books, 1990) tells a moving story in which a region's human and natural features combine to create tragedy. Through the language used by the narrator (a ten-year-old white boy), Taylor's description of the weather and natural environment, and the events that unfold, the reader is immersed in the rural South of 1931. Discussion questions that will help students explore the regional concept include:

- What caused this incident?

- Could it have happened in another region? Why or why not?

- Could in happen in the South today? Why or why not?

- Have we gained more control over the human or natural causes of the incident in the years since 1931? Explain your answer.

Students could contrast the problems of the rural South in the early years of the Depression with those in another region, the Dust Bowl, by reading *Moxie*, by Phyllis Rossiter (New York: Four Winds, 1990). Set in rural Kansas during the Depression, the book tells of a family's struggle to stay on their farm despite drought, dust storms, hobos, and fire. The people in this region share with those in the South the problems of poverty and unemployment, and intolerance is also evident, although it is not racially based. The natural hazards faced are almost directly opposite those of the South, however—drought instead of too much rain. The questions given above for *Mississippi Bridge* could be adapted for use in discussing this book.

All Themes

As mentioned earlier, all five themes are interrelated; although it is useful to separate the themes as we are becoming familiar with their meaning and use, in reality separating them is nearly impossible. The interrelationships among the themes are well illustrated in a unique book created by Steve Lowe. Lowe has selected passages from Henry David Thoreau's *Walden* (New York: Philomel Books, 1990); accompanied by Sabuda's linoleum cut illustrations, the excerpts provide young readers with a vision of Thoreau's life on Walden Pond. The introduction indicates where Walden Pond is and why Thoreau chose to spend "two years, two months, and two days" there. The text and illustrations contribute to a strong sense of place, as well as to an understanding of the special aspects of forest regions in the Northeast. Thoreau's interactions with the environment are well described, as are his observations of trains passing by his cabin and the visitors he receives, both indicators that even people in remote areas are connected to people in other places.

One of the cautions to keep in mind as you are beginning to use the themes is not to teach the themes themselves, but to use them to decide what to teach. One useful way to do this is to teach students to ask geographic questions about places they encounter in books (and in reality!). These questions might include the following, which are based directly on the themes:

- Where did this story take place?

- According to the story, what is this place like? Where else could I find information about what this place is like?

- How did the environment in the story affect people? How did people affect the environment?

- What examples of movement among various places on earth can be found in the story?

- Are any regions—areas that share special characteristics—mentioned in the story? If so, what characteristic made each region special?

Students could share geographic information about story settings through picture cubes. Picture cubes are made from six equal-sized squares; information or pictures are put on five of the squares before they are taped together (the sixth square is left blank so string for hanging the cube can be taped to it). Each one of the five squares could be devoted to one of the themes. For example, on one square the student could draw a map showing the location of the story setting; on the second square, he/she could make a chart showing physical and built features as well as personal reactions to the place; on the third, he/she could draw two pictures showing how the place has been changed by humans; and so on. A finished cube would look like the following:

Geographic questions could also be used in simulated press conferences with characters from stories students have read. While the main story written by "reporters" after the press conference might focus on the character's achievements or some other aspect of the story, the geographic information could be used to write a "sidebar," or accompanying article, on where the character lives.

References

Garreau, Joel, *The Nine Nations of North America* (New York: Avon, 1981).

Guidelines for Geographic Education: Elementary and Secondary School (Washington, DC: American Association of Geographers, and Macomb, IL: National Council for Geographic Education, 1984).

Hill, A. David, and Regina McCormick, *Geography: A Resource Book for Secondary Schools* (Santa Barbara, CA: ABC-CLIO, 1989).

K-6 Geography: Themes, Key Ideas, and Learning Opportunities (Washington, DC: Geographic Education National Implementation Project, 1987).

Model Guides for Using Children's Literature in Geography

Characteristics of a Good Guide

Susan Hepler, on whose work the format used for the guides in this section is based, has identified the following characteristics for quality guides for literature-based programs:

- "A good guide should improve the quality of the reader's experience with the book....Through talk, readers should be able to say, 'I hadn't thought about that before,' or 'I had, but I couldn't put it into words'."

- A good guide should include questions that cause students to "examine why people act as they do" and, in moderation, "encourage readers to identify with whatever aspects of the text match their perceptions."

- A good guide should include activities that "serve the reader and the book first, not some particular curriculum area." The activities should help students examine and expand on the ideas in the book and their reactions to those ideas.

In developing the guides included in this section, we have attempted to apply these criteria. Thus, although we emphasize geographic understanding, we do not focus exclusively on geographic questions or activities; we *do* try to ensure that students consider the environment and humans' relationship to it as they develop meaning from each story, poem, or book. For each book, we include at least some questions that, like those used in the Great Books program, have "no single right answers, requiring both the student and teacher to examine factual information, assess motivation, and make inferences" (Gallagher 1991). We do sometimes include extension activities that take the geographic understandings beyond the book; these activities are separated from the activities more directly related to the book with several asterisks. With some of the guides we include black-line masters for handouts needed to do the activities.

Many of the books will be most effective if read aloud, especially with younger students. Having several copies of the book may still be a good idea, however, to allow close examination of the illustrations, which often include a great deal of geographic information. For primary grade students, we include a relatively small number of questions in each guide. For older students, the number of questions increases, and we divide the list of questions by chapter or section of the book.

The chart on the next page lists the books for which guides are provided, appropriate grade levels, and geographic themes addressed.

References

Gallagher, Arlene F., ed., *Acting Together: Excerpts from Children's Literature on Themes from the Constitution* (Boulder, CO: Social Science Education Consortium, 1991).

Hepler, Susan, "A Guide for the Teacher Guides: Doing It Yourself," *The New Advocate* (Summer 1988).

Title	Grade Level	Themes				
		Location	Place	Relation-ships	Movement	Region
My Cousin Katie	K-1		X	X		
Is Anybody Up?	K-2	X				
The Empty Lot	K-3		X	X		
Time To Go	K-3		X	X	X	X
Tar Beach	K-3		X		X	
Mojave	K-6		X	X		X
My Grandmother's Journey	2-4				X	
Stop the Presses! Nellie's Got a Scoop	3-4				X	
Trucker	3-4	X			X	
Go Fish	3-4		X	X		
Morning Girl	3-5	X	X	X	X	
Brother Eagle, Sister Sky	3-6			X		
Jayhawker	4-6	X	X		X	
Grasshopper Summer	4-6	X	X	X	X	X
The Ups and Downs of Carl Davis III	4-6		X	X	X	X
John Muir	4-6	X	X	X	X	
The Star Fisher	5-6		X			
Downriver	6	X	X	X	X	X

My Cousin Katie,
written and illustrated by Michael Garland
(New York: Crowell, 1989).

Summary

In anticipation of a visit, a child describes the farm where Cousin Katie lives. The rich colors of Garland's paintings draw the reader into Katie's daily activities.

Initiating Activities

1. Ask students to name some relatives they like to visit. Why do they like to visit these relatives? Where do they live? What special things can students do there? What is special about the relatives themselves?

2. Show students the cover of the book and explain that the child telling the story is going to visit Cousin Katie. Where do students think Cousin Katie lives?

Discussion Questions

1. Make a schedule of Katie's daily activities. How are they similar to or different from your own? Which of Katie's activities would you most like to do? Which would you least like to do?

2. Imagine that you are having a picnic with Katie and her mother. What foods do you think might be in your picnic lunch? What can you see, smell, hear, and feel as you sit by the pond? Why is this a special place for a picnic?

3. Where do you think the child who is telling the story lives? Why do you think that?

Follow-up Activities

1. Have students draw simple maps of Katie's farm, using the two-page drawing early in the book as the main source for their maps but adding other features (such as the pond) according to their own ideas.

2. Have students create picture books about the homes of relatives they like to visit. Display the books on a table on which you have placed an outline map of the United States. Help students locate and label the places their relatives live.

Is Anybody Up?,
written and illustrated by Ellen Kandoian
(New York: Putnam's Sons, 1989).

Summary

A little girl named Molly gets up at 7:30 on Saturday morning to have breakfast. While no one else in her family is up yet, many people and animals who live in Molly's time zone are eating breakfast—an Inuit woman in Baffin Bay, Molly's grandpa in Miami, a parrot in Columbia, and others. As each is joined ·by someone else, they say hello in their own language. Delicate watercolor drawings illustrate the text.

Initiating Activities

1. Ask students when they would hear or ask the question, "Is anybody up?" What do they think a book with that title might be about?

2. Display the cover of the book and ask students to observe as many details as they can. If necessary, guide them with such questions as:

   • What is the little girl wearing?

   • How can you tell what time it is?

Discussion Questions

Read the book through from beginning to end, stopping before the last page, which explains time zones.

1. What are some things all the people in the story have in common? (They eat breakfast at the same time, they wear clothes, many of them have a clock, many of them have families they live with in some kind of house.) What do the animals have in common with the people? (They need food.) How are the people different? (They eat different foods, they wear different styles of clothing, they speak different languages.)

2. What do you think the people were saying when they spoke in their own languages near the end of the story? Why do you think that?

3. How does it make you feel to know that other people in faraway places are doing the same thing that you are doing at the same time?

Follow-up Activities

1. Post a large world or Western Hemisphere map. With younger students, you may want to highlight each of the locations mentioned in the book. Divide the class into eight groups, assigning each group one of the people or animals eating breakfast at the same time

as Molly. Each group is to create a symbol for that person or animal and place that symbol on the map in the appropriate location. When all the locations have been marked, ask students if they notice any pattern to the locations. They should note that the locations form a rough vertical line on the map. Review the cardinal directions with students, pointing out which places the book indicates are north of where Molly lives and which are south. Then ask students to decide where Molly lives. They should choose a location on the east coast of the United States, between New York and Miami.

2. Read the last page of the book to students. Using a balloon, draw wedges on the surface, as shown below, to indicate how the time zones are drawn. If feasible, divide the class into groups of five, and give each group an orange to dissect. The sections of the orange are similar to time zones. Ask students how many sections an orange would need to have a section for every time zone. (24)

3. On a world map or globe, locate the time zone in which your community is located. Help students identify other locations north or south of them in the same time zone. Ask students to write or draw a story showing what people in the locations they identified might be doing while they are at school, playing after school, or getting ready for bed.

* * * * *

4. On subsequent days, you may want to greet students with the phrases meaning "hello" from the book. Challenge students to identify what language you are speaking and locate the country where that language originated on a map.

The Empty Lot,
by Dale H. Fife, illustrated by Jim Arnosky
(Boston: Little, Brown, and San Francisco:
Sierra Club Books, 1991).

Summary

Harry Hale decides to sell a two-acre lot that was once part of his grandfather's farm. Visiting the lot, he is surprised by how the city has grown up around it, as well as by the abundant wildlife living in this "empty" space. He decides not to sell the lot. Pastel drawings illustrate the substantial text, which makes the book a likely read-aloud candidate for the early grades.

Initiating Activities

1. Read the title of the book aloud to students, making sure they know what a *lot* is. Ask students what *empty* means to them. If they drew a picture to show what *empty* means, what would be in the picture? Show students the cover illustration and ask them if it looks like they imagine a picture of *empty* would look. What questions does the picture raise in their minds?

2. Read the dedication aloud to students, explaining what a dedication is (a statement explaining to whom or what the author is paying respect with the work). What or whom is Ms. Fife dedicating the book to? (The Ohio countryside and the Stieber woods) Ask: Does this give you any ideas about the book?

Discussion Questions

1. What was the "tap...tap...tap" sound Harry heard? How would hearing this sound in a lot you thought was empty make you feel? How do you think Harry felt when he heard this sound?

2. Why do you think someone would want to build a gas storage tank, a bigger factory, or a parking lot? Why are these things needed in cities?

3. Why do you think the children liked to play in the empty lot? Do you have a similar place where you like to play? Do you think Dale Fife had such a special place when she was young? Could a place that had many objects made by humans also be special? Why or why not?

4. Why did Harry change the sign? Do you like the new sign? Can you think of a better sign to put in the lot?

Follow-up Activities

1. With the students, discuss what made the lot a special place. One of the factors that made the lot special was that it remained relatively unaffected by humans in the middle of a city, where human influence often seems to totally obscure the natural environment. Some communities set aside undisturbed "greenbelt" areas to ensure that some of the natural environment is preserved. Students should not, however, come away with the idea that human-made places cannot be very special also. With the students, you may want to create a bulletin board display showing special places in your community, both natural and human-made.

2. Have students draw pictures showing the "occupied" lot as it existed during Harry's visit. Then have them draw pictures showing what it would look like if Harry sold the lot. What would change? What would stay the same? How would the changes be helpful? How would they be harmful?

3. Take students to a greenbelt area, an empty lot, a backyard, or an open area of the schoolyard. Ask them to sit quietly and listen for evidence that the area is not empty. What sounds tell them that wildlife lives in this area? What other evidence can they find that wildlife is present?

* * * * *

4. Encourage students to give Harry's empty lot a name based on one or more of the types of wildlife living there. You may need to provide an example or two (e.g., Quail Quarters or Dragonfly Den) to get students started. Each student should design a sign for the lot showing the name. To provide practice with map work, give students copies of Handout 1 and challenge them to use a classroom atlas to figure out the names of the cities and counties shown on the map with picture symbols. Students should give the name of the city or county and the state in which it is located.

Answer Key: 1. Eagle Point, Oregon; **2.** Big Bear City, California; **3.** Aspen, Colorado; **4.** Beaver City, Nebraska; **5.** Los Coyotes, Texas; **6.** Deer Park, Illinois; **7.** Buffalo, New York; **8.** Turtle Creek, Pennsylvania; **9.** Palm City, Florida; and **10.** Bass Harbor, Maine.

WILDLIFE NAMES

   Each place shown on the map below is named for a plant, animal, bird, or fish. Use an atlas and the clues below to figure out the name of each city. Write its name and the state in which it is located on the map.

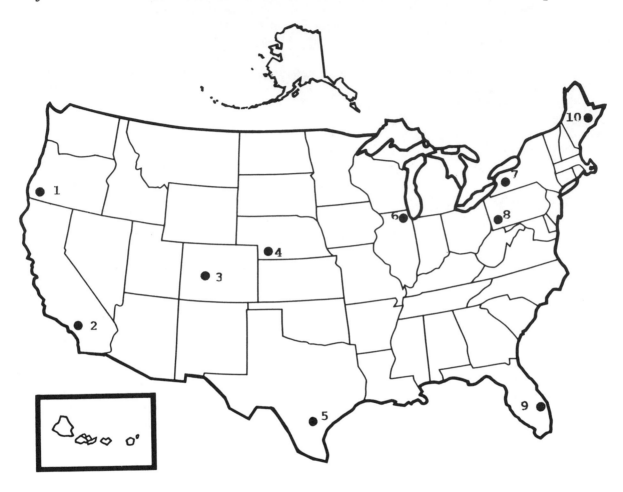

1. I'm the national bird.

2. I'm not a teddy.

3. My leaves turn yellow in fall.

4. I'm known for being busy.

5. I howl at night.

6.   My antlers look heavy.

7.   I really lived on the Great Plains, not in the East.

8.   I stay in my shell.

9.   My fronds sway in the breeze.

10.   Go fishing to find me.

Time to Go,
by Beverly and David Fiday,
illustrated by Thomas B. Allen
(Orlando, FL: Gulliver Books, 1990).

Summary

A young boy whose family is leaving their farm walks around his home one last time. Facing pages present drawings in charcoal and colored pencil, one showing how the farm looks as the boy makes his final tour and the other showing how it looks in his memory. As the family leaves, the boy vows to return. Although there is not a great deal of text, the vocabulary makes the book best suited for reading aloud.

Initiating Activities

1. Ask students to describe situations when they have heard the phrase "time to go." When does hearing it make them happy? When does it make them sad?

2. Display the cover of the book and let students speculate on whether the phrase has a happy or sad meaning for the boy shown on the cover. What about the cover makes them think so?

Discussion Questions

1. Where did the boy in the story live? (On a farm; the illustrations seem to suggest that it was in the Midwest)

2. Describe a day in the boy's life as a farm child. What would be fun about this life? What questions do you have about it?

3. Why do you think the boy was leaving his home? (The book does not say for sure, but the illustrations seem to indicate that dry weather, or a drought, made it impossible to continue farming.) How did he feel about moving? Can you think of other examples of people who have had to leave places and ways of life they loved?

4. How does this story make you feel? Why?

Follow-up Activities

1. Ask students to write a diary entry for the boy based on the information provided in the book. Then tell students to imagine that the boy moves from his farm to your community. How would the events of a typical day change? Have students write a second diary entry reflecting life in the new setting.

2. On the chalkboard, construct the outline for a chart comparing farm, small town, and city life. Ask students to think of both good and bad aspects of living in each place. Post their responses on the board. Based on the chart they have compiled, where would students rather live? Did they choose a place similar to or different from where they now live? Why?

3. Primary students are unlikely to understand why a family would lose its farm because of bad weather. To give them some rudimentary understanding, ask them to list all the things that a farmer needs in order to grow crops. A simplified list would include good soil, enough warm weather for the crop to grow to maturity (growing season), enough rain (but not too much!), machinery to work the land, and seed. Write the list on the chalkboard and ask students to identify which items depend on nature; put check marks next to those items. Draw a simplified flow chart on the chalkboard, similar to the one below. Briefly explain each of the steps in the chart. Then erase the box saying "rain falls" and ask students what would happen to the rest of the boxes. Erase the "crop grows," "farmer harvests the crop," and "farmer sells the crop" boxes. What is the result? (The farmer won't have money to pay bills and buy seed for the next year; if the farmer owes money to a bank and cannot pay, the bank may take the land and sell it to get its money back.) Since students are likely to think this process is quite unfair, allow them time to express their feelings.

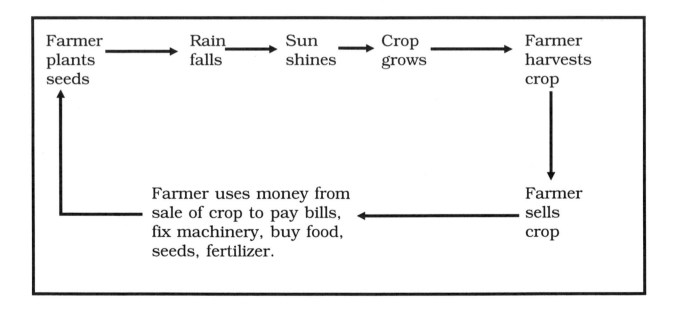

Tar Beach,
written and illustrated by Faith Ringgold
(New York: Crown Publishers, 1991).

Summary

Eight-year-old Cassie Louise Lightfoot imagines soaring over New York City as she lies on Tar Beach (the roof of her apartment building). As she flies over such landmarks as the George Washington Bridge, she claims them for her own, allowing her to redress past injustices against her family. Lively paintings bordered with patches from the author's original story quilt illustrate the book.

Initiating Activities

1. Ask students if they have ever imagined flying through the air with no airplane, balloon, or other device—that is, simply flying with their bodies. Let students share their thoughts on the question. Then ask them to reflect on why people dream of flying. What do they think flying would help them do? How would it make them feel?

2. Show the cover of the book, and let students speculate on such things as where Tar Beach was, where the people shown on the cover lived, why a girl is shown in the air above the bridge in the background, and what the patches on the top and bottom of the page signify.

Discussion Questions

1. What or where was Tar Beach? What did Cassie like about sleeping on Tar Beach? What were some of the other things the Lightfoots and the Honeys did on Tar Beach? If you were on Tar Beach with the two families, what would you see, hear, taste, smell, and feel?

2. Why did Cassie want the George Washington Bridge for her own? How did she claim it in the story? What did the bridge look like when she flew over it?

3. Why wasn't Cassie's father allowed to join the union? Do you think that was fair? What would you have done if you were Cassie? Did you like her response of claiming the building by flying over it? Why or why not?

4. Why did Cassie decide to take her brother Be Be with her when she flew over the union building? Have you ever had an experience like this with a younger brother or sister (or have you done something similar to an older brother or sister)?

5. Do you agree that all you need in order to fly is "somewhere to go that you can't get to any other way"? Why or why not? Have you ever taken a flight like Cassie's in your imagination? Describe the experience.

Follow-up Activities

1. Read selected items about the author and her artwork (described at the end of the book) with the students. Help students locate New York City on a map of the United States. If possible, show photographs of the George Washington Bridge. Discuss the use of quilts as both a storytelling and folk art medium. Be sure students understand that the patchwork border on the book's cover and pages was taken from the original story quilt. Interested students may want to try their hand at producing a story quilt, either for an original story of their own or for a story they have read.

2. Read the African-American folktale, *The People Could Fly*, aloud to students (an excellent version recorded by a storyteller is available from Childrens Press). In this story, slaves escape from the physical and psychological violence imposed on them by flying away, much like Cassie flew in *Tar Beach*. Lead students in a discussion of the similarities between the two stories and of the power of the human imagination.

3. Many students have probably not had the experience of being on a building's roof. Lead students in a discussion of the kind of roof that is needed in order to enjoy a rooftop garden, picnic area, and the like. Encourage students to speculate on why many tall buildings in large cities have such roofs while homes in smaller towns often do not. Consider such factors as rainfall and snowfall, as well as open yard space available and appearance. Tell students that rooftops are still used for many purposes, including gardening, sunbathing, and even playgrounds for day care centers in large cities. Let students design a rooftop that they would like to visit.

* * * * *

4. Show two photographs of the same place, one taken from ground level and one from an aerial perspective. Have students compare and contrast the two pictures. Which shows a view that is similar to the view shown in a map? Challenge students to imagine that they are flying over the school and draw pictures showing the view they would have from above the school. If possible, compare these with ground-level pictures, noting the similarities and differences.

Mojave,
by Diane Siebert, illustrated by Wendell Minor
(New York: Crowell, 1988).

Summary

This book-length poem vividly evokes the Mojave Desert through its language and illustrations. The text and pictures touch on the landforms of the desert, its wildlife, people's history in the desert, and seasons in the Southwest. The beauty of the language deserves oral reading, but students should also be given time to reread passages and study the paintings in detail.

Initiating Activities

1. Show students the book and ask: What kind of place do you think the Mojave is? Find the Mojave Desert on a wall map of the United States. Have students list what they know or can deduce from the map about deserts in general and the Mojave in particular. Also have them describe their feelings about deserts.

2. Read the first three lines of the poem to the students and ask them what it sounds like. (A poem and an invitation) Ask students to speculate on what kind of book this is going to be.

Discussion Questions

Read the book through once and allow students to share their responses to the words and pictures. Reread the book more slowly, using the following questions to analyze the text in greater detail.

1. What does the poet compare the desert to? What parts of the desert does she equate with parts of the face? What effect does the comparison have on you as a reader?

2. Imagine you are the raven flying over the desert. What do you see?

3. What do you think the poet meant by "their crumbling walls personify/The dreams of those who used to be." What happened to the miners who tried to live in the Mojave? Do you think anyone could live in the Mojave? Why or why not?

4. Which words help you feel what each season is like in the desert? In which season would you rather visit the desert? Why?

5. How much of what we said we knew about deserts was correct? How much was incorrect? (If students wonder why sand was not mentioned in the book, share with them the information that only 10-20 percent of the North American Desert is covered by sand.) What new information about deserts did we gain by reading this book?

Follow-up Activities

1. Divide the class into four groups and give each a copy of the book. Have each group make a poster on one of the following topics: desert wildlife, desert landforms, weather in the desert, and people in the Mojave Desert. With older students, you may want to require that they provide illustrated definitions of words from the poem related to their topic.

2. Have students create a travel brochure for a desert vacation. The brochure should describe the beauty of the desert, as well as the items that a vacationer would need to bring along, depending on the season.

3. Assign students to work in small groups to make a chart showing similarities and differences between your region and the Mojave desert. They may want to use the categories listed in activity 1 to organize their charts.

4. Make Siebert and Minor's later collaborations, *Heartland* (New York: Crowell, 1989) and *Sierra* (New York: HarperCollins, 1991), available for students to read and compare with *Mojave*. The comparisons may focus on the language Siebert uses, Minor's artwork, or the factual information conveyed about the regions. All of these comparisons will help students grasp the special character of each region.

* * * * *

5. Have students complete Handout 2, which provides practice in locating the world's deserts.

Answer Key: See completed map below; **1.** Sahara Desert, **2.** Australia; **3.** middle latitudes; **4.** the Patagonian Desert is located in the middle latitudes in Argentina in southeastern South America (highly motivated students may identify specific latitudes and longitudes); **5.** Great Basin Desert, Great Salt Desert, Death Valley, Painted Desert, Mojave Desert, Sonoran Desert, Chihuahuan Desert.

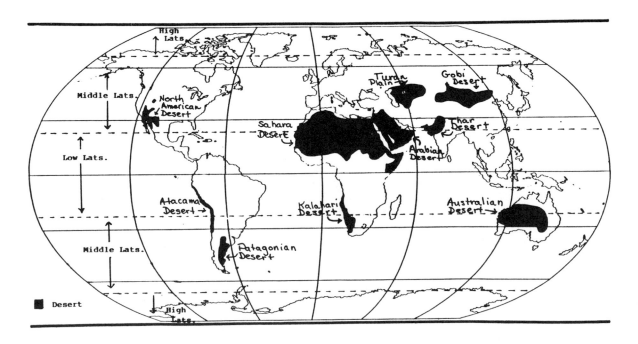

THE WORLD'S DESERTS

Deserts are created because of the way air moves around the earth. In some places, moist winds do not blow across the land; instead they blow across the ocean. In other places, moisture is lost as air travels over mountains. The land behind the mountains thus is dry.

The map on the next page shows the world's desert regions. Use the table below to identify each desert. Then answer the questions.

Desert	Location
Sahara	Northern Africa
Australian	Central and Western Australia
Arabian	Arabian Peninsula (East of the Sahara)
North American	Southwestern North America
Gobi	Central Asia
Patagonian	Southeastern South America
Kalahari	Southern Africa
Atacama	Western South America
Thar	Northwestern India
Turan Plain	East of the Caspian Sea in the former Soviet Union

1. Which desert is the largest? _____

2. Which continent has the largest percentage of its land in desert?

    _____

3. In Asia, are more deserts located in the low, middle, or high latitudes?

    _____

4. Describe the location of the Patagonian Desert as accurately as possible.

    _____

    _____

5. (Extra Credit) The North American Desert is actually made up of several smaller deserts. Use an atlas to find the names of these deserts.

    _____   _____

    _____   _____

THE WORLD'S DESERTS

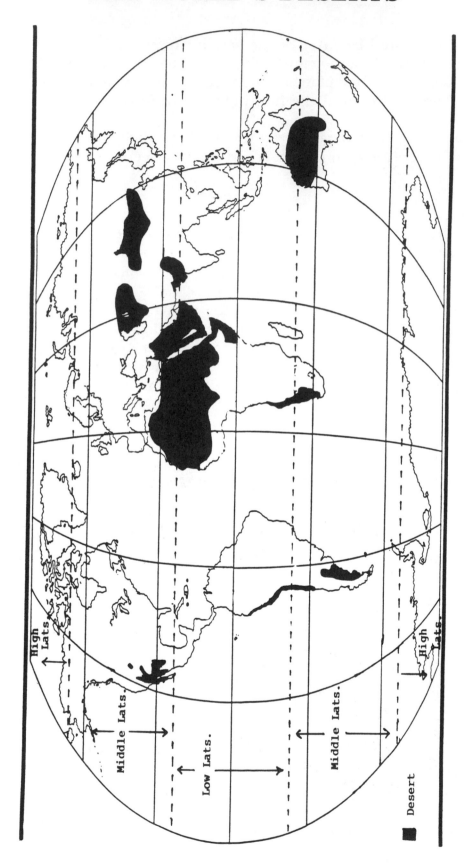

My Grandmother's Journey,
by John Cech,
illustrated by Sharon McGinley-Nally
(New York: Gulliver Books, 1991).

Summary

Korie's grandmother, as a bedtime story, tells of her early life in Russia, including experiences with two gypsy fortune tellers and the incredible losses her family suffered during the Russian Revolution and its aftermath. With her baby in her arms, she and her husband tried to escape Russia during World War II but were captured and forced to work in German factories. After the war, the family finally arrived in the United States. The vivid bordered illustrations give the story, based on the recollections of the author's mother-in-law, the feeling of a folktale.

Initiating Activities

1. Ask students if they ever talk to parts of their body. Allow some time for silliness, as well as serious responses. Next, ask them to think about what a grandmother might say if she were to talk to her body.

2. Display the cover of the book and draw students' attention to the title. Tell them that the grandmother in this story talks to her feet, saying things like "Feet, where haven't you been?" Why do students think she would say something like that? (Because she went on a long journey by foot.)

Discussion Questions

1. Why do you think Korie wanted to hear her grandmother's story again?

2. Gypsies were people who moved around, not settling in one specific place. People who had settled in a town or on a farm often feared the gypsies or thought they had special powers. Why do you think they might have feared the gypsies? How do you think the grandmother felt when the Gypsy told her there would come a time "when your every footstep will be a pain"? How would a prediction like this make you feel, even if you did not believe that gypsies could see into the future?

3. A revolution is a war in which the people of a country or a group of the people throw out the government and set up a new one. They usually do this because they don't think the old government is helping the people. Did the revolution help or hurt the grandmother's family? How do you think people could disappear "like a tree would or a horse does"? What do you think helped the young couple to keep going when so many bad things had happened to them?

4. Are you surprised that, during World War II, people were more willing to help the couple when they saw the baby? Do you think people are usually more willing to help children than older people? Think of some examples that support your view.

5. Imagine that you were the grandmother on the day she sailed by the Statue of Liberty. How would you have felt?

Follow-up Activities

1. Tell students to imagine that the grandmother's feet could talk. What would students say to her feet? What might the feet say back? Let pairs of students role play a conversation between a student and the grandmother's feet. Again, you will need to allow for some silliness at first, but encourage students to ask serious questions and give serious answers.

2. Read "A Note on the Story" with students. On the chalkboard, draw a timeline showing the events in Feodosia Ivanovna Belevtsov's life. Then ask students to identify on the timeline each time she moved in response to these events. Extend the timeline to the present and add some major events that have occurred in recent years, such as the Persian Gulf War. Point out that events like these still cause people to move, just as events in Feodosia's life caused her to move. Ask students to clip newspaper or magazine articles about people who move because of traumatic events in their homelands. Create a bulletin board display of these clippings around a world map, using string or yarn to link each article with the proper location on the map.

3. Invite a refugee to your class to discuss his or her experiences. Before the visit, work with students to develop a list of questions to ask the visitor. You may want to include "What would you say to your feet?" to follow up on the theme of the story.

4. Encourage students to read *How Many Days to America?* by Eve Bunting, illustrated by Beth Peck (New York: Clarion, 1988). This book tells the story of modern refugees, fleeing their Caribbean home by boat. The dangers of this journey can be compared to those in *My Grandmother's Journey*. What motivates people to face such dangers?

Stop the Presses, Nellie's Got a Scoop! A Story of Nellie Bly,
written and illustrated by
Robert Quackenbush
(New York: Simon and Schuster, 1992).

Summary

Nellie Bly was one of the nation's first female reporters, getting many of her stories by going "undercover" in such settings as asylums and prisons. This biography, written in Quackenbush's typically amusing style, covers the events of her life from her childhood in Pennsylvania, to her early days as a reporter in New York, to her years of running a business, and finally to her return to journalism. A set of illustrations showing two children examining items ostensibly from Bly's valise—rendered in pen and ink with color highlights, like the illustrations showing Bly's life—provide a story within a story.

Initiating Questions

1. Ask students to name the jobs they would like to have when they are adults. Have they been interested in the same jobs for a long time, or do their interests change often? If you wanted to be a teacher from an early age, share that fact with students. Tell students that they are going to be reading about a person who wanted to be a writer from the time she was very young.

2. Show students the cover of the book and ask them to speculate on the kind of writer that Nellie Bly became and on when she lived. If students are not familiar with the words *presses* or *scoop*, define them as a class.

Discussion Questions

1. Who are the two children shown at the bottom of page 9? Why are they in the book?

2. Why do you think writing was considered a man's profession? What jobs were considered suitable for women? Why do you think that was true? What could someone do if they were discouraged because they were excluded from interesting jobs?

3. Would an article about divorce cause as much excitement today as it did in Nellie's time? Why or why not? What kind of article would cause an uproar? Why did Nellie change her name? Do you think writers today ever change their names to protect themselves or their families?

4. Why did Nellie go to Mexico? Why do you think so little was known about Mexico in 1886? Are there places today that we know so little about that articles written there would help boost newspaper sales?

5. What kind of a place was Blackwell's Island? Do you think Nellie was courageous or foolish to go there to write a story?

6. How did becoming famous for her tour around the world change Nellie's life? Do you think she enjoyed the fame? Why or why not?

7. What did Nellie do when she got married? Did any of these actions surprise you? Why or why not?

8. What were Nellie Bly's two most important achievements? Give reasons for your choices.

Follow-up Activities

1. Provide students with a world map and have them plot not only Nellie's around-the-world trip (which is shown on a map at the front of the book), but also her many moves and reporting trips in the United States. Along the lines connecting places, have them write the reasons for the various moves; for example, on the line between Cochran, Pennsylvania, and Pittsburgh, students might write, "lost wealth and needed work."

2. A game based on Nellie's trip around the world is provided at the back of the book, but it does not accurately portray the conditions Nellie experienced on the trip. For example, Nellie's ship did not collide with another ship, causing her to spend time on a raft. Thus, if students play the game, you should make clear to them that the events are things that *could* happen to someone on such a journey, not things that *did* happen to Nellie. If students play the game, be sure they follow their own progress on a world map as they proceed toward their destination.

3. Have students consider the question asked on the last page of the book—what would Nellie write about if she were a reporter today? Provide newspapers and magazines that students can look through for possible topics. They can then create posters that answer the question; these posters might be given such titles as "Nellie's News for the Nineties."

4. The question of what makes someone decide at an early age what career they want to have and then pursue that career with great energy, often at considerable personal cost, is one that is worth students' consideration. As a continuing activity, students might interview adults in a variety of careers, finding out when they decided that was the job they wanted to have, why they picked that job, and what they have had to do to become successful in it.

Trucker,
written and photographed by Hope Herman Wurmfeld (New York: Macmillan 1990).

Summary

Black-and-white photos and text give readers a view into the lives of a Tennessee trucker and his family. Phil Marcum hauls lumber, steel, and produce between Canada and Florida. While his job keeps him from taking part in some family activities, he tries hard to be present for as many as he can, including a visit to his daughter's school to show her classmates his truck and talk about his job as a trucker.

Initiating Activities

1. Ask each student to jot down something he/she could not get along without today; it may be an item of clothing, something they ate for breakfast or plan to eat for lunch, a school supply, or something else important to students. Next, ask them to decide whether they think the items they listed were made in your community. If not, how might the items have gotten to your community? List all of their answers on the board.

2. Display the book and ask students to identify the form of transportation the book focuses on. Let them speculate on the kinds of products that are transported by truck.

Discussion Questions

1. What kinds of products did Phil Marcum haul on his truck? What factors determined the products that he hauled? (The kind of truck he has, where he wanted to drive)

2. Based on the story, what are the challenges of trucking? What are its rewards? What parts of trucking do you think you would dislike?

3. How does Phil's job affect his family? What advantages or disadvantages would having a parent who is a trucker have?

4. How do truckers help each other when they are on the road? Do you think this help is important? Why or why not?

5. What questions would you like to ask a trucker if one visited your class?

Follow-up Activities

1. Have students plot as accurately as they can the routes Phil took on the two trips described in the book. How far did he travel? According to the information provided, about how much did he spend on fuel? What other expenses did he have on the trip?

2. Invite a trucker to visit your class and describe his/her job to students; if possible, a family member might accompany the trucker to class to discuss effects on families. Allow students to ask the questions they generated in response to question 4 above.

* * * * *

3. Distribute copies of Handout 3 to students and allow them time to complete it. Most will need to consult with their families in order to finish the exercise.

YOUR CIRCLE OF NEEDS

Draw a circle of needs like the one shown below on the map on the next page. Put the circle of needs on the state in which you live. With your family's help, identify some items that help you meet your needs in each area. Write these items below. If items come from outside your state, write their states of origin, too. Circle the items you think could have come to your state by truck. On the map, draw an arrow from these items' states of origin to the proper section on your Circle of Needs.

For example, if your mother drives a car to work each day, you might list "car" below, along with "Michigan." Since cars are transported on special trucks, you would circle the word "car" and draw a line on the map from Michigan to the "Work" section of your Circle of Needs.

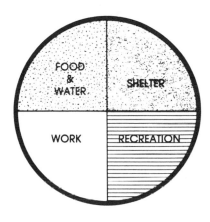

Things that help me meet my needs:

Food and Water   Shelter

Work   Recreation

YOUR CIRCLE OF NEEDS

Go Fish,
by Mary Stolz, illustrated by Pat Cummings
(New York: HarperCollins, 1991).

Summary

Thomas pesters his grandfather until he offers to take Thomas fishing in the Gulf of Mexico near their Florida home. They discuss the wildlife they observe while fishing and then cook their catch for dinner. Before bedtime, Thomas's grandfather tells him a story like the African folktales pased through their family for many generations. Charming black-and-white illustrations accompany the text.

Initiating Activities

1. Ask students what the words "Go fish" mean to them. Probe until they mention both the card game and actually going fishing.

2. Ask students what they like about playing card games. List their responses on the chalkboard. Then ask them what they like about fishing and list those responses on the chalkboard. (If none of the students have been fishing, let them speculate on what they would enjoy about the experience.) Do playing cards and going fishing have anything in common? If students do not mention the fact that they can both be enjoyed with friends and family, point this out and tell students that they are going to be reading or listening to a book in which family members go fishing and play "Go Fish" all in the same day. Show the cover of the book and let students guess who the family members are.

Discussion Questions

(After reading Chapters 1 and 2)

1. Name all the things Thomas did to get his grandfather's attention. What finally worked? Why do you think saying "It's snowing" finally caused Grandfather to put down his book?

2. What did Grandfather collect and make? Do you think these hobbies had anything to do with where he and Thomas lived? Would these hobbies be easy to participate in where you live?

3. What do you think Thomas meant when he said that there were some things he knew but couldn't make himself believe? Are there things you know but can't believe?

4. List the kinds of wildlife that Thomas and his Grandfather saw while they were fishing. Which kinds of wildlife could you also see where you live? Which kinds could probably only be seen near a large body of water?

5. Why is patience important in fishing? Do you think Thomas was a patient boy? When do you find it hard to be patient?

(After reading Chapters 3 and 4)

6. What did Thomas and his Grandfather catch? Why did they put some of the fish they caught back into the Gulf?

7. Why did Thomas tell Donny's mother a fib when he went to their house to play? Do you think the fib was a good idea? Why or why not?

8. Why was Ringo acting strange after dinner? What else might make a cat act this way? What did Grandfather mean when he said Ringo's "instinct was stronger than his judgment"? Is a person's instinct ever stronger than his or her judgment? Give an example.

(After reading Chapters 5 and 6)

9. Imagine that you are sitting on the porch with Thomas and Grandfather. What can you see, hear, smell, feel, and taste? What would be different if you were sitting on your own porch? What would be the same?

10. What was the *ibeji*? What is the cult of *ibeji*? What natural phenomenon did Grandfather use the cult of *ibeji* to explain?

11. What did Thomas think he would tell his grandchildren when he is older? How do you think handing stories down from one generation to another helps young children? Does it also help the older people who tell the stories? If so, how?

Follow-up Activities

1. Invite someone who fishes in your area to come to the class and display his/her equipment and talk about the fish that are common.

2. Encourage students to read African folktales that explain natural phenomena, such as the *Just So Stories*. After students have read several such stories, challenge them to write their own folktales explaining some aspect of weather or animal behavior and appearance.

Morning Girl,
by Michael Dorris
(New York: Hyperion, 1992).

Summary

Told in the voices of 12-year-old Morning Girl and her younger brother, Star Boy, this book re-creates Taino life in 1492. The children's family faces conflicts, the loss of a baby, and a hurricane, but the family members' love for one another and the island on which they live is obvious. Then, one day during her morning swim, Morning Girl sees a canoe full of "fat people" who speak a language not known to her. To be polite, she invites them ashore. The book ends with an excerpt from Columbus's diary.

Initiating Activities

1. Tell students that they are going to be detectives as they read a book called *Morning Girl*. They are going to look for clues as to where Morning Girl and her family lived. Encourage students to think of the kinds of information that will be useful as clues (e.g., information about weather and landforms).

2. Ask students to speculate on why someone would be named Morning Girl. What might a girl with this name be like?

Discussion Questions

(After reading Chapters 1-3)

1. Were we right about why Morning Girl was given that name? Do you think it is a good name for her?

2. Do you think Morning Girl is right that parents don't know their children the way that brothers and sisters do? Why or why not?

3. How did Star Boy get his name? Was it a good name for him? Who are you more like—Star Boy or Morning Girl? Explain your answer.

4. Describe Morning Girl's emotions when she found out about the new sister and then again when their mother left and came back without the new baby. How were Star Boy's feelings similar to and different from Morning Girl's?

5. What clues did these chapters give to where the children and their family lived? What ideas do these clues give you?

(After reading Chapters 4-6)

6. Why was Star Boy pretending to be a rock? How did he think like a rock? Do you think you could pretend to be a rock for an entire day? Why or why not?

7. Why did Morning Girl want to know what people saw when they looked at her? How did her parents help her?

8. Describe the storm in which Star Boy was caught. What did he think about and do to survive the storm?

9. What clues did this chapter give us as to where the family lived? Where do you think they lived?

(After reading Chapters 7-9)

10. What happened to Star Boy at the celebration after the storm? Why do you think Morning Girl did what she did? Would you have done the same thing? Why or why not?

11. Why was Star Boy mad at everyone after the celebration? What do you think he meant when he told his mother he learned that "at night you must be your own friend"?

12. What happened when Morning Girl went to swim the morning after Star Boy disappeared again? Who do you think the "fat people" in the canoe were? Why? Where do you now think Morning Girl and Star Boy lived?

(After reading the Epilogue)

13. Were we right about who the people in the canoe were? Were we right about where Morning Girl and Star Boy lived?

14. After reading Columbus's words, do you think Morning Girl should have welcomed the people ashore as she did? Why or why not?

15. Do you think Columbus was right in saying that the people of the island were "poor in everything"? Why or why not? How does the excerpt from Columbus's diary make you feel?

Follow-up Activities

1. Have students draw illustrations for the book, showing the special aspects of the place where Morning Girl and Star Boy lived and the relationships between people and the land.

2. Based on the story and the excerpt from Columbus's diary, have students write additional chapters for the book, describing what might have happened to Morning Girl and Star Boy after the Europeans arrived. Encourage older students to do research to find out what did happen to the Taino people following Columbus's arrival.

Brother Eagle, Sister Sky,
a message from Chief Seattle,
illustrated by Susan Jeffers
(New York: Dial, 1991).

Summary

Chief Seattle's words, adapted from a speech given at a negotiation with government agents in the 1850s, eloquently describe the relationship of humans and the environment. His plea to protect the environment is beautifully illustrated by Susan Jeffers' paintings depicting Native Americans in various natural settings.

Initiating Activities

1. Show students the cover of the book, the frontispiece, and the title page. Ask them to free write about their responses to these pages for three minutes. At the end of three minutes, ask students to share their responses in groups of three. They do *not* need to read what they wrote; instead they should discuss the thoughts that the free writing exercise helped them articulate.

2. Let students speculate on what they think the book is about. Then read the introductory page and talk briefly about the tradition of public speaking that was common in many Native American groups. Because many groups made decisions through a consensus process, the ability to explain one's position eloquently and persuasively was an important aspect of leadership. Students can evaluate Chief Seattle's skill as you read the book aloud to them.

Discussion Questions

Read the book aloud once to let students enjoy the language and paintings. Allow some time for general responses to the book, including students' ideas about Chief Seattle's oratory. Then reread the book more slowly, using the questions below to guide students' analysis.

1. Why do you think Chief Seattle began with a question? Is that an effective technique for gaining an audience's attention? Why or why not?

2. What does it mean to be *sacred*? How do you treat things that are sacred? Do most of us treat the earth as though it were sacred?

3. Chief Seattle compares people and other parts of the natural environment to members of a family. Do you think this is a good comparison? Why or why not?

4. Chief Seattle points out some ways that people depend on the natural environment. What are they? According to Chief Seattle and the illustrations, how do people change the environment?

5. Chief Seattle says "Teach your children what you have been taught...what befalls the earth befalls all the sons and daughters of the earth." What do you think he meant by this? Do you think children can ever teach their parents about protecting the environment? If so, how?

6. What do you think Chief Seattle meant when he said, "It will be the end of living, and the beginning of survival"? If Chief Seattle could see our community, do you think he would say people are living or surviving here? Explain your thinking.

7. Which drawing is most meaningful to you? Why?

Follow-up Activities

1. Divide the class into several small groups, assigning each group an activity such as the following:

- To present the book to the class in a Reader's Theatre format. Reader's Theatre is an approach to shared reading in which the text of a book is turned into a script for several narrators. If your students do not have any experience with Reader's Theatre, you may want to consider scripting the book for them. However, if they have experience with the approach or are skilled readers, they may enjoy experimenting with the scripting process themselves.

- To write an illustrated letter to Chief Seattle about the condition of the environment today.

- To create a mural illustrating the idea of a "web of life" in which humans are "merely a strand."

2. Distribute Handout 4, which presents a quotation from the Sioux chief Tatanka Yotanka (Sitting Bull). Ask students to illustrate this quotation in the way that Jeffers illustrated Chief Seattle's speech. Interested students might do research to create a book of illustrated quotations related to humans and the environment; the quotations need not be restricted to Native Americans, but could be from people of many nations and ethnicities.

THE LITTLE GROVES OF OAK TREES

The following is a quotation from a Sioux Chief named Tatanka Yotanka (Sitting Bull). Read the words and create a series of illustrations to show what the words mean to you.

> I wish all to know that I do not propose to sell any part of my country, nor will I have the whites cutting our timber along the rivers, more especially the oak. I am particularly fond of the little groves of oak trees. I love to look at them, because they endure the wintry storm and the summer's heat and—not unlike ourselves—seem to flourish by them.
>
> Tatanka Yotanka

Jayhawker,
by Patricia Beatty (New York: Morrow, 1991).

Summary

In the turbulent years before the Civil War, young Lije Tulley and his father are Jay-hawkers, Kansas abolitionists. After his father is killed on a raid of a Missouri plantation, Lije goes undercover with a group of Confederates. His job is to inform the Kansans of planned bushwacking activity, raids on Kansas territory. After several years undercover, Lije returns to his family shortly before the infamous raid on Lawrence.

Initiating Activities

1. Ask students to imagine that they lived in the time before the Civil War. They did not believe in slavery. However, in a neighboring state are many people who own slaves. What attitude would they adopt toward their neighbors—would they "live and let live" or try to do something about slavery in the neighboring state? If so, what would they do? Allow time for discussion of various options.

2. Tell students that the book they are going to read is about a group of people who lived in Kansas before the Civil War. Review some of the history of "bloody Kansas" with students. Tell students that the group they are going to be reading about called themselves the Jayhawkers, after a mythic bird, the Jayhawk, which is said to be a cross between the powerful hawk and the mischievous blue jay. Encourage students to speculate on the characteristics of a group of Kansans who would name themselves for this mythic bird. Some students might want to draw pictures showing what a Jayhawk and a Jayhawker would look like and/or do.

Discussion Questions

(After reading Chapters 1-4)

1. On a map, locate the area where the Tulley's lived. With an arrow, show the route they probably took when they moved to Kansas. Use arrows to show the routes that South-erners might take into Missouri. What would all of this movement into Missouri and Kansas likely result in along the border? Are there any places you think similar conflicts might have arisen in the years before the Civil War? Explain why you picked these locations.

2. Why was Chapter 1 called "The Blessing"? What clues does Chapter 1 give you that religion played a part in abolition? Do you think the slaveholders were also religious? Why or why not?

3. How did the Tulleys get word of John Brown again? Why do you think information was passed in this way? When news about Brown arrived at the newspaper, how did it come? How long had it taken for this news to arrive? What was the news? How did Lije and his family react to the news?

4. Describe some of the weather the Tulleys experienced in 1859. How did the weather affect their activities? Why didn't the Tulleys leave when some other families did? Would you have left Kansas if you were Mr. Tulley? Why or why not?

5. Describe the land Lije, Mr. Tulley, and Mr. Cousins traveled through on the raid. Why did Mr. Cousins say "it's a good night for our kind of work"? How do you think Lije managed to get back home to Kansas after the two men were killed? Were you surprised by anything Lije thought or did?

(After reading Chapters 5-9)

6. Describe some of the national events in 1861 that had an impact on Kansas. Which events do you think had the most significance for Kansas? Why?

7. What did Mr. Montgomery ask Lije to do for the Jayhawkers? Based on what you have read so far about the Jayhawkers and bushwackers, describe the dangers of this job. What would your answer have been if you were Lije?

8. Describe the surroundings of the house with the gray shutters, as well as the people that Lije met there. How were the times Lije spent alone at the house with Lotta similar to and different from being at home with his family? How do you think Lije felt living at Lotta's house?

(After reading Chapters 10-13)

9. What effect did Missouri's not seceding have on the bushwackers? How did most people that Lije met in Missouri seem to feel about the Union? Why do you think the state had not seceded yet?

10. Describe the site where the Wilson's Creek battle occurred. What preparations were made for the battle? How did Lije react when he saw the man who had burned his home lying dead?

11. How did Lije find out that Jim Hickok wasn't a real bushwacker? How did this make him feel? Do you think it would be especially important for a young person doing a very difficult job to feel that they were not alone among the enemy? Why or why not?

12. What do you think Jim Hickock meant when he said "Bushwacking is the worst kind of war work there is. For that matter, Jayhawking is, too, though there's a better cause behind it"? Do you think a better cause makes jayhawking right and bushwacking wrong? Explain your answer.

(After reading Chapters 14 and 15)

13. Why did Quantrill decide to ride on Lawrence? Why do you think Quantrill hated Lawrence so much?

14. What happened in the cellar of the house in Lawrence? Why do you think Prentiss did what he did? How do you think he felt about the war?

Follow-up Activities

1. Have students construct a timeline. On one side, they can show national events, including those mentioned in the book but also others that they have learned about from other sources. On the other side, they can show events in Kansas and specifically those events in which Lije took part. How does their timeline help show how one state, such as Kansas, is part of events in the larger nation?

2. Ask students if they can think of any other conflicts in which location has been extremely important. If no one mentions the Persian Gulf War, ask how location was important to this conflict. Assign students to find articles that describe the Persian Gulf region. What important resources are found in this region? How did these resources play a role in the conflict? Students may also suggest other conflicts that can be researched in the same way.

Grasshopper Summer,
by Ann Turner (New York: Macmillan, 1989).

Summary

When Sam White's father decides to leave Kentucky for Dakota Territory shortly after the Civil War, Sam is sad. The difficult journey west and the first few months on the plains do nothing to convince him that leaving home was the right thing to do. Yet, by the time locusts wipe out their crop and threaten to send them back to Kentucky, Sam is willing to take desperate measures to stay in his new home, including asking his grandfather for money against his father's wishes.

Initiating Activities

1. Ask students to think about a memorable summer they have spent. If they were going to write a book about that summer and call it _____ Summer, what word would go in the blank? Allow time for students to share various titles. Post them on the chalkboard. Then go through the list asking students to classify the titles according to whether students think they describe a good summer or a bad summer.

2. Share the title of the book, Grasshopper Summer, with students. Ask: What does this title suggest? Do you think the summer referred to was a good one or a bad one? Encourage students to be as descriptive as possible in their responses—for example, what do they imagine the weather was like in "grasshopper summer"?

Discussion Questions

(After reading Chapters 1-5)

1. Where did Sam live? Based on the clues in the first chapter, when do you think the story took place?

2. How are Sam and his brother Billy different? How do their activities at home and school show these differences? Do you think the brothers are alike in any way? Which brother do you think you would have liked better?

3. Why was Sam's father so upset about Sam playing in his army jacket? Describe the disagreements between Sam's father and grandfather. Do you tend to agree with one more than the other? Why?

4. Look in a history book to find out where Dakota Territory was. What do you think the land and climate were like there? How would Dakota Territory be different from Kentucky? If you were Sam and Billy, would you be excited about moving? Why or why not?

5. Make a list of the things the Whites took in the covered wagon with them. What were some of the things they left behind? How would leaving things behind make you feel?

(After reading Chapters 6-10)

6. What did the White family members think about former slaves and Indians? How can you tell that they did not all think the same way? Why do you think people who grew up in the same place and knew the same people might have different ideas about people?

7. Describe the scene at the Mississippi River. What made the crossing so difficult? How did the boys feel about the Mississippi afterwards? Who or what do you think was responsible for the colt's death?

8. The Whites met several other families in Missouri. Where were these families going? On a map, trace possible routes from Missouri to their destinations. Which family would probably have the hardest trip? Why?

9. How did Billy get left behind? How did the family members react? How would you have reacted in a similar situation?

10. Describe Dakota Territory or draw a picture of what you imagine it was like. Why do you think Sam said "Something about that wide blue sky, that endless grass, made us quiet"? Why do you think he couldn't stop shivering when he looked at the prairie?

(After reading Chapters 11-15)

11. How did Sam "make his mark" on Dakota Territory? Why do you think this was important to him? Are you surprised that he still thought of himself as a "Kentucky boy"? Why or why not?

12. What things were important to the Whites in deciding where to put their house? Draw a picture that shows what made the location they picked a good one.

13. Describe the work the family did to build their house. Why do you think Mr. White didn't want Mrs. White to help? Do you think she was right to help anyway?

14. How did the members of the family change the little house to make it seem like home? Can you develop a definition for *home* that accounts for their actions?

(After reading Chapters 16-20)

15. How did Sam find his friend Allan? How did Sam's attitude change after he found out Allan was living near him? Why do you think Sam "felt at home for the first time" on the Fourth of July?

16. Describe what happened when the grasshoppers came. How did the grasshoppers change the Whites' farm? How did the Whites react to the grasshoppers?

17. Why didn't Mr. White want to ask the boys' grandfather for money? Do you think he was right? Why or why not?

(After reading Chapters 21-25)

18. What actions did Allan and Sam decide to take? Do you think their letters were a good idea? What do you think the responses will be?

19. What did the White family decide to do? What did the Grant family decide to do? Why do you think they reached different decisions? What would you have done if you were making the decision whether to stay in Dakota Territory or leave?

20. What new experience did Sam have near the end of the book? Why hadn't he seen snow before? What did he think about as he walked around in the snow? Based on his thoughts, how do you think Sam had changed since moving to Dakota Territory? Were the changes positive or negative ones? Explain your answer.

Follow-up Activities

1. Have students construct a diorama showing how the Whites located their home to take advantage of positive aspects of the prairie and to minimize negative aspects.

2. Interested students could research the climate, wildlife, and other natural features of Kentucky and the Dakota Territory and make a chart showing the similarities and differences. In which setting would students rather live now? In the 19th century? Are the two answers different? Why or why not?

3. Ask students to imagine that they are government officials who receive Allan's letter. How would they respond?

4. Encourage students to find out more about the life cycle and habits of the grasshopper. How do scientists account for such large infestations as the one described in the book?

The Ups and Downs of Carl Davis III,
by Rosa Guy (New York: Delacorte Press, 1989).

Summary

Because his parents fear Carl Davis III will become involved with drugs in New York City, they send him south to live with his grandmother. As he struggles for acceptance in a school with few other black students, he learns that some problems reach beyond the city. By presenting the story in the form of letters from Carl to his parents and friends back in New York, the author is able to convey the 12-year-old's confusion, worries, and increasing maturity.

Initiating Activities

1. Show students the book's dust jacket and ask them whether they think it shows one of Carl Davis III's ups or downs. Allow them to share similar "down" experiences they have had.

2. Read the "To Whom It May Concern" statement on p. 1 of the book. Ask: What do you think this means? What form do you think the book is going to take?

Discussion Questions

(After reading Carl's letters of March 1, 3, 13, and 15)

1. What is happening in Carl Davis III's life? How does he feel about it? Are his feelings justified? Why or why not?

2. Find Carl's original home on a map. Pick a location in South Carolina where you think Spoonsboro might be. What do you think the differences between these two places would be? Why do you think Carl's parents sent him south?

3. Based on the evidence in his first four letters, do you think you would like Carl if he enrolled in your school? Why or why not?

(After reading Carl's letters of March 17, 19, and 21)

4. What problems is Carl having with his grandmother and Deacon Jones? How are their ideas different from his? What advice would you give Carl about dealing with the two older people?

5. What differences does Carl see between his new school and schools in New York? What similarities? Would a teacher like Miss Attaway be hired to teach in your school? Why or why not?

6. What does Carl mean by "Terrifying Energy"? Have you ever sensed something like "TE" when you were around older students?

(After reading Carl's letters of March 24 and 26 and April 1)

7. What point was Carl trying to make in his "wide-open discussion" of Columbus? Why do you think it made Reggie so angry?

8. Why doesn't Carl have anyone to play with near his grandmother's house? What friend does he tell his mother about? Why was this friend so important to him?

9. Why did Carl call Inge names? Did calling her names help him solve his problems? Did it make him feel better?

(After reading Carl's letters of April 3, 5, 8, and 16)

10. Why does Carl say that life is getting better in South Carolina? Do you think he means it? Why or why not?

11. According to Carl, how are Johan and Russell alike? What do you think Carl and Johan have in common?

12. What new activity does Carl have that he could not do in New York? What does he enjoy about it?

13. Why does Carl say that "the memory of black folks isn't any longer than that of whites"? Do you agree or disagree?

(After reading Carl's letters of April 18, 22, and 27 and May 3, 5, and 10)

14. What happened to Carl in the woods? Why was he so frightened? Would you expect something like this to happen in a small town?

15. Why did Carl suddenly start writing to his father? Why is he so angry with his mother? How are his letters to his friend Russell different from his letters to his parents? Are his letters to his mother different from his letters to his father?

16. What made going to town to do his grandmother's shopping such a great experience for Carl? Why does he say his friend Russell should have such an experience? Why didn't he have as good a time at the party his grandmother's friend gave for him?

17. What did Carl mean when he said that "The war which gave us our independence in 1776 merely set the stage for a continuing revolution"? Why didn't the teacher, other students, and principal seem to agree with him?

(After reading Carl's letters of May 12, 15, 16, and 24)

18. Why did Carl's parents send him to South Carolina? What are Carl's reasons for saying that they were wrong to do so? Do you think that the place in which a person lives can determine whether they decide to use drugs? Why or why not? Try to use the experiences of Russell and Johan to support your reasoning.

19. What does Carl mean by calling himself "the Vegetable"? What does he mean when he says Miss Attaway is a "most negatively perceptive teacher"? Why did Miss Attaway object

to Carl's paper on Malcolm X? Why do you think Reggie Owens had never heard of Malcolm X? Why do you think Inge wanted to know more about Malcolm X? Have you heard of Malcolm X? Do you think he was a man "significant in changing the course of human events"?

(After reading Carl's letters of May 28, 29, and 30)

20. What was Grandma's "one lesson in history"? Do you agree that students should learn about what makes a country great, as well as the reasons it isn't as great as it could be? Why or why not?

21. What did Johan mean when he said "I must go back home to disagree"? Are you afraid to disagree with people where you live? If not, do you think there are parts of the United States where you would be afraid to disagree?

(After reading Carl's letters in June)

22. What does Carl say in his letters early in June about staying in South Carolina? Are you surprised? Why does he change his mind by the end of the book?

23. What shocking thing does Carl learn about his father? How does Carl deal with this information?

24. What do the changes in the way Carl signs his letters tell you about how Carl has changed while living in South Carolina? How would you describe Carl as of June 30? If he had never left New York, what do you think he would be like on that date? How important do you think where people live is to the ways in which they mature?

Follow-up Activities

1. Conduct a class study of the things that parents worry about concerning their children. Each student should ask his/her parents or guardians to identify their top three worries, special features of the community where you live that contribute to those worries, and whether they think the worries they listed would be different if they lived somewhere else. If your community is a large city, students may want to ask about possible worries if they lived in a rural community in another region, and vice versa. Appoint a group of students to compile the class data and report on it to the class. Use the data as a basis for discussing the relationship between place and certain kinds of human problems; be sure students understand that while specific problems may be more common in certain areas, that does not mean that everyone who lives in the area has the problem or that people who live in other areas do not have the same problem.

2. Create a "Hall of Fame" in the classroom, beginning with the people mentioned by Carl's Grandma in her history lesson to Miss Attaway's class but expanding to include your students' own heroes and "sheroes." For each person included, students should prepare a poster with a picture of the person, a description of his/her work, and a map showing locations that were important in the person's life.

3. Encourage students to write Carl Davis III letters in which they describe their own community, school, and experiences similar to those Carl went through. In their letters, students should comment on whether they believe where they live influences the choices they make.

John Muir,
by Eden Force
(Englewood Cliffs, NJ: Silver Burdett Press,
1990).

Summary

John Muir's life as recounted in this biography provides an opportunity for students to see what one person can do when he has a love of place and is concerned about the relationship between people and the environment. Born in Scotland, John Muir emigrated to Wisconsin as a boy, but found his true home in the Yosemite Valley of California. Among his many accomplishments are serving as the first president of the Sierra Club and educating people around the world about the importance of preserving the wilderness.

Initiating Activities

1. Tell students they are going to be reading a biography in a series called *Pioneers in Change*. What do they think that phrase means? What are the implications of putting the words *pioneers* and *change* together? Show the cover of *John Muir* and ask students to speculate on what kinds of change Muir worked on. (They should be able to guess that his work had something to do with nature.)

2. Ask students to write down some of the characteristics that they think a person would need in order to be a "pioneer in change" related to nature. What would such a person need to know? What would they need to be able to do? What would be important to that person—that is, what would their values be? Encourage students to keep their lists to compare with what they learn as they read the biography of John Muir. They may want to add things to the list or delete items from it as they read.

Discussion Questions

(After reading Chapters 1 and 2)

1. What were the "scootchers" referred to in the title of Chapter 1? Give some examples from the book. Have you ever done a scootcher?

2. Describe John's home life as a young boy. How did living in Scotland affect the way he lived? How did his father's beliefs affect the way John lived? Do you think you would have enjoyed being a member of the Muir family? Why or why not?

3. Where did the family decide to move to? How did John feel about the move? What were the reasons for his happiness?

4. What different forms of transportation did the Muirs use in moving to their new home? Which section of the journey seemed most dangerous or difficult? the most fun?

5. What were the "wonderful glowing lessons" John learned in Wisconsin? Who taught these lessons?

6. Describe John's life after the move to Wisconsin. Do you think he still thought living in Wisconsin was better than living in Scotland? Why or why not?

(After reading Chapters 3-5)

7. Describe some of the inventions John took to the state fair in Wisconsin. Why do you think people were so fascinated with these inventions? What do you think John's inventions tell us about him?

8. How did the Civil War change Madison? How did it change the Southern towns that John traveled to? Can you think of other examples of events that would change a place? Can you think of any events that have changed your own town? If so, how?

9. Define *botany*. How did John live while he was studying botany in Canada? What kind of knowledge and skills would it take to live this way? Do you think you could live this way? Why or why not?

10. What evidence did John find to support his ideas about how Yosemite Valley was formed? Draw a picture to show some of John's ideas about glaciers. How did people react when John said that the ideas of a famous geologist were wrong?

(After reading Chapters 6-7)

11. What did Muir mean by his "other self"? What did his "other self" help him do?

12. What made Muir worry about the wilderness, especially Yosemite? What evidence did he see that Yosemite was "doomed to perish"?

13. Why did Muir begin writing about his experiences? What "moved" through John's writing? What effect did he hope his writing would have on other people? Do you think that it did? Why or why not?

14. Describe some of Muir's experiences in Alaska. How was working and studying in Alaska similar to his work in Yosemite? How was it different?

(After reading Chapters 8-10)

15. Why did Muir and his friends start the Sierra Club? What was its purpose? Do you know if the Sierra Club still exists today? If not, how could you find out?

16. In his book *The Mountains of California*, Muir wrote about "ecology," depicting nature as a "dynamic system." Define each of these terms. Is Muir's way of thinking about nature and humans' relationship to nature similar to or different from the way we think about nature today? Explain your answer.

17. Do you think Muir was right when he said that only Uncle Sam could save trees from fools? What evidence can you give to support your answer?

18. Describe the Hetch Hetchy controversy. What were the different positions on this controversy? Which side eventually won? Do you agree with the decision that was reached? Why or why not?

19. What were some of the last new places Muir visited in his life? Why do you think he wanted to go there? Why did his friends think he was crazy to do so? Would you want an older person that you cared for to go on such a trip? Why or why not?

20. Name some of the people influenced by John Muir's work. Do you think his influence on other people is an important aspect of his life? Why or why not?

Follow-up Activities

1. Have students prepare a world map that shows John's travels from boyhood until his death. They may want to use a special symbol or color to show places that John lived and another symbol or color to show places that he visited. They might also create special symbols to show what John studied or was looking for at particular locations.

2. Ask students to refer to their original lists of the characteristics they felt a person would need in order to be a "pioneer for change" related to nature. How would they change their lists after reading John Muir's biography? Encourage students to use their lists of characteristics to write a job description for a "pioneer for change."

3. Assign students to use the "Important Dates" at the back of the book as the basis for an illustrated timeline of John Muir's life. Each entry on the timeline should include an indication of the location in which it occurred, either through a small map showing the location or a picture showing what the place was like.

4. Interested students might research the work of the Sierra Club today. What kinds of materials do they produce? What kinds of projects do they undertake? What groups oppose their work? Why?

5. John Muir's work supported the creation of specific national parks and monuments but also encouraged the overall national parks program. Ask each student to select one national monument or park that is of interest to him/her and create a mobile that shows what makes the park or monument a special place. The locations of all the parks and monuments selected by students should be shown on a classroom map.

6. Have students create a "Jackdaw" for John Muir. A Jackdaw is a box containing artifacts and documents that represent the person and the culture in which he lived. The Jackdaw could be shared with other classes to give them a sense of John Muir's life and times.

Summary

Joan Lee's family is moving from Ohio to West Virginia to start a new business. Because the West Virginia of 1927 is unaccustomed to Chinese-Americans and not terribly accepting of "outsiders" of any kind, the Lees struggle to gain acceptance in the community. Caring friends, both adults and young people, help the family triumph over the town's backward views and their own differing opinions about how family members should act.

Initiating Activities

1. Write the words *star fisher* on the chalkboard and ask students to speculate on the possible meaning of the phrase. Accept all responses, posting them as well. Explain that students will be reading a book titled *The Star Fisher* and should keep their possible definitions in mind as they start reading.

2. Read the "Author's Preface" with students and discuss Yep's background. What did Yep mean by "West Virginia has always been more real to me than China"? Why would a person keep a file of his/her family's "escapades"? Does this seem like unusual behavior for the average person? for a writer? What outline for the story does the preface hint at? Does it provide any clues as to what the star fisher of the title is?

Discussion Questions

(After reading Chapters 1-3)

1. Describe the five members of the Lee family. What conflict between the parents and children becomes obvious very early in Chapter 1?

2. Imagine that you are with the Lee family at the railroad platform in Clarksburg. What would you see, smell, hear, and feel in those first moments in your new home town? Free write for three minutes, being as descriptive as you can. Now look at what you have written. Which words describe the natural features of the place, like the wildlife, landforms, and so on? Which words describe the human aspects of the place, such as the buildings, the people and their actions? Which words describe your imagined reactions to the place? Did the natural or human-made features of the place most affect your imagined reactions to it? Why?

3. Now imagine that you are with Joan and Emily having tea at Miss Bradshaw's house. Do the same free writing exercise you did above. How was your response to this place different from your response to the railroad platform? This time which affected your reactions most—natural or human-made features?

4. Describe the parents' and children's reactions to Miss Bradshaw and to the men who wrote a nasty message on the fence. Why do you think the parents and children reacted differently? What did father mean by "The nail that sticks out gets hammered"? Can you think of a proverb that means the same thing? Can you think of a proverb that means the opposite? Which view do you think most Americans take?

(After reading Chapters 4-7)

5. Try to draw a picture of the Lee's new home using the description in Chapter 4. How accurate do you think your drawing is? Explain your judgment.

6. What is a star fisher? What is the importance of the story of the star fisher?

7. Why did Joan and her mother argue the first day of school? Do you think one of them was more right than the other? Why?

8. Describe what happened to Joan at the store. Why was it such an embarrassing experience for her? Do you think Mr. Edgar was nice to her or do you think he made her feel worse? Why do you think their mother told the children to eat separately?

9. What happened at school to make Joan wish she still lived in Ohio? Why do you think she compared herself to the star fisher's daughter?

10. What stories about her early life did Miss Lucy tell Joan? Why do you think she told Joan those stories? How did the stories make Joan feel?

(After reading Chapters 8-10)

11. What caused the fight between Joan and her mother on the evening of the first day of school? Can you understand Joan's feelings? Her mother's? If you had been there, how might you have kept them from quarreling?

12. What did Joan discover about Bernice's family? Why did she decide not to tell her mother about Bernice's family? Do you agree that "holding back" the truth from her mother was all right? Why or why not?

13. When Joan thinks of moving to China, she again compares herself to the star fisher's daughter. What similarities does she see? Is there anywhere that really feels like "home" to Joan? How do you think that makes her feel?

(After reading Chapters 11-14)

14. Why was it important for the mother to take a pie to the social? Why didn't the children want her to? What ways of preventing her from doing so did they come up with? Can you think of others?

15. Describe the scene at the social. What roles did Bernice, the Reverend, and Miss Lucy play in making the evening a success? How did the evening change life in West Virginia for the Lees?

16. What worries does Mrs. Lee have about her children growing up in America? What did Joan mean when she said, "in our hearts we'll always be Chinese"?

17. What is the last comparison Joan makes between herself and the star fisher? How does this comparison show that Joan's thinking has changed since she first arrived in West Virginia?

Follow-up Activities

1. Have students draw paired illustrations showing the comparisons between Joan and the star fisher's daughter. Do their illustrations show why *The Star Fisher* was a good title for this book?

2. Encourage interested students to find out more about Chinese-Americans and their history. Students could construct a timeline map, showing patterns of Chinese immigration to the United States and within the United States.

Downriver,
by Will Hobbs (New York: Atheneum, 1991).

Summary

Seven teenagers enrolled in a wilderness education program not-so-affectionately known as "Hoods in the Woods," hijack the leader's equipment and attempt to raft the Colorado River without a map, a permit, or adult assistance. While some days on the river are idyllic, the teens also experience injuries and betrayal by two of their group. The narrator, a 15-year-old girl named Jessie, emerges from the experience with new friends, new confidence, respect for the wilderness, and a willingness to work out the problems with her family.

Initiating Activities

1. Ask students if they know what *downriver* means; if not, help them develop a definition. Read the dedication page to students. Based on the Roosevelt quotation and the book's title, what do students think the book is about?

2. Have students locate the source of the Colorado River on a U.S. map and describe its location as accurately as possible. With students, trace the river's route through the Grand Canyon to the mouth. Ask them to describe the location of the mouth as accurately as possible.

Discussion Questions

(After reading Chapters 1-4)

1. Where are the San Juan Mountains? What do you think the weather in October would be like in the San Juans? On what do you base your answer? What challenges would the weather pose in climbing a mountain, such as Storm King Peak?

2. What do you think Al meant when he said society lacks a ritual by which "young people can decisively achieve adulthood"? Can you think of any examples of such rituals, either in this culture or another? Do you think such rituals are important? Why or why not?

3. Describe what happened on Storm King. What caused the problems? Who do you think was responsible?

4. After the problems on Storm King, Freddy was friendlier toward Jessie. How did she respond? Why do you think she rejected his overtures? What feelings do you think she was having at the time?

5. Describe the eight teenagers on the "Discovery Unlimited" program. Try to predict some of the problems they may face as a group. Which teens do you think might be the cause of problems? Who do you think is most likely to "blow the whistle"? Which of the eight would you most like to be friends with? least like to be friends with? Why?

(After reading Chapters 5-9)

6. Trace the group's route from where they left Westwater Canyon to where the teens left Al. Describe the scenery they saw on the drive.

7. Describe what Troy did in the convenience store. Why do you think he did it? How might you explain the different reactions to his offer to buy everyone whatever they wanted? How would you react to such an offer?

8. Why do you think the teenagers decided to drive off without Al? If they had seen Al walking across the parking lot just as they were starting the van, do you think they still would have gone? Do you think they would have gone if one of the kids had made a strong argument against going? Why do you think they decided to raft down the Colorado River instead of going to one of the other places that was suggested?

9. Find Lee's Ferry, Arizona, on a map. How accurate was the group's estimate that they would travel 200 miles between there and Lake Mead? Do you think traveling 20 miles a day is reasonable? Why or why not?

10. Describe some of the problems the group had in their first two days on the river. What skills helped them solve these problems? What could they have done to avoid the problems?

(After reading Chapters 10-13)

11. Why do you think the water in the side canyon the group explored was so much warmer than the water in the Colorado?

12. What did Jessie learn about Freddie when they found the Hopi shrine? Does this information help you better understand Freddie and some of his actions? Explain your answer.

13. What did the mile-by-mile guide contain that kept Jessie up reading it at night? How did all the information make her feel? Why do you think Troy threw the guide away? If you had been with the group, what would you have done after Troy threw the guide away? Why do you think Jessie continued being friendly with Troy?

14. Why did Jessie tell Star things like she had "pudding in her head"? Do you agree with Jessie? How might you account for Star's behavior?

(After reading Chapters 14-17)

15. Describe the conflict over what to do at Crystal. Which side had the better arguments? Why couldn't the others be convinced?

16. What health problems did the group develop near the end of the book? Who or what do you think was responsible for these problems?

17. After Al picked up the kids, Rita looked at Lava Falls from the helicopter and said she still thought they could have "kicked its butt." Do you agree? Why or why not?

18. Imagine that it is four months later. What do you think Jessie and her friends will be doing?

(After reading Chapter 18)

19. How accurate were your predictions? How much of the change in group members' lives do you think was attributable to their experience on the river? Explain your answer.

Follow-up Activities

1. The book referred to several ways in which flashfloods affected the river. Ask students to draw pictures that illustrate how a flashflood could change the river in some way.

2. Encourage students to write a journal or diary of the trip on the river from the standpoint of a character other than Jessie. The journal should describe the character's reaction to events, as well as the physical surroundings.

3. If possible, obtain a mile-by-mile guide of the Colorado River and allow students to explore it, learning more about the awesome power of the river and the beauty of its sur-roundings. Interested students could create a large wall map of the river's course, showing the various side canyons and rapids that the group encountered.

4. Will Hobbs has experienced all of the adventures he includes in his books. Students could write to him in care of the publishing company to ask questions about rafting that the book prompted.

THEMATIC UNITS, CHILDREN'S LITERATURE, AND GEOGRAPHY

Introduction

The three units that follow are intended to serve as models for ways in which works of children's literature can serve as the centerpiece or the stimulant for interdisciplinary units that develop the five themes of geography. The units require from one to three weeks of class time to complete, blending literature/language arts and geography with other subject matter and skills.

The first unit, "Just a Dream: How People Change the Environment," is suitable for use with students in grades K-3. Centered around Chris Van Allsburg's book, *Just a Dream*, the unit focuses on the geographic theme of relationships within places, but also can be used to address objectives in science and art.

The second unit, designed for students in grades 3-5, is based on a photographic essay by Joan Anderson and George Ancona, *The American Family Farm*. The unit addresses all the geographic themes while developing students' research and reporting skills.

The work at the heart of the third unit is Holling Clancy Holling's *Paddle-to-the Sea*. This unit, appropriate for grades 4-6, also addresses all five geographic themes. It also engages students with ideas from history and economics and includes a major art project.

JUST A DREAM:
HOW PEOPLE CHANGE THE ENVIRONMENT

Unit Overview

Just a Dream, written and illustrated by Chris Van Allsburg (Boston, MA: Houghton Mifflin, 1990), is the story of Walter, a young boy who is insensitive to the environment until he dreams of a future with huge garbage dumps, gross air pollution, and a hotel on top of Mount Everest. The charming story provides a perfect vehicle for helping young students explore the ways in which people change the environment, as well as ways in which they can protect it. The unit begins with the story, branching out to a variety of related activities for examining people's relationship with the natural environment. The unit could serve as an interdisciplinary art, language arts, social studies, and science unit.

Objectives: At the end of this unit, students will be able to:

1. Describe ways in which humans change the environment, using examples from the story and from real life.

2. Describe ways in which humans can protect the environment, using examples from the story and from real life.

Grade Level: K-3

Time Required: 5-7 class periods

Materials: For **Parts 1** and **2** of the unit, you will need a copy of *Just a Dream*, as well as drawing materials, which will also be needed for **Part 4** of the unit. For **Part 3**, you will need copies of Handout 5, a scale, and a used grocery bag for each student. For **Part 4**, you will need an index card for each student, plus petroleum jelly or transparent tape; several magnifying lenses would also be useful. For the demonstrations, you will need a clear quart (or larger) jar and jar lid, water, matches, four to six strips of paper 4 inches x 1/2 inch, and an ovenproof glass bowl (clear or white will work best).

Procedure

Part 1 (1 class period):

1. Hold up your copy of *Just a Dream* and ask students when they have heard the title phrase. (It is usually used after a child has had a bad dream.) What does the title suggest that the book may be about? (Someone who has a bad dream)

2. Read the first six pages of the story with students, pausing to let them comment on Walter's actions.

3. Have students use their predictions about the book, combined with what they have learned about Walter in the first few pages of the story, to draw a picture showing what Walter might see or do in the future. Display the finished pictures around the classroom.

Part 2 (2 class periods):

1. Show students the first picture from Walter's dream. What does it show? (A huge pile of trash so high that it covers the bottom half of several houses) Where do students think the trash came from? Write "trash" on the chalkboard under the heading "Problems of the Future."

2. Next, read the explanation of the dump. What did Walter discover about the houses covered by trash? (They were on the street where he lived, but no one lived there anymore.) How did he feel?

3. Continue in the same manner, letting students speculate on the pictures of Walter's dream and then reading and discussing the accompanying text. Continue creating the list of "Problems of the Future" on the chalkboard. Discuss the colors that are most noticeable in Walter's dream. (Browns and yellows) What effect do these colors have on students when they look at the pictures? Stop reading before Walter wakes up.

4. Briefly let students comment on whether any of their predictions about Walter's dream were accurate. Did their pictures generally show a future that was better or worse than it appeared in Walter's dream?

5. Ask students to think about how Walter might feel after waking up. Have them write an ending for the book telling what Walter might do as a result of his dream. If time allows, let them share their endings in small groups.

6. Read the remainder of the book with students. How accurate were students' predictions about Walter's behavior after the dream? How was the future of Walter's second dream different from the future of the first dream? (There was no pollution; people were using machines that seemed old-fashioned because they did not pollute or use energy resources.) What colors did the author use in the pictures of this future? (All colors, including greens, blues, browns, and so on) What made the two futures so different? (People's actions)

7. Ask students to think about the following question: Do you think the future of our town will be more like Walter's first dream or his second one?

Part 3 (1-2 class periods):

1. Tell students they are going to conduct some investigations to learn more about how they can influence the kind of future their community will have. They are going to look at the problems of trash and air pollution. Their first assignment is to begin saving everything they would normally throw away in the next 24 hours. Give each student a used grocery bag and ask them to put their trash—at home, school, and any other activities—into the bag for one entire day.

2. Begin the following day by telling students that the average American throws away 4 pounds of garbage every day! Weigh each student's bag of trash and record the results. You may wish to expand on this activity by having students construct a human graph. To do this, designate three areas on the chalkboard. Students who threw away 0-3 pounds of trash should stand in one area, those who threw away 3-5 pounds of trash should stand in the second, and those who threw away more than 5 pounds should stand in the third.

Students can compare the lengths of the three lines of students to determine whether, based on one day's worth of trash, more of the students in their class are below average, average, or above average in trash production.

3. Explain to students what happens to trash in your community. Point out that many communities put their trash in landfills. Many of these landfills will be filled up in the next few years; some are already full.

4. With the class, brainstorm ways to reduce the amount of trash. Be sure the 3 R's—reducing the amount of trash generated, reusing items that would be thrown away, and recycling material—are mentioned. Which of these methods did Walter use after his dream? (Recycling, since he sorted the trash) Discuss reuse of trash items, using Handout 5.

5. Divide the class into groups of four or five students and ask one student in each group to spread his/her trash on the floor. (You may want to put down a piece of plastic for each group to spread their trash on.) Challenge the groups to see how much of the trash they could eliminate by using the 3 R's. Allow time for groups to report to the class.

Part 4 (2-3 class periods with several days after the first class period to collect pollution):

1. Tell students that their next investigation is going to be about air pollution. With the class, develop a simple definition of air pollution: soot, smoke, or dust in the air. Take a quick poll to find out whether students think air pollution is a problem in your community.

2. Explain that students are going to have a chance to see whether the air in your school or classroom contains any pollution. Give each student a 3 x 5 card. Students should write their names on their cards. They should then either apply a thick layer of petroleum jelly on the card or three strips of transparent tape across a hole cut in the card.

3. Have students tape their cards near doors, windows, heat ducts, on the floor, or on windowsills in the classroom. Allow the cards to collect air pollution for several days.

4. After several days, have students inspect their pollution cards, using magnifying lenses if available. Let students compare their own cards with those of classmates to find out who has the most and least pollution. Discuss what students observe, trying to determine reasons for differences in the amounts of pollution caught.

5. Ask students where they think the pollution comes from. (Possible answers include car exhaust, wood stoves, factories, fireplaces, and so on.) Explain that you are going to do some demonstrations to show where pollution comes from.

6. Pour a little water in the quart jar. Swirl the water around so all sides are wet. Pour out the water. Fold each strip of paper in half lengthwise. Hold all the strips together and light one end. Drop the papers and the match into the jar. Quickly put the lid on the jar. Have students observe the smog forming in the jar. Discuss the following questions:

- What happened in the jar? (Smoke filled the jar) What color is the smog? (Gray or white)

- Try to look through the smog. How does it change what you see? (It makes seeing things difficult.)

- How is our air like the air in the jar? (It has water vapor and smoke in it.) What did the smoke and water vapor make when they mixed in the jar? (Smog) What do you think happens when smoke and water vapor combine in the air around us? (They create smog.)

Let a student take the lid off the jar. Then ask:

- How does the smog smell? (Bad)

- What happened to the smog when it went out of the jar? (It went into the air in the classroom.)

7. Tell students you are going to show them another way that pollution is created. Light a match and hold it directly under the bowl. Move the match around so that soot is deposited over a wide area. Discuss what would happen to this soot if there wasn't a bowl above the fire. (It would go into the air.)

8. Discuss with students some of the ways that people can help prevent air pollution: by riding bicycles or walking instead of driving cars, by car pooling or taking public transportation, by building factories that have special equipment to clean the air that goes out their smokestacks, by burning clean fuels in fireplaces (natural gas instead of wood, for example), by not smoking, by preventing forest fires and other large fires.

9. As a culminating activity, ask each student to draw a poster that will convince people to adopt one of the ways of protecting the environment students learned about in this unit.

10. As a follow-up, students might enjoy reading *The Great Kapok Tree: A Tale of the Amazon Rain Forest*, written and illustrated by Lynne Cherry (New York: Gulliver Books, 1990).

REUSING HOUSEHOLD OBJECTS

Here are some ideas for using plastic milk bottles:

- Fill with sand and use as exercise weights.
- Store cereal in the bottles. It will stay fresh and pour easily.
- Use as a watering can for hanging plants.
- Cut off the top and set the bottle on the kitchen counter. Use for wet garbage or trash.
- Cut off the bottom. Use the top as a funnel for putting oil in the car.

Can you think of any ways to use cardboard tubes from wrapping paper, paper towels, or toilet paper?

Can you think of any ways to reuse coffee cans?

Can you think of any ways to reuse plastic pop bottles?

THE AMERICAN FAMILY FARM: COMPARING WAYS OF LIVING ON THE LAND

Unit Overview

Farming aptly demonstrates the geographic theme of the interrelationships between people and their environment, a fact well illustrated in Joan Anderson's text and George Ancona's photos in *The American Family Farm: A Photo Essay* (San Diego: Harcourt Brace Jovanovich, 1989). Ancona and Anderson, concerned about the demise of family farms, began the book as "a tribute to a dying way of life." In their profiles of three farm families, however, they show that while the business is a difficult one, it is one that many families will go to extraordinary lengths to maintain.

This unit built around the book begins with shared reading and discussion of the book. Students then work in small groups to answer questions they have about farming in the three regions represented in *The American Family Farm*. They present their research findings to the class and complete a written evaluation exercise.

Geographic Understandings: At the end of this unit, students will be able to:

1. Locate the MacMillan, Adams, and Rosmann family farms on a map of the United States.

2. Describe each of the farms, including information about natural features (e.g., landforms, soil, climate, growing season) and human features (e.g., farm buildings and equipment, types of crops grown), as well as the farm family's and their own responses to the place.

3. Give examples of the ways in which farm families depend on the environment, as well as examples of ways in which they change the environment.

4. Explain how products from the farms get to markets.

5. Identify the region of the United States in which each farm is located and describe how the farm is representative of farms in that region.

Grade Level: 3-5

Time Required: 8-9 class periods

Materials: You will need a colorful picture of a family farm and a wall map of the United States. Ideally, you should have three copies of the book, *The American Family Farm*. You will also need a variety of materials for student use in making presentations on their assigned families. Make arrangements with the librarian or media specialist to assist the

students during the two research periods in **Part 2** of the unit. You will need copies of Handout 6 for all students for the evaluation exercise.

Procedure

Part 1 (3 class periods):

1. Display the picture of a family farm and ask students to imagine that they were in the location shown in the picture. What would they be doing? What would they see, hear, taste, touch, smell, or feel? Ask students to free write for three minutes in response to these questions about the picture.

2. After the three minutes are up, have students share some of the words and phrases they wrote; post their responses in three columns on the chalkboard—one column for words related to the natural environment, one column for words related to the human characteristics of the place, and one column for emotional responses to the place. Focusing particularly on students' personal responses to the farm picture, encourage them to give reasons for why they responded as they did.

3. Read the quotation from Willie Nelson that appears on the back cover of the book: "If we lose the family farm, we lose more than a piece of land or individual growers of food. We lose a self-sufficient spirit that has made our country great. We just can't let that happen." Clarify with students what is meant by a "self-sufficient spirit" and ask for their thoughts on why a self-sufficient spirit has made our country great.

4. Have students focus on why we might lose the family farm, as indicated in Nelson's quotation. Show students *The American Family Farm* and read the "Introduction" with the class. Discuss such questions as: Are you surprised at the number of family farms going out of business? Do you think this is a "dying way of life"?

5. Anderson and Ancona point out that the three families they studied, even though they live far from each other, have a lot in common. Point out that there are differences, too. Using information from the "Introduction" and students' own knowledge, develop a class list of five to eight categories on which farms could be compared. These might include the following: location, climate/growing season, size of farm, type of soil, crops/livestock raised, farming techniques used, special issues/problems. Make a chart on the chalkboard in which the class can fill in information on the three farms.

6. Read the section of the book on the MacMillans aloud with the class, stopping to locate the farm on a U.S. map and to fill in sections of the chalkboard chart as appropriate. Mark the location of the farm in some semi-permanent way, such as with a pushpin or piece of self-adhesive notepaper, so that you can refer back to the map later in the unit. After you have read the selection, let the class discuss what they would and would not like about living on the MacMillan family farm. What questions do they still have about the farm? For example, they might wonder why people who do not know much about farming are trying to change regulations governing how farmers take care of their animals. Post their questions on the chalkboard.

7. Repeat the above procedure for the Adams and Rosmann families.

8. When you have completed the chalkboard chart for all three farms, conduct a large-group discussion of the similarities and differences among the three farms. Focus particular attention on the similarities in terms of dependence on the environment and the differences in terms of ways in which the families adapt to and change the environment.

Part 2 (3 class periods):

1. Draw students' attention to the map where they marked the locations of the three farms. Have them identify the regions of the United States in which the three farms are located (Northeast, Southeast, and Midwest). Do they think these farms are typical of other farms in their regions? Why or why not?

2. Divide the class into thirds, assigning each group one of the three regions identified in step 1. Each group's task is to conduct research on farming in the assigned region, determining in what ways the farm described in the book is typical of farms in the region and in what ways it is atypical. The groups may also want to address the unanswered questions that the class posted in discussing its assigned family. Groups will have two days to research farming in their region.

3. The third day of this part of the unit should be used to plan group presentations. The presentations **must** answer the question of how the farm covered in the book is and is not typical of other farms in the region, but it can also present other information students have gathered. Encourage students to be creative in planning their presentations. For example, they might want to create a mural, collage, or photo essay; present a skit; prepare a panel discussion or television news special; etc.

Part 3 (2-3 class periods):

1. Before groups begin their presentations, tell the class that each student will need to be able to answer questions about the other regions following the presentations, so they should listen carefully and take notes.

2. Have each group make its presentation, followed by questions from the other groups and general discussion.

3. Distribute Handout 6 and ask students to complete it as an evaluation exercise for the unit.

WRITING A LETTER TO THE EDITOR

Imagine that you are a member of the MacMillan, Adams, or Rosmann family. (You must choose one of the families you did not research in the small-group activity.) You have just read an article in a newsmagazine that says the family farm is out of date and cannot be successful.

Write a letter to the editor of the magazine. In the letter, explain why you believe in the family farm, as well as some of the problems you face as a family farmer. Your letter should answer the following questions about your farm:

- Where is it located?

- What is your farm like? Why do you like it?

- How does the environment influence the way you live? How does your family adapt to and change the environment?

- Who buys your products?

- How is your farm similar to and different from other farms in your region?

When you have finished your letter, exchange letters with a classmate. Give each other suggestions for improving the letters.

PADDLE TO THE SEA: A JOURNEY ON THE GREAT LAKES

Unit Overview:

Paddle-to-the-Sea, written and illustrated by Holling Clancy Holling (Boston: Houghton Mifflin, 1941), is one of the all-time best-sellers in children's literature. In the story, a Chippewa boy carves Paddle to resemble his father's canoe when it was loaded with packs and supplies for a long journey. The boy then places Paddle on a snowy hillside in the Nipigon region of Canada, just north of Lake Superior. The spring thaw carries Paddle down into Lake Superior, beginning a four-year journey through all of the Great Lakes, out the St. Lawrence, and into the Atlantic.

Through its story, its beautiful language, and its evocative illustrations, *Paddle-to-the-Sea* takes children out into the world. Unit activities focus on mapping, Great Lakes shipping, pollution, and the historic French and Indian trade. The major classroom activity is the creation of a Great Lakes collage.

Geographic Understanding: At the end of this unit, students will be able to:

1. Trace Paddle's route on a map of North America.

2. Describe several of the places Paddle visited on his journey.

3. Give examples from the story of how people depend on and destroy or change the environment.

4. Explain the importance of the Great Lakes as a transportation route for products from the region.

5. Describe how the Great Lakes region is similar to and different from the region in which they live.

Grade Level: 4-6

Time Required: 10-15 class periods

Materials: You will need the book *Paddle-to-the-Sea* each day of the unit. You will also need a wall map of North America and a base for constructing the collage (a piece of homosote is recommended but a large bulletin board or cardboard carton can be used), as well

Unit adapted from *Paddle-to-the-Sea*, developed by the Match Program of the Boston Children's Museum (Boston, MA: American Science and Engineering, 1974). Used by permission of the Boston Children's Museum.

as materials for constructing the collage; numerous ideas are presented in the **Procedure** section. Other materials you will need are listed below:

Part 2: Items representing a beaver pond, such as a beaver-cut log or twigs chewed by beavers.

Part 3: Waterways templates, enlarged to twice the size provided using a photocopying machine or the overhead projector; cardboard and construction paper to make the waterways for the collage; scissors.

Part 6: A collection of castaway items, such as a feather, a sandal, a grain of sand, a wooden slat, and a bottle.

Part 8: Multiple copies of the Possession Cards, cut apart for the groups. Enough copies of the role cards for one-third of the class to have each.

Part 9: Resource materials with articles on the Great Lakes and St. Lawrence Seaway today.

Procedure

Part 1 (1 class period):

1. Gather the children around you and read the first three chapters of the book aloud, allowing time for students to examine the pictures.

2. Encourage students to share their reactions to the book before focusing the discussion more specifically on dreams and what they mean to people. The book says a lot about dreams, on which some Indians placed great emphasis. Many Indians believed dreams were clearer views of life than one can normally attain. Dreams often serve as guides. The dreamer has a direction, an inner force that organizes actions, makes them meaningful, and gives them impetus.

3. Tell the class about the unit you are about to undertake, explaining that every day you will read further in the story. With each reading there will be some special things for the class to do. Use a wall map of North America to locate the starting point of Paddle's journey, and allow students to speculate on where he will end up.

Part 2 (1 class period):

1. Pass around a beaver-cut log or other items you have found to represent the beaver pond. Caution them that the quills are very sharp. Explain that these things are found in the next chapter and invite the students to imagine from these clues what's going to happen. Don't tell them what the objects are if they don't guess.

2. Read Chapter 4 and allow some talk about how the predictions compared to the story. Read Chapter 5 to the students.

3. Ask students if they know what a lumberjack is. (Someone who cuts trees for use by people) What are some of the ways in which people use wood? Allow time for students to name some of the uses. Then ask each student to compile a list of all the products or materials made from wood or wood fiber that they have used in the last 24 hours.

4. Compare students' lists. What are some items that all students used? What were some unusual items made of wood? Point out that one of the common wood products used in schools is paper. On average, every American uses 600 pounds of paper per year.

5. Ask students to consider what may happen to forests if people continue using a large number of wood products. (The forest could be lost.) How can people prevent the loss of forests? (Be sure that lumber companies replant and do not overharvest, prevent forest fires, recycle forest products, etc.)

6. Return students' attention to the story. What do they think will happen to Paddle now that he is in the river with the lumber? Allow time for them to made predictions.

7. Read Chapters 6 and 7. Lead a discussion about what the students thought was going to happen and how the students feel about the lumberjack who rescued Paddle.

Part 3 (2-3 class periods):

1. Read Chapters 8, 9, and 10 and allow time for discussion.

2. Explain the idea of the collage and show students how it will be set up. This can be done in a number of workable ways, each with its own characteristics. Generally speaking, a 4' x 8' sheet of homosote resting on boxes or on the floor is best. The children can work around all sides of it; it readily takes paint, tacks, tape, nails, and staples; things can be built up on it; and it can easily be moved. A big sheet of heavy cardboard (such as from a refrigerator carton) or a piece of plywood on the floor or on a table also work well. Vertical surfaces, such as bulletin boards, are not recommended but are acceptable; however, gravity limits thc things you can attach.

3. Divide the class into six groups of approximately equal size. Assign one group to each of the Great Lakes and one group to the St. Lawrence Seaway. Set the groups to work creating the waterways using the templates provided (you may wish to enlarge them on a photocopying machine or using an overhead projector). Thc crucial task is positioning them on the map surface in proper relation to each other. Make sure a map of the area is available for reference and check the layout before the lakes are drawn in finally. The St. Marys, St. Clair, Detroit, and Niagara Rivers can be drawn in free-hand. Ask students to find the origins of the place names.

Look at the illustration in *Paddle-to-the-Sea* on the lower left of Chapter 2. The Great Lakes really do pour into one another, and showing the relative elevations of the lakes does add to the understanding of canals. It's a lot of work, but we recommend that the lakes be mounted onto the collage to illustrate elevations. If you intend to show the elevation, it is helpful to know each lake's approximate elevation above sea level. You can simulate these differences by using each lake template to cut additional layers of backing material and then building up various thicknesses.

Lake	Elevation (feet above sea level)	Thicknesses
Superior	600	2 corrugated cardboard and 6 construction paper
Michigan	578	2 cardboard and 3 paper
Huron	578	2 cardboard and 3 paper
Erie	570	2 cardboard and 1 paper
Ontario	245	1 cardboard and 1 paper

4. Once the lakes have been placed on the base, students can begin adding other information, such as pictures of Paddle's adventures, things he saw, people or animals he met, and so on. The pictures can be attached near where the adventure took place, with a string or line connecting the picture to the exact location. A lot can be done with the collage, and it's important for students to see its potential. Remember, though, that everything doesn't have to be done at once. The collage can be added to in regular sessions or in free time. All contributions are welcome, but if there are many, they should be small or be put on top of each other. By superimposing and intermingling all kinds of subjects in an array of forms, the collage models the real world as we encounter it; it reflects the complexity and richness of experience. The more thickly the students encrust the collage with honest and understood things, the more Paddle's journey and the Great Lakes will mean to them.

Ideas for things to include in the collage will occur to you and the students as you work on it, but here is a list of things to start with:

- Paddle's own path with place names, distances traveled, length of time underway, season of the year, etc.
- A map showing many Indian nations.
- American state outlines and Canadian provinces.
- location of your school relative to the lakes.
- topographic features (plains, mountains) and vegetation (forest, grasslands).
- other waterways.
- lore associated with various places.
- children's drawings related to the story.
- areas where different people lived and traveled (Indian nations, voyageurs, French, English, etc.), accompanied by dates.
- objects such as ore samples, grain, trinkets, leaves, shells, cars, boats, etc.
- people and places meaningful to the children: where relatives live, where famous people were born, etc.
- trips the students have taken in this region.
- routes of explorers with dates.
- shipwreck sites with dates and other statistics.
- Coast Guard stations and lighthouses.
- cities and landmarks.
- locks.
- animals and sea life.
- lake depths.

Part 4 (2 class periods):

1. Ask students what they would think if they saw a lake or river that was red. What could make the water this color? Allow them to speculate on reasons for the water to be red. Then explain that Paddle will experience red water in today's episode.

2. Read Chapters 11, 12, and 13 to the class, being sure they recognize that iron ore made the water red.

3. Take students outside to show them how enormous the freighters are. Have the children form the outline of a freighter out in the school yard or in a nearby park or along the street. The outline should be about 730 feet long by 65 feet wide. Pacing off 730 feet with approximate 3-foot strides is accurate enough.

Doing this on a straight street should be fairly easy: the children forming the sides can stand along both sidewalks; if it is a closed street, a few may stand in the street for the bow and stern sections. Another teacher can watch for traffic on one end while you handle the other. Use some children to form two lines across the ship to mark off the center cargo hold from the fore and aft cabin areas.

When everyone is in position, have them all wave their hands or, better yet, have them wave flags that they have made in advance.

The idea of all this is simply to make the great ships of the lakes as real for the students as possible.

5. Allow time for students to work on the collage.

Part 5 (2 class periods):

1. Begin by reading Chapters 14, 15, 16, and 17.

2. The crew at the Coast Guard station estimated that Paddle had gone between 700 and 2000 miles depending on the winds he'd encountered. On the collage, complete Paddle's path as far as Whitefish Bay lighthouse. It should look something like the picture at the bottom of p. 13 of the book. Have each group try to compute the distance Paddle has traveled and compare their computations.

Stress that the distance cannot be measured exactly, just as Paddle's course cannot be charted exactly.

3. Use the diagram with Chapter 17 to discuss how a lock works. Note that Great Lakes canal locks are not filled by pumping. Water flows by gravity through valves from the upper lake into the lock. When the lock is emptied, lower valves are opened and water drains into the lower lake.

4. Allow time for students to add to the collage.

Part 6 (1 class period):

1. Read Chapters 18 and 19.

2. Spread out a collection of castaway items on the floor. These might include a feather, a sandal, a grain of sand, a wooden slat, a bottle. Gather the class to see the items, pick them up, and talk about the stories these items might have to tell—where they have been, how they were made, etc.

3. Have each student write a story for one of the objects, telling of its journey and how it came to be cast up on the shore. The idea is to remind students that items found on beaches are not "just there"; they came from somewhere. Encourage students to let their imaginations run free.

Part 7 (1-2 class periods):

1. Read Chapters 20, 21, 22, and 23 and allow time for general discussion. Challenge students to think about the following questions, which they will be exploring in this lesson:

The father sent Paddle on his way because "Somewhere, someone who had faith in currents, in winds, and also people put thought and careful work into this carving. And I'll not be the one to stop his Paddle-to-the-Sea." What did he mean by that? Why did other people help Paddle on his voyage?

2. Tell students they will try to find answers to these questions by acting out some of the situations in which people had to make a choice about Paddle. Keep the children in their same groups—they should be communicating well with each other.

3. Let the students begin with the situations described in the book and then go on to invent new ones involving other characters, places, and circumstances. Make sure they understand, though, that each situation has to have an essential tension, or conflict of interest between the characters, to reveal values and choices.

The museum man situation is ready-made. Three children can play the father, museum man, and daughter, enacting the scene in which the museum man offers to buy Paddle. The two main roles are clear, but the daughter can be played in various ways, each with its own flavor:

"Whatever Daddy says is all right with me."

"I don't care what either of you say; I want to keep it."

"Oh, Daddy, sell it to him and give me the money."

Two or three "casts" can come up to enact this scene. In general, each will offer an interesting variation. The father's comment about faith in currents and winds and people can, of course, be incorporated. Initially, you may have to help the characters to establish their positions."

The lumberjack made a decision, too. First he was going to take Paddle home to his son, Henri. Then when he read Paddle's message, he changed his mind. Two children can enact the talk the lumberjack must have had with himself: one "pro-Henri" and one "anti-Henri."

Other situations are only alluded to in the story. We know the places where Paddle was picked up and the kinds of people who found him. From this information, let the children create for themselves situations that might have taken place. Small groups can huddle to plan their skits and then present them.

The key is to find a good reason for *not* sending Paddle on and to build a situation around that. If the children have trouble thinking of situations, these can be used to get them going:

- Port Colborne. A sailor finds Paddle and plans to put him back, but a friend says it would be better to keep Paddle because he'll never make it through the Welland Canal locks or survive the Niagara Falls.

- A railroad engineer in Cleveland debates with himself whether to take Paddle to the ocean by train or let him keep going by water.

- Ashtabula. Four Cub Scouts find Paddle. The boy who found him wants to show him to his father and then send him on. Another claims Paddle belongs to the Scout den because a member found him. The third boy says the boy who found Paddle should give him to the den because many of its activities involve learning about Indian culture. A fourth boy says he only belongs to whoever made him and that he should be put back right away.

- A mechanic in Erie who carves ships as a hobby admires Paddle and is tempted to keep him with his collection of fine carvings.

4. When the skits are over, round up the thinking on the original question: What was it about Paddle and the people he met that kept him on his journey? The idea of this lesson is to make the children see that, with respect to Paddle, people made choices that reflected what they believed in. Students, too, should think about what they value when they make decisions.

Part 8 (2 class periods):

1. Read Chapter 24. Tell the class about some of the items over the mantle in the picture. You don't have to go into detail: see what interests the children. Make the point, however, that before the old lady got these items, they were probably traded many times, most recently, of course, for money. The things the old lady has collected make a collage of the Great Lakes, much like the one the children have been making. In their variety, the objects stand for the clutter of people and events that have gone before.

Looking from left to right, the items include:

- Canoe paddle.

- Ball-headed club.

- Rapier.

- Tomahawk.

- Rifle and powder horn.

- Beaded pouch—beadwork and fringed leather, probably made by the Iroquois.

- Birchbark basket—decorated with Huron designs.

- Ceremonial pipe—the Plains Indians carved out of red stone.

- Beaded Chippewa mocassin.

- Knife in porcupine quill sheath—quills were dyed, flattened, and stitched into intricate designs.

- Bear trap.

- Arrows in quiver.

- Plains Indian spear—with a metal point held on by thongs and a feather.

2. Conduct a bartering session. Take half of the class and divide it into three groups—Chippew Indians, Huron Indians, and Frenchmen. Give each group its possessions and appropriate role cards, and allow them time to look these things over. Have one member of each group read the role card aloud so all positions are clear.

3. Explain the ground rules:

The Chippewas and the French are too widely separated to trade with each other, so both groups must trade with the Hurons, who live between them. The Hurons should trade first with the French and then with the Chippewas, after which they can go back and forth. Of course each group should try to make the best deals it can.

4. Let the trading begin. When they have finished, have the groups note what they wound up with. Ask them how they feel about the trades they made. Are they satisfied? Could they have done better?

Return all the items and let them try again. How did they fare this time?

5. Now let the other half of the class have two bartering sessions. They will already know how to proceed and will probably drive harder bargains.

6. Have the class as a whole look at the record of transactions and talk about what happened. Use the following questions to stimulate discussion:

- What changes took place between first and second bartering sessions or across all four?
- Why do people trade things anyway? Why don't they do it much anymore—or do they?
- Were the Indians treated fairly? What does *fair* mean to you?
- What risks were there for the French?
- How is the worth of a thing determined?
- What is a dollar worth? What is a person worth? What is a dream worth?

The idea, quite obviously, is not to recreate a specific event in history, but rather to characterize a relationship in which the children can glimpse the people who went before, how they lived, and what they valued.

Part 9 (2 class periods):

1. Read Chapters 25, 26, and 27. Talk about the story, gathering students' thoughts with such questions as:

- How must the Chippewa boy have felt at the end?
- What did Holling mean when he wrote, "For four years he had been what he was supposed to be, a Paddle-to-the-Sea. And he had done what he was supposed to do. And so he showed no surprise, even at crossing the ocean"?
- What has the story meant to you?

2. Remind students that *Paddle-to-the-Sea* was written more than 50 years ago. How do they think Paddle's journey would be different today? What might he see that he didn't see 50 years ago? What might be gone? What new problems or adventures would he have? What would be the same? Allow time for general discussion of these questions.

3. Ask each group to find several articles on their assigned body of water that give some ideas about what Paddle's journey would be like today. Allow time for the groups to share their findings with the class. Try to reach some general conclusions about the changes that Paddle would see if his journey occurred today.

4. Work on the collage can continue as long as there is interest. When it's finished, take the class on a "grand tour." Perhaps the collage could be displayed so others in the school can examine it and marvel at its complexity.

Lake Superior

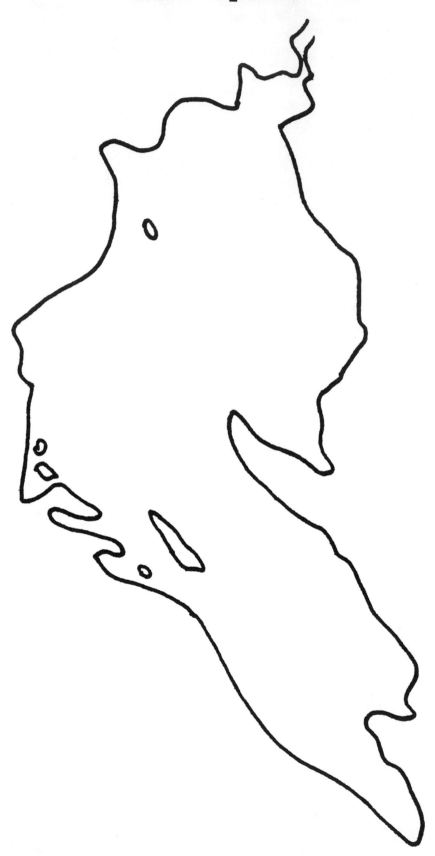

Lake Michigan

Lake Huron

Lake Erie

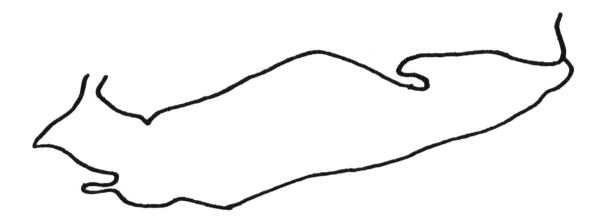

Lake Ontario

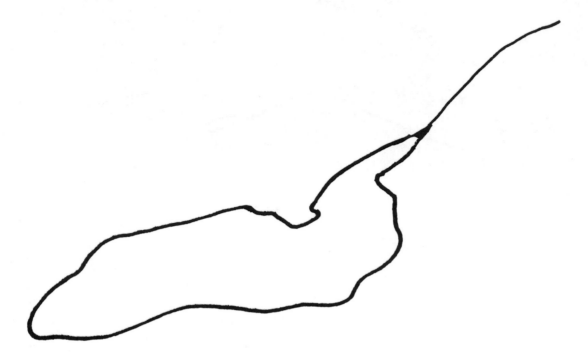

St. Lawrence Seaway

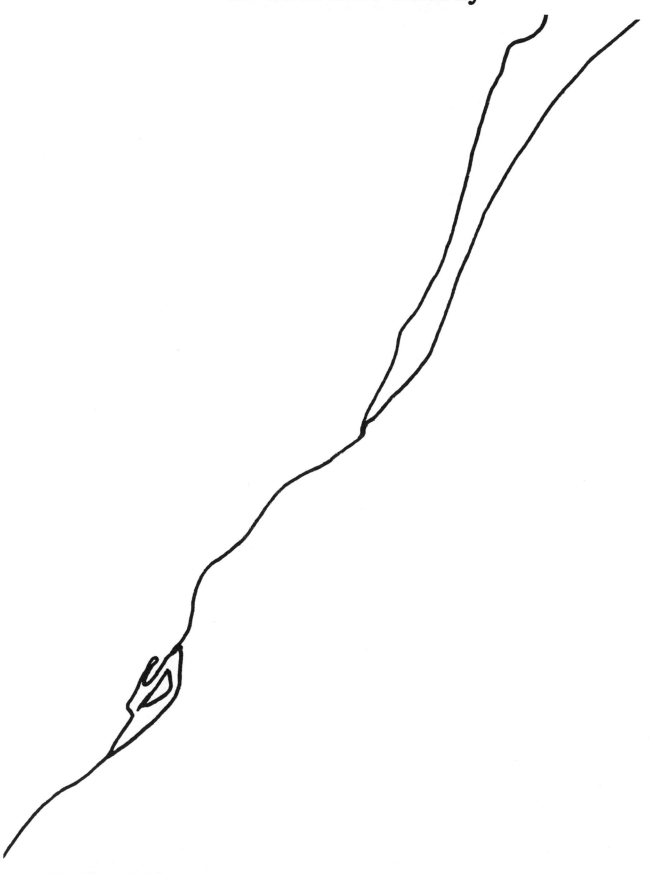

Chippewa

mini-pelts	basket
club	canoes

Huron

corn	mini-pelts
sunflower seeds	bear-claw necklace
wampum necklace	

French

trade cloth	brass beads
ribbon	seed beads
mirrors	mocotaugen (knife)

Hurons

The first time that the French traders came to our village we were frightened. Their ship was so large that we thought it was strange. But the Frenchmen brought us many things: a metal knife, a kettle, and many tiny colored beads. And in exchange they wanted beaver pelts. We wondered if the Frenchmen would come back.

Their ship will hold many men. The Frenchmen told us that there are more people in France than there are stars in the sky—and they all want beaver to wear on their heads.

It is good that our Chippewa neighbors will come soon with many beaver skins in their canoes. Beaver no longer come to our traps because we have trapped too many for the European traders.

We can get many strings of wampum from our neighbors who live near the ocean. We have so much wampum that maybe we can trade some to the Chippewa for their beaver skins.

Our women have many baskets filled with corn and sunflower seeds to trade for still more beaver.

And if we do not get beaver skins from the Chippewa, the Frenchmen will not want to trade with us. The coming of the Europeans has changed our lifestyle. Trading is now necessary because game is scarce. Before, game provided food, clothing, and tools.

Items we will trade for:

- cloth
- ribbon
- seed beads
- brass beads
- tomahawk pipe
- mirrors

Frenchmen

Now that our ship has arrived safely from France, we can travel to the land of the Huron Indians, who will trade beaver skins for the cargo of our ship.

Because all the beavers in Europe have been trapped, there is a great demand for beaver skins. When we left France, all the fashionable aristocrats wanted beaver hats and fur trimming on their clothes. The Indians can trap more beaver than our ships can hold.

We brought many colorful things to trade for beaver skins. There are so many guilds and factories in Europe that we were able to purchase a large supply of trade items for little money.

This time we should take home some baskets and other items made by the Indians.

We will need a canoe so we won't have to carry the furs on our backs like the last time. The canoe trip will be dangerous for us because we are not experienced at traveling by canoe. Of course, the return to France with a ship load of beaver skins will make us very rich. That is, if everything goes well.

Items we will trade for:

- beavers
- baskets
- clubs
- canoes

Chippewas

Many beaver have come to my traps. By trading with the Huron, we will have enough food for the winter.

Now it is time to pack our canoes and travel to the land of the Huron. Because our women are busy scraping fur skins, they do not have time to grow much corn. Our fields are small because we live in the forest. The Huron women grow large fields of corn and sunflowers.

The Huron nation has things that came from the French traders who want many beaver skins. Our woods are so full of beaver that it is easy to trap them. We do not understand how the Frenchmen could wear so many beaver. The Huron will trade many strings of wampum for our beaver skins.

The Europeans have not come to our lands, but we are changing our ways of living out of necessity to survive.

Items we will trade for:

- seed beads
- mirrors
- sunflower seeds
- cloth
- ribbon
- brass beads
- corn
- wampum
- mocotaugen (knife)

Recently Published Folktales, Myths, Legends, Fables, and Stories
(Organized by Continent or Region of Origin)

Northern America

The Adventures of High John The Conqueror, by Steve Sanfield, illustrated by John Ward (New York: Orchard, 1990). African-American; intermediate.

Dream Wolf, by Paul Goble, illustrated by the author (New York: Bradbury, 1989). Native American; primary and intermediate.

Earthmaker's Tales: North American Indian Stories about Earth Happenings, retold by Gretchen Will Mayo, illustrated by the reteller (New York: Walker, 1990). Native American; intermediate.

How Glooskap Outwits the Ice Giants: And Other Tales of the Maritime Indians, retold by Howard Norman, illustrated by Michael McCurdy (Boston: Joy Street, 1990). Native Americans of Nova Scotia and New England; intermediate.

Jump on Over! The Adventures of Brer Rabbit and His Family, by Joel Chandler Harris, adapted by Van Dyke Parks, illustrated by Barry Moser (New York: Harcourt Brace Jovanovich, 1990). African-American; primary and intermediate.

The Rainbow People, by Laurence Yep, illustrated by Patrick Benson (New York: Holt, 1990). Chinese-American; intermediate.

Raven's Light: A Myth from the People of the Northwest Coast, by Susan Hand Shetterly, illustrated by Robert Shetterly (New York: Atheneum, 1991). Native American; primary and intermediate.

Raw Head, Bloody Bones: African-American Tales of the Supernatural, selected by Mary E. Lyons (New York: Scribners, 1991). African-American; intermediate.

The Sea Lion, by Ken Kesey, illustrated by Neil Waldman (New York: Viking, 1991). Native American; primary and intermediate.

The Talking Egg: A Folktale from the American South, retold by Robert D. San Souci, illustrated by Jerry Pinkney (New York: Dial, 1990). Creole; primary and intermediate.

The Three Brothers: A German Folktale, adapted by Carolyn Croll, illustrated by the adaptor (New York: Putnam, 1991). German-American; primary.

The Three Little Pigs and the Fox, by William H. Hooks, illustrated by S.D.Schindler (New York: Macmillan, 1990). Appalachia; primary.

Latin America

All of You Was Singing, by Richard Lewis, illustrated by Ed Young (New York: Atheneum, 1991). Aztec; primary and intermediate.

Borreguita and the Coyote: A Tale from Ayulta Mexico, by Verna Aardema, illustrated by Petra Mathers (New York: Knopf, 1991). Mexico; primary.

Feathers Like a Rainbow: An Amazon Indian Tale, by Flora, illustrated by the author (New York: Harper, 1990). Amazon River Basin; primary.

Llama and the Great Flood: A Folktale from Peru, by Ellen Alexander, illustrated by the author (New York: Crowell, 1990). Peru; primary and intermediate.

Songs of the Chirimia, retold and illustrated by Jane Anne Volkmer, translated into Spanish by Lori Ann Schatschneider (Minneapolis: Carolrhoda Books, 1989). Guatamala; primary and intermediate.

Why There Is No Arguing in Heaven: A Mayan Myth, by Deborah Nourse Lattimore, illustrated by the author (New York: Harper, 1990). Mayan; primary and intermediate.

Europe

The Arrow and the Lamp: The Story of Psyche, retold by Margaret Hodges, illustrated by Donna Diamond (Boston: Little, Brown, 1990). Greece; intermediate.

Buried Moon, retold by Margaret Hodges, illustrated by Jamichael Henterly (Boston: Little, Brown, 1989). England; primary.

Elfywyn's Saga, by David Wisniewski, illustrated by the author (New York: Lothrop, 1989). Iceland; primary and intermediate.

Hershel and the Hanukkah Goblins, by Eric Kimmell, illustrated by Trina Schart Hyman (New York: Holiday House, 1990). Ukraine; primary.

The Little Snowgirl: An Old Russian Tale, adapted and illustrated by Carolyn Croll (New York: Putnam, 1990). Russia; primary.

Oom Razoom or Go I Know Not Where, Bring Back I Know Not What, by Diane Wolkstein, illustrated by Dennis McDermott (New York: Morrow, 1991). Russia; intermediate.

Tam Lin, retold by Susan Cooper, illustrated by Warwick Hutton (New York: McElderry, 1991). Scotland; primary.

Tony's Bread: An Italian Folktale, by Tomie dePaola, illustrated by the author (New York: Putnam, 1990). Italy; primary.

Middle East/North Africa

The Diamond Tree: Jewish Tales from Around the World, retold by Howard Schwartz and Barbara Rush, illustrated by Uri Shulevitz (New York: HarperCollins, 1991). Turkey, Iraq, and other nations; intermediate.

The Egyptian Cinderella, by Shirley Climo, illustrated by Ruth Heller (New York: Crowell, 1990). Egypt; primary.

Fortune, by Diane Stanley, illustrated by the author (New York: Morrow, 1989). Persia (Iran); primary and intermediate.

Subsaharan Africa

Awful Aardvark, by Mwalimu, illustrated by Adrienne Kennaway (Boston: Little, Brown, 1990). Africa; primary.

How the Guinea Fowl Got Her Spots: A Swahili Tale of Friendship, retold and illustrated by Barbara Knutson (Minneapolis: Carolrhoda, 1989). East Africa; primary.

How Many Spots Does a Lepard Have? And Other Tales, by Julius Lester, illustrated by David Shannon (New York: Scholastic, 1990). Africa; intermediate.

The Orphan Boy, by Tololwa M. Mollel, illustrated by Paul Morin (New York: Clarion, 1991). Maasai; primary.

Rabbit Makes a Monkey a Lion: A Swahili Tale, retold by Verna Aardema, illustrated by Jerry Pinkney (New York: Dial, 1990). Africa; primary.

Tower to Heaven, retold by Ruby Dee, illustrated by Jennifer Bent (New York: Holt, 1991). Africa; primary.

Traveling to Tonda: A Tale of the Nkundo of Zaire, retold by Verna Aardema, illustrated by Will Hillenbrand (New York: Knopf, 1991). Zaire; primary.

Asia

The Bird Who Was an Elephant, by Aleph Kamal, illustrated by Frane Lessac (New York: HarperCollins, 1989). India; primary and intermediate.

The Dwarf Giant, by Anita Lobel, illustrated by the author (New York: Holiday House, 1991). Japan; primary.

The Empty Pot, by Demi, illustrated by the author (New York: Holt). China; primary.

Grandfather Tang's Story, by Ann Tompert, illustrated by Robert Andrew Parker (New York: Crown, 1989). China; primary.

The Greatest of All, by Eric A. Kimmel, illustrated by Gloria Carmi (New York: Holiday House, 1991). Japan, primary.

A Letter to the King, by Leong Va, illustrated by the author, translated by James Anderson (New York: HarperCollins, 1991). China; primary.

Lon Po Po: A Red-Riding Hood Story from China, translated and illustrated by Ed Young (New York: Philomel, 1990). China; primary.

The Seven Chinese Brothers, by Margaret Mahy, illustrated by Jean Tseng and Mou-sien Tseng (New York: Scholastic, 1989). China; primary.

The Shining Princess and Other Japanese Legends, retold by Eric Quayle, illustrated by Michael Foreman (Berkeley, CA: Arcade, 1990). Japan; intermediate.

Three Strong Women, by Claus Stamm, illustrated by Jean Tseng and Mou-sien Tseng (New York: Viking, 1989). Japan; primary.

Australia and Oceania

Land of the Long White Cloud: Maori Myths, Tales and Legends, by Tiri Te Kanawa, illustrated by Michael Foreman (Berkeley, CA: Arcade, 1989). New Zealand; intermediate.

Index of Book Titles

Seeking Diversity

Seeking Diversity

Language Arts
with
Adolescents

Linda Rief

With a Foreword by Nancie Atwell

HEINEMANN
Portsmouth, NH

Heinemann Educational Books, Inc.
361 Hanover Street   Portsmouth, NH 03801-3959
Offices and agents throughout the world

Every effort has been made to ensure that copyright holders and all other writers—including teachers and students—have given their permission to reprint borrowed material. We regret any oversights that may have occurred and would be happy to rectify them in future printings of this book.

Page vii: Demott, Benjamin. 1990. "Why We Read and Write." *Educational Leadership*. (March) 47 (6), 6.

Page 9: From *One Child* by Torey L. Hayden. Copyright 1980 by Torey L. Hayden. Published by Avon Books. Used by permission of The Putnam Group, Inc., New York.

Page 12: Farley Mowat letter reprinted by permission of the author.

Page 67: Reprinted by permission granted by Ann Landers and Creators Syndicate.

Page 67: "The little boy and the old man" from *A Light in the Attic* by Shel Silverstein. Copyright 1981 by Shel Silverstein. Used by permission of HarperCollins.

Page 90: From *A Girl From Yamhill* by Beverly Cleary. Copyright 1988 by Beverly Cleary, William Morrow Co. Used by permission of William Morrow Co., Inc./Publisher, New York.

Chapter 8 first appeared in a shorter version in *Educational Leadership*, Vol. 47, No. 6, March 1990.

Library of Congress Cataloging-in-Publication Data

Rief, Linda.
    Seeking diversity : language arts with adolescents / Linda Rief : with a foreword by Nancie Atwell.
        p.   cm.
    Includes index.
    ISBN 0-435-08724-X. — ISBN 0-435-08598-0 (pbk.)
    1. Language arts (Secondary)—United States—Case studies. I. Title.
    LB1631.R53   1992
    428'.0071'2—dc20   91-34800
                                                              CIP

Cover photo of Mandy Tappan by James Whitney

Designed by Maria Szmauz

Printed in the United States of America

92 93 94 95 96   10 9 8 7 6 5 4 3 2

For
Donald H. Graves
and
Donald M. Murray
teachers, writers, colleagues, friends
who asked me to tell them more
and trusted me to be able to do it

**We should be seeking diversity,
not proficient mediocrity.**

Donald M. Murray

We do not write and read primarily in order to ensure that this nation's employers can count on a competent, competitive work force. We write and read in order to know the human world, and to strengthen the habit of truth-telling in our midst.

Benjamin DeMott

Contents

Foreword

In *Turning Points: Preparing American Youth for the 21st Century*, the Task Force of the Carnegie Council on Adolescent Development notes that for many students, early adolescence "represents their last best chance to avoid a diminished future" (8).* Middle schools and junior highs are described as "potentially society's most powerful force to recapture millions of youth adrift and help every young person thrive during early adolescence."

"Yet all too often," the report continues, "these schools exacerbate the problems of young adolescents. A volatile mismatch exists between the organization and curriculum of middle grade schools and the intellectual and emotional needs of young adolescents."

In Linda Rief's classroom at Oyster River Middle School the intellectual and emotional needs of her eighth grade students are at the crux of the language arts curriculum. These adolescents thrive—as readers and writers and thinkers, as citizens, as individuals, and as members of an extraordinary classroom community. Every September Linda creates a context for students to engage with her in acts of genuine literacy, and they seldom disappoint her. *Seeking Diversity* is the blueprint for this creation, a book about how one teacher draws on her experiences as a reader and a writer and her understanding of adolescence to build a literate future for her students.

I wonder if any junior high English teacher has taken his or her students more seriously than Linda Rief or, in turn, been taken more seriously by kids. Her own literacy is the key to theirs. She writes with them and for them, reads with them and to them, and speaks her mind always about what she thinks and knows. Because they see her as a reader and writer, students pay attention. In a self-evaluation at the end of eighth grade, one student wrote to Linda:

*Carnegie Council on Adolescent Development. 1989. *Turning Points: Preparing American Youth for the 21st Century*. Washington, DC: The Carnegie Corporation.

I like to hear what you say. There are so many things that you pull out of us in our writing that I'm glad you share your writing with us. I would feel stupid sharing my writing with you if you never had anything to share with me. I'm glad you write. That encourages me.

And another student concluded:

You helped me by not being a teacher. Not telling me what I did wrong or write. (Ha, Ha. A joke.) You told me what you *liked*.

Linda rejects the posture of the English teacher as a neutral authority who objectively "presents" books and "corrects" writing. Instead, she speaks, reads, and writes in her classroom based on her own experiences and tastes and expects her kids to do the same. Students quickly understand that Linda makes it her business to learn about their literature and can be counted on to recommend books they will love. They also understand that she is a trustworthy responder to their writing because Mrs. Rief the writer knows what she is talking about.

The range and excellence of writing included in *Seeking Diversity* is so astonishing it is easy to forget that these are thirteen-year-old authors. Appendix A consists of excerpts from the writing and reading portfolios of ten eighth graders, from the articulate and focused to the scattered and hard to reach. Those of us who have taught middle school students will recognize that while the writing and reading documented here and elsewhere in the book are extraordinary, the kids are not. They exhibit all the enthusiasm, petulance, bravado, insecurity, sincerity, and cynicism of adolescents anywhere: Linda Rief's students are everyone's. The range of tastes, styles, and idiosyncracies represented in their writing and reading suggests what is possible for all junior high students and engenders new respect for adolescents, written language, and the influence of a teacher who believes in both.

Teachers will find *Seeking Diversity* to be particularly helpful in its discussion of the mechanics of writing and reading workshops in a traditional junior high setting. Linda skillfully adapts the particular methods that make sense to her until they fit her theories, classroom, teaching load, and schedule, rather than blaming the methods or her students when things don't work out as she expected. She knows that teaching is always a struggle and that there is no one right way, and she rejects instant answers and quick fixes in favor of a slow growth model of teacher development that increases the likelihood that she and her kids will find success. Linda invites other teachers to change by showing how she concentrated on just one or two things at a time in transforming her classroom, solving the problems of the workshop as they arose.

Secondary teachers will especially appreciate Linda Rief's solutions to the difficulties of teaching process writing and reading to 125 students a day, five days a week. She describes how she makes time for writing and reading in a schedule that allows just one period a day for language arts instruction, how she manages 125 dialogue journals, handles individual writing conferences that can't be daily or prolonged because of numbers yet still keeps the response personalized, builds and maintains a classroom library and keeps track of her books, and communicates with parents in order to inform them and enlist their help. She also shares titles of her students' favorite books, lists of books to use with specific units of study, classroom organization techniques, forms that she and peer responders use in writing conferences, and questions that students ask of themselves in their trisemester and final evaluations.

Perhaps most significantly, *Seeking Diversity* shows how Linda Rief's students have used portfolios to research and demonstrate their writing and reading abilities. Amidst all of the recent attention to portfolio assessment in educational testing circles, Linda's work shows that portfolios can only have value as an evaluative measure when students are truly immersed in writing and reading, make big choices as writers and readers, and have regular, sustained time to write, read, and analyze their progress. Portfolios are not a magical new method; they are hard and time consuming work. But when literacy assessment is the goal, rather than the micromanagement of lists of predetermined subskills, portfolio evaluation gives teachers and parents the richest, most accurate picture possible of what students can do.

Every year Linda Rief expects the world of 125 eighth graders. She imagines for them an undiminished future as readers and writers, and she puts her trust in the irresistible power and pleasures of written language. *Seeking Diversity* is her vision of how the language arts can create great opportunities for all adolescents and for a society that needs them to grow up articulate, capable, and whole.

NANCIE ATWELL
The Center for Teaching and Learning
Edgecomb, Maine

Acknowledgments

Seeking Diversity grew out of my attempts to make sense of my teaching. It is as much about my learning as it is about my students' learning. I am most grateful to my finest teachers, Donald Graves and Donald Murray, for helping me discover in myself what I know, and allowing me to take risks with all that I don't know. They, along with Nancie Atwell and Tom Romano, have taught me to trust myself as a learner and to have confidence in my students. The passion these four teachers have for reading, writing, and learning is contagious. Their teaching, their learning, their writings have enriched my life and the lives of my students.

Al Rocci, my first principal at Oyster River Middle School in Durham, NH, and now principal of the Emerson School in Bolton, MA, trusted me with my first classroom, before I ever took an education course. When I asked, "How can I become a teacher?" after being a teachers' aide for more than a year, Al supported and advised me all along the way. Every time he stepped into my classroom to evaluate me, he left me a better teacher.

Anne Heisey, principal at Stratham Memorial School in Stratham, NH, and Richard Tappan, English teacher at Oyster River High School, were my first mentor teachers. I wanted to be like them. Now I know, I have to be myself; yet, they are each a piece of who I am as a teacher.

I am grateful to so many teachers who have taught me by their words and their examples that teaching is really life-long learning:

* Mr. Webb—my high school English teacher in Hingham, Massachusetts,

* Jane Kearns, Jane Hansen, Paula Fleming, Jack Wilde, Tom Newkirk, Elizabeth Chiseri-Strater, Ruth Hubbard, Mary-Ellen Giacobbe, and Susan Stires—from the New Hampshire Summer Writing Program,

* Gary Lindberg, Mekeel McBride, Tom Carnicelli, Les Fisher, Anne Diller, John Carney, and Grant Cioffi from the University of New Hampshire English and Education Departments,

* poets Jean Nordhaus, Roland Flint, at Georgetown University, Gigi Bradford, from the Folger Shakespeare Library, and Catheryne Flye, who provided positive support and extensive advice about my writing and speaking while on a Kennedy Center Fellowship in Washington, DC,

* my colleagues at the Oyster River Middle School, especially Dolly Bechtell, Joan Savage, Beth Doran, Fran Mooradian, Paula Ickeringill, and Michele McInnes, for sharing their knowledge of adolescents and their years of expertise in the classroom,

* Dori Stratton, Judi March, and Lisa Noble, interns who supported and enhanced my teaching with their ideas and their questions,

* science fiction writer James Patrick Kelly, playwright Paul Mroczka, and poet Elizabeth Kirschner, who have worked with my students (and me) through the Artists in Residence Program sponsored by the NH Council on the Arts. Several of the pieces in this book were written under their tutelage,

* and especially to all my students, who have taught me more than anyone else over the last ten years. The whole purpose of this book was to carry their unique voices into other classrooms. The following students have portfolios in Appendix A and additional writings throughout the book: Tricia Crockett, Karen Gooze, Janet Moore, Gillian Nye, Matt Platenik, Sandy Puffer, Andy Reiff, John Roy, Jay Seger, Scott Troxell.

The students involved with the art of literature murals are, for the *Go Ask Alice* mural: Charlie Bryon, Anne Sisk, Ben Stamper, Amy Ulrich; for *The Outsiders* mural: Angie Dufour, Jeff Halldorson, Jennifer Hamel, Todd Whitford; for the *Night* mural: Sarah Gooze, Nahanni Rous, Jay Seger. Jeff, Sarah, and Nahanni also have writings elsewhere in the book.

Additional students whose writing appears in this book are: Becky Antonak, Lisa Bergeron, Craig Blakemore, Abby Burton, Chris Cluke, Dorn Cox, Kyle Downey, Monica Dufour, Graham Duncan, Lizzie Firczuk, Matt Fitzgerald, Kate Freear, Melissa Geeslin, Patty Grass, Craig Gwinn, Mark Hegarty, Steve Heick, Kristan Hilchey, Amanda Jones, Kate Keefe, Donaldo LaFerriere, Scotti Long, Holly MacKay, Toby Mautz, Lindsay Oakes, Marie Andre Roy, Jill Shapiro, Shannon Spencer, Jennifer Stevens, Stacey Sweet, Sara Weidhaas, Scott Wells.

Whether their writing appears or not, all of my students voices are here, between the stories, between the pages. They will continue to teach all of us.

I am so lucky to be in an environment where teachers are trusted as professionals who know what works best for their students. A special thanks to my superintendent John Powers, my principal Donald Wilson, and assistant principal Steve LeClair, for their continual support of me in and out of the classroom. Thank you, too, to Andrea Ross and Jackie Hardy, our school secretaries, and Don Mone, our custodian—all of whom have no idea how their smiles, kind words, and wonderings ("How's the book coming?") kept me going.

I am also grateful to all the teachers and administrators I've met and worked with in workshops and writing programs throughout the country—from Texas, Colorado, Michigan, Illinois, Arkansas, Washington, New York, Pennsylvania, California, Georgia, Nebraska, Iowa, Tennessee, Vermont, Maine, Connecticut, Rhode Island, Massachusetts, and New Hampshire . . . Your questions, your concerns, and your insights have helped shape my classroom and the format of this book. My classroom has been a laboratory in the search for making sense of teaching and learning.

Three people have had a direct influence on the final version of this book. I am so grateful for the time they took in giving my words such a close read, for their positive support, and for their on-target advice:

* Philippa Stratton, editor-in-chief at Heinemann, for taking full advantage of Post-it notes,

* Michael Ginsberg in Louisville, Kentucky, whose positive comments made me believe adolescents have something to teach even university folks,

* and my dearest friend and colleague Maureen Barbieri, a seventh grade teacher at Laurel School in Shaker Heights, Ohio, who read every word of this book through every draft, who always answered her phone no matter how many times a week I called with revisions ("How does this sound?"), and who loves books, kids, teaching, writing, and elegant dining as much as I do.

Jim Whitney, a media specialist and photographer at Lewis and Clark University in Portland, Oregon, planted the seed for this book. I am grateful for his gift of being able to capture the essence of reading and writing on film, as he has done so many times. After developing the picture of Mandy reading (on the cover), he blew up the print, labelled it "Linda's Book," and presented a copy to Philippa Stratton at Heinemann. Jim knew my kids had important things to share with teachers before I had the first words on paper.

Thank you to Cheryl Kimball, my production editor at Heinemann, for her editorial expertise.

Every teacher needs a good bookstore. Elizabeth Nowers and Linda Armirotto at the Little Professor Bookstore in Dover, New Hampshire, have introduced me to, and chased down, the numerous books which fill my room.

Finally, a special thanks to my husband George, and my sons Craig and Bryan, who knew when to leave me alone, when to bring me coffee, and when to listen. In the three years it has taken me to write this book, they have learned to cook (far better than I can), do their own laundry, and vacuum the house. If I'd only known, I would have started this writing business twenty years ago.

To all of you, for your support, advice, patience, and love—thank you!

Introduction

I am not the same person who started teaching ten years ago. My classroom this year is very different from the way it was last year. I expect that—I want that. Several things have changed my thinking about what teaching and learning are all about.

My first teaching job was as a substitute. I had had no education courses. I had never taught before. I had spent a year in the school as a teachers' aide. When the principal couldn't find a sub for an ill teacher, I asked to be given the chance to try it. I taught the way I'd been taught—read the next story in the anthology, answer the questions at the back of the book, and talk about the right answers. Do that for three days and then spend two days on exercises from the grammar text. I loved the control. I loved being the one who knew all the right answers. I felt incredibly smart. I had the teacher's manual with the answer key.

I have to admit, however, that in less than a week I was feeling quite uncomfortable. Anyone could do this. There was no special training needed to sit on a stool at the front of the room, call out kids' names to read out loud (I didn't even need to know the kids—I had a seating plan), give them a few silent minutes to answer the questions, and then call on them for the right answers. The only training I needed was how to keep from falling off that stool—falling out of sheer boredom. This was not fun.

I was in a classroom divided by a cardboard curtain. It took me less than a week to know enough to sidle up to that curtain and listen to every word the teacher beside me said. The strange part was, it was the students who were doing the talking—about their writing, about their opinions of their reading. I listened intently to this passionate man, Dick Tappan, challenge his students. I spent weeks like a spy huddled next to that room divider writing down everything he did or said. He was having fun.

I wanted to teach like Dick Tappan. But I needed the background. I took two courses at the University of New Hampshire. In one class the professor strode to the front of the room, dumped a pile of books on his desk, and asked each of us to come forward and arrange the books in a way that made sense to us. I was dumbfounded. Come up to the front of the room? Everyone would see I was the oldest person who had ever gone back to school. I was also convinced that no matter how I arranged those books, it would be the least sophisticated of all. Everyone would know how stupid I was. Other students moved to the front. They arranged the books by topic, by author, by size, by color, by likes/dislikes, and so on. I saw nothing. I was petrified of what this professor wanted. I had become blind to learning because nothing made sense to me. I didn't understand what had happened to me until I read Frank Smith (1978): "Nothing can be taught unless it has the potential of making sense to the learner . . ." (p. x). When students are anxious about what the teacher wants, they can't make sense of what they are doing.

I have carried away two things from Tom Carnicelli's simple exercise that I try to remember every day in my classroom. When students live in confusion about what I want, they too become blind to learning. I want them to be fully aware of what we are about as learners day after day. No hidden agendas. No crafty games.

Secondly, no matter what I present in that classroom, we all see things differently. Some students arranged the books by subject; others saw colors, size, and so on. We are all different—all bringing our own learning experiences and learning styles into that room. No matter what I present, each student sees it differently and takes his or her own meaning from that experience. It is this diversity that I try to foster in my classroom. I want to hear all the diverse voices of my kids. I want them to hear each other. We are all learners/teachers.

The second course that has stayed with me over the years was a writing course. As usual, I was late. The only empty chair was next to the professor. He beckoned me in. Immediately I was shocked. He expected us to write. I had taken several writing courses, but no one had asked me to write—we just talked about writing. I wrote. We got into pairs. I was with the professor. I could barely talk as he said, "Why don't you read me your piece?" I read. He smiled, laughed, repeated some of my words, and said, "Tell me more." I did—because for the first time, someone wanted my words, reacted to what I had to say, and made me feel like a writer. I took another look at what I'd written. I wanted to make it better.

I carry Don Graves's words with me into my classroom like a stack of well-loved books. "Tell me more," I say to my students. This book is about what they've had to tell me—what they've had to teach me. It's about the diversity of voices in one classroom and how those voices have come to be.

My classroom has evolved slowly. I didn't change everything overnight. I focused on one or two things at a time. I listened to my students, trying constantly to figure out what was working and what wasn't working. Forrest taught me about *real* writing. He wrote a powerful response to an editorial I shared from the newspaper. He asked, "Where do I send this?" I said, "You can't. That editorial was several years old. It was just an exercise to get you to write a persuasive piece."

"That's the stupidest thing I ever heard," he said. "I want to send this to the editor."

Forrest was right. From then on I concentrated on making the writing real—for genuine purposes. Not simply a "stupid" exercise.

Another year my focus was on conferences. A professor taught me what not to do. "This is garbage," he said, throwing my master's thesis across the desk. With his comment all the constructive, positive responses from the other four readers went right out of my head. I couldn't revise the paper for months. I carry that lesson with me every time I kneel down to talk with kids about their writing. What response helps the students best? Who gives that response? How can I best model constructive response? My first question to the students has become, "How can I help you?"

Mary Ellen Giacobbe opened my eyes to reading. In a summer course she taught at the University of New Hampshire she asked each of us to bring five favorite books we were currently reading, or had recently read, to the first class. I couldn't find five recently read books. I realized I wasn't reading. I thought I didn't have time. That scared me. Reading was part of my curriculum. How could I have neglected it so badly? That next year I worked hard at integrating reading along with the writing: What helps kids to become better readers? How could I best help them? How could we share what we know about books and authors in the classroom?

I am constantly asking myself: What works for learners? What doesn't work? What can we do to make this a richer learning environment? It is not a static, quiet classroom. When I began looking over the folders of writing, the logs and journals, and the letters I've saved from my students over these ten years, I began to realize how much we have changed. My expectations have changed. I expect more and I get more. The quality and quantity of writing and reading have risen steadily. Each year I let go of more and more and the students take more and more responsibility for their own learning. Their thinking is deeper, richer, and more diverse. I am constantly amazed at what they are able to do and how they are able to do it.

Part of the challenge for me is to keep up with them. I have to learn beyond my classroom. I have to put myself in situations that challenge my thinking, my comfort. I take courses that push my knowledge. I find myself hiding behind other students, hoping the professor won't call on me because I'm having trouble understanding the vocabulary and the concepts. But I push myself to figure it out. I listen hard. I reread. I rewrite what I think. And I try to relate it all to my own experiences. I have to be a learner in and out of my classroom so I won't lose sight of what it's like for my students—so I will continue to hear their voices. I don't ever want to set myself up in the front of that classroom again sitting on a stool with all the answers. Like Byrd Baylor in her book *The Other Way to Listen,* I want always to remember, "If you think you're *better* than a horned toad you'll never hear its voice—even if you sit there in the sun forever" (1978).

Like Bill Childs, an art teacher at Oyster River High School, I want to be a practitioner of my craft for professional *and* personal reasons. "Even as I am a teacher of art," Bill says, "I can't imagine not being an artist and a student as well. Certainly as I work I am reminded of the challenges, frustrations, and achievements that are experienced by my students. As it is important to feel a sense of satisfaction through the growth of my students, it is important to me to feel that I am achieving something on a more personal level as well."

What do I want for my students? I want them to leave my classroom knowing they *are* readers and writers, wanting to learn more, and having a number of strategies for that learning in any field. I want them to like learning and to like themselves. I want them to know they have important things to say and unique ways of saying them. I want them to know their voices are valued. I want learning to be fun. Most importantly, I want them to gain independence as learners, knowing and trusting their own choices. Like Aunt Addie Norton (Wigginton, 1985) I want my students to learn by doing: "I tell you one thing, if you learn it by yourself, if you have to get down and dig for it, it never leaves you. It stays there as long as you live because you had to dig it out of the mud before you learned what it was." I want the same things for me.

Writing this book has not been easy. I want to do justice to what thirteen- and fourteen-year-olds are capable of doing. Getting that clearly on paper while saying "this is where we are all at now, and it may not be right yet" is difficult. This is a draft of my teaching, my learning. It's scary sharing a rough draft of one's work. I don't want anyone to take what I write as a final draft—the last word. This is where my learning and teaching are at the moment.

I even thought of making this a blank book with only an invitation to visit my classroom. Teachers could make their own discoveries as they learned with all of us. They could fill the blank pages with their own ideas as they adapted ours. They could list book titles, jot quotes from my students, make sketches of the organizational layout of the room—anything. They could take what they needed and wanted from this classroom and change it on those blank pages to suit their own needs and purposes. Most likely they would see and learn things I never saw. They would give me new insights into my own classroom.

I hope that is what happens anyway—that as teachers read the pages of this book, they will jot notes to themselves that fit their situations, their students, their philosophies. Some teachers may take whole ideas; others may take just bits and pieces. I hope they will send me their ideas, their book titles, because that sharing is how we all learn best—from each other.

I certainly learned a lot from Nancie Atwell and her book *In the Middle* (1987). I learned from Tom Romano and his book *Clearing the Way* (1987). They taught me to trust the voices of my students, to be a learner myself, and to take risks. They taught me to be passionate about learning.

But I am not Nancie Atwell or Tom Romano, and I can never do exactly what they do in the same way they do it. We all carry our own personalities, histories, and agendas into a room the minute we step in. I adapted and changed their ideas, their structures, their strategies to fit me and my kids. What I do today, I may not do tomorrow. One thing will remain constant: I will always have questions.

What I share with you in this book is a glimpse into my classroom: what works and doesn't work, the successes and the failures. I don't believe any one system of learning works for every child. Everyone learns differently. As a teacher I believe my job is to find out what works best for each child. There are kids I never reach. Sometimes I don't work hard enough to find out why. Other times there are outside forces over which I have no control. No system is perfect or is *the* answer. All I'm sharing is where I am in that learning process right now. From year to year it is never the same.

At the end of the book are a number of appendices. Appendices B–S include handouts. Appendix A is a glimpse into the portfolios of a variety of my students: their self-selected, most effective pieces of writing; examples of their responses to literature; and self-evaluations of their learning. This look into portfolios shows the intensity of emotions with which adolescents today are struggling, and their honesty and determination in dealing with some of life's toughest questions. It also shows just how sophisticated and articulate thirteen- and fourteen-year-olds can be when they are given choice, time, and positive response to their writing and reading.

"Letting students use their personal language is perhaps the most powerful educational tool a teacher can put in their hands," says Tom Romano. "Using language makes us think, and writing makes us use language" (1987). Writing is thinking. Look at the reading and writing of these students and you will get to know them personally—you will get to know what they are thinking.

The students (I have five sections of students, approximately twenty-five students per class, heterogeneously grouped) whose writing I have chosen to share come from a variety of backgrounds. Some are academically successful at everything; others are successful at little. Some are poor; some are rich. Some are gifted; some are learning disabled. Some have parents who value learning; others have parents who don't even subscribe to a newspaper. I believe all kids *can* learn, if they *want* to learn.

During the summer after eighth grade, I received several letters from Nahanni. One postcard from London said: " . . . We saw Tintern Abbey. My mom says Wordsworth wrote a poem about it. It's a ruin of a monestery built in medieval times. The sunlight pours in through stone walls that are only half there. There is no roof and I think this is the way a building is meant to be. Why shut out the sky? I thought of you there. We could spend a day sitting with pen and paper. I can see poems floating in the spaces between open staircases . . . "

I want these pages, these glimpses into the literate lives of adolescents to be like "poems floating" into other classrooms. For like Nahanni I want all middle-school students to have the opportunity to create these literate occasions (Graves, 1990, 16) "in the spaces between open staircases." Listen to these voices—and know how literate human beings come to be.

Author's Note: The examples of student writing which appear in this book have been left the way the students wrote them. Log entries and self-evaluations are first drafts, often containing spelling, punctuation or usage errors.

Final pieces have been edited to the best of each student's ability. When final pieces go beyond the classroom, I act as an editor and show the students what they missed. In the same way the copy editor of this book has saved me from embarrassment over a misspelled word or a usage problem, I do the same for my students.

Works Cited

Atwell, Nancie. 1987. *In the Middle*. Portsmouth, N.H.: Boynton/Cook.

Baylor, Byrd. 1978. *The Other Way to Listen*. New York: Macmillan.

Graves, Donald. 1990. *Discover Your Own Literacy*. Portsmouth, N.H.: Heinemann.

Romano, Tom. 1987. *Clearing the Way*. Portsmouth, N.H.: Heinemann.

Smith, Frank. 1978. *Reading Without Nonsense*. New York: Teachers College Press.

Wigginton, Eliot. 1985. *Sometimes a Shining Moment: The Foxfire Experience*. Garden City, N.Y.: Anchor Books.

From the Middle
to the Edge

> enuine, independent reading and writing are not the icing on the cake, the reward we proffer gifted twelfth graders who've survived the curriculum. Reading and writing are the cake. Given what we know about adolescents' lives and priorities, can we afford to continue to sacrifice literate school environments for skills environments?
>
> Nancie Atwell, *In the Middle*

On the first day of school in my eighth-grade classroom I explain what we can expect of each other as learners, how the room is organized, and why I do what I do. A hand shoots up and Shawn asks, "Lemme get this straight. You mean we get to *read* during reading, and *write* during writing?"

Within three days Shawn has read *Wild Cat* (1975) by Robert Newton Peck ("kinda gross when the babies get eaten"), *The Cat Who Went to Heaven* (1958) by Elizabeth Coatsworth ("I think I gotta read this one again. I was kinda confused."), and *Of Mice and Men* (1984) by John Steinbeck ("If the smart guy didn't have the dumb guy taggin' along, I think he coulda gotten a job easier . . . I think he feels bad for him though.").

What do these books have in common? They are all thin. At this point it is Shawn's only criterion for choosing books. At fifteen, he is in eighth grade for the third time and confesses to me that he has never read a book in his life.

"Yo, Mrs. Rief, I'm three [books] for three [days]. Better get more [thin] books," Shawn smiles as he fingers through the shelves of paperbacks.

At our team meeting that afternoon, the resource room teacher reads from Shawn's record at previous schools. She tells me that Shawn *hates* reading and writing and has real problems with both, and that I should not expect much. If he is backed into a corner, she reports, he becomes belligerent. I thank her for the information, wondering what Shawn's former teachers considered reading and writing.

September 9, the fourth day of school. Sitting at round tables or on large pillows scattered throughout the room, we are all reading or writing. Shawn sits at a table on my left. I watch as he bends over his reading log, his thick hands writing furiously. He finishes, leans over, pokes me in the arm, and says, "Read this. This guy's pretty good."

I tuck my finger into the page I'm reading from One Child by Torey Hayden (1980). Shawn hands me a Great Brain (Fitzgerald, 1972) book. The part he wants me to read explains how the Great Brain solved the mystery because he realized that glasses fog up in a warm room after being out in the cold, so the person couldn't have seen what he claimed he saw. Shawn laughs. "That happens to me all the time. This guy's pretty good." I chuckle in agreement.

Sandy sits on my right. She turns the book she is reading upside down on the table, opens her reading log, and writes:

> Sept. 9
> 30 min.
> p. 13–36
> *Toward a Psychology of Being*

What I think he is saying is that a healthy person [who] wants to grow wants to be more independent and more self-reliant. Growth-motivated people aren't directed by their environment or social conditions. I haven't really understood what Maslow has been talking about cuz he keeps mentioning existentialism and I have no idea what that is. Usually, though, by the first eighty pages everything begins to fall into place and I can read a lot faster because I know what the author is talking about and I don't have to keep reading paragraphs over and over. If I still don't understand what's being said I read on and sooner or later it falls into place.

Sandy and Shawn may be at very different ends of the ability spectrum, but they have one important thing in common. They are genuine readers and writers bringing meaning to, and taking meaning from, their chosen texts. Soon they understand clearly what I'm after: teach me what you know, how you've come to know that, and what you are able to do with that knowledge. Sandy and Shawn will challenge me throughout the year—Shawn, to keep him reading, writing, and thinking about his learning; and Sandy, to keep up with her intellectually and to nudge her to teach me more. It doesn't matter what level these students appear to be. My job is to guide them beyond what they can already do, to challenge them to challenge themselves.

Could I address Sandy's and Shawn's needs with a teacher's manual or workbooks aimed at the "average" eighth grader? Of course not. Workbooks don't address the unique learning styles, the extraordinary ideas, the honest thinking, or the prior knowledge each child brings to the classroom. My students are my curriculum. I want to nurture that uniqueness, not standardize my classroom so that the students become more and more alike, their only aim to pass minimum competency tests.

With Sandy and Shawn engrossed in their own reading and writing I return to One Child. I read it because a student recommended it to me. Hayden is a teacher of severely disabled children. Sheila, a six-year-old child abandoned by her mother and abused by an alcoholic father, was placed in Hayden's classroom after tying a two-year-old boy to a tree and setting him on fire. Hayden tries to reach her. I read:

One night I brought in a copy of *The Little Prince*. "Hey Sheil," I called to her. "I've got a book to share with you."

• • •

"Come and play with me," proposed the little prince. "I am so unhappy."

"I cannot play with you," the fox said. "I am not tamed."

" . . . What does that mean—'tame'?"

• • •

"It is an act too often neglected," said the fox. "It means to establish ties."

" 'To establish ties'?"

"Just that," said the fox. "To me, you are still nothing more than a little boy who is just like a hundred thousand other little boys. And I have no need of you. And you, on your part, have no need of me. To you, I am nothing more than a fox like a hundred thousand other foxes. But if you tame me, then we shall need each other. To me, you will be unique in all the world. To you, I shall be unique in all the world . . . "

• • •

The fox gazed at the little prince, for a long time.

"Please—tame me!" he said.

"I want to, very much," the little prince replied. "But I have not much time. I have friends to discover, and a great many things to understand."

"One only understands the things that one tames," said the fox.

"What must I do, to tame you?" asked the little prince.

"You must be very patient," replied the fox. "First you will sit down at a little distance from me like that—in the grass. I shall look at you out of the corner of my eye, and you will say nothing. Words are the source of misunderstandings. But you will sit a little closer to me every day . . . "

Sheila put her hand on the page. "Read that again, okay?"

I reread the section. She twisted around in my lap to look at me and for a long time locked me in her gaze. "That be what you do, huh? . . . Tamed me . . . It be just like this book says, remember? I do be so scared and I run in the gym and then you come in and you sit on the floor . . . And I peed my pants, remember? I be so scared. I think you gonna whip me fierce bad 'cause I done so much wrong that day. But you sit on the floor. And you come a little closer and a little closer. You was taming me, huh?"

I smiled in disbelief. "Yeah, I guess maybe I was."

"You tame me. Just like the little prince tames the fox. Just like you tamed me. And now I be special to you, huh? Just like the fox."

"Yeah, you're special all right, Sheil." (Hayden 1980, 101–3)

Hayden creates a safe, loving reading environment where a relationship forms between a student, a teacher, and books. She confirms what Sheila knows, letting her know that what she thinks is important. To this teacher, each student is unique. Like the fox, she knows the importance of patiently establishing ties with her students.

In Hayden's classroom, Sheila is pondering, wondering, thinking, connecting . . . things that literate people do.

I try to create this same literate environment in my own classroom— the literate environment that Nancie Atwell describes as a place where people read, write and talk about reading and writing. Where everybody is student *and* teacher (1987).

Neither can I separate reading, writing, speaking, and listening. They are integrated processes finely woven into a tapestry of literacy. The components of language have to be taught as a whole for learning to be meaningful. Each aspect of language use enriches another. I immerse my students in good literature and offer continual opportunities for writing, talking, listening, and thinking so they can begin to answer their biggest questions—Who am I? Why am I here?—through their reading and writing.

Sandy and Shawn are only 2 of the reasons we are all learners in the classroom: there are 123 others in five classes. We all have names, and like Sandy and Shawn we all have our own, unique voices. Don Murray once suggested that we should "seek diversity in writing, not proficient mediocrity." I'm seeking that same diversity in reading, too.

For learning to be meaningful for each of us, I have to see my students as individuals. I can only hear the diverse voices by offering them choices, giving them time, and responding positively to their reading and writing (Giacobbe 1985). Most importantly, I must model my own process as a learner. So I write with my students and read with them. They value what I ask them to try if I value the same process, and I understand the process of learning best when I'm a participant. I share my failures as well as my successes—my weaknesses as well as my strengths.

At the end of every year I ask students, "If you had to pick the one thing I did that helped you the most as learners, what would it be?" They inevitably say, "You write with us, and you read with us."

All teachers should be readers and writers, but teachers of language arts *must* be writers and readers. How many schools hire home economics teachers who do not cook or sew, industrial arts teachers who will not use power tools, coaches who have never played the game, art teachers who do not draw, Spanish teachers who cannot speak Spanish? Yet, how many prospective English teachers are asked in their interviews, "What are you reading? What are you writing?"

When I first began teaching, my students read and wrote in my voice, because they answered my questions. Then I asked myself what made writing and reading easy for me, and I realized that to express themselves effectively all learners have to read and write and speak in their own voices. But they can discover their voices only by answering their own questions, their own inquiries. Now I constantly ask my students, "What do *you* think?" as they read and write.

Richard Rodriguez, author of *Hunger of Memory* (1983), once told an audience at a National Council of Teachers of English luncheon in Texas, "Listen to my voice and wonder how it comes to be." We must do the same with our students: listen to their unique voices and ask how they have come to be. It is not easy to encourage such diversity in our classrooms. It takes time for students to find their own voices and their own texts. And it requires us to trust that students do have important things to say.

I expect good reading and writing in which one process enriches the other, in which students' ideas and wonderings and questions invite risks, taking them to the outer edges of what they know and what they can do. I expect good reading and writing, in which process and product are woven tightly into literate tapestries of wonder and awe.

My students seldom disappoint me.

While we were reading from *The Diary of a Young Girl* (Frank 1972) together, I read aloud from Elie Wiesel's *Night* (1986), and from several children's books, including *Rose Blanche* by Innocenti (1985) and *Hiroshima No Pika* by Maruki (1980). In my reading log I wrote:

Just finished Night by Wiesel. I felt tears lumping in my stomach, in my throat, tightening my whole body. I don't think I would have been so upset by this book if Mom hadn't died a year ago. I felt Wiesel's relief when his father died, a burden relieved, yet he loved his father so much. After taking care of Mom with my sisters all summer I felt a burden relieved. Yet, in the moment it was over I wanted her back. Wiesel's last two sentences haunt me: "From the depths of the mirror, a corpse gazed back at me . . . The look in his eyes, as they stared into mine has never left me." (p. 109) I wonder if the look of his father has ever left him?

In response to *The Inner Nazi* (Staudinger 1981), which she chose to read on her own, Sandy wrote in her log:

I wonder why I am so drawn to Hitler's insanities, and not the Jews' suffering? It's not that I sympathize with the Nazis, but I think it has something to do with "evil" always being more interesting than "good" . . . After WW II, everyone said "Oh, that won't happen again, we won't let it" but it is happening . . . all over the world (S.Africa). It really makes me wonder, when Americans come together, we write letters, make phone calls, demonstrate, petition and protest when Coke changes its formula, but we can't end apartheid in South Africa.

Steve, seldom enticed into whole books, chose to respond to a dirt-bike magazine:

the CR 500 is kick but the whole way it has a superer strong motot, awsome clutch, good brakes, and the suspention is as smooth as a babeys butt

Sandy and Steve, like Shawn, are the extremes. Sandy's writing reflected a strong connection with her reading:

Television Pogrom
Numb with indifference
I watch
rows and columns of goose-stepping figures
move across the screen
in the comfort of my own home
the horrors
of the camps
flicker on the mocking glass bubble
Then
as if a dose of reality might be
too much
it's a break to suburbia
where a trim woman
tells me to use lemon-brite on my
no-wax floors.

Sandy played with poetry while she made the connections between what she read and what she knew. Steve wrote about his dirt bike adventures. I tried poetry because Sandy taught me how to take the risk. In one poem, both Wiesel's *Night* and my mother were still on my mind.

Waiting for Her to Die

I change
dressings cleanse wounds
measure medicines mash foods
launder sheets plump pillows
vacuum rugs scrub floors
scour tile dust knick-knacks
mend curtains repair furniture
mow grass trim hedges
plant shrubs arrange flowers
and tiptoe
tiptoe so I won't
wake her "don't
tiptoe just because my eyes are closed"
she whispers from the hospital bed in her livingroom
so I don't
but I do
wash windows so that
she and I
can see clearly.

I wrote a letter to Elie Wiesel and shared it with the class.

Dear Mr. Wiesel,

I just finished reading your book *Night*. It made me weep inside. It is so clear how much you loved your father. You took me back with you as you gazed at his face. I reached out to touch him as I was reading your words. For this book alone you deserve the Nobel Peace Prize, for you are not letting people forget.

I could not help but think of my Mom as I read the page on your father's death. My mom battled cancer for five years. By the time she died there was little left to her body. Her eyes were those of a corpse—one who fought too long. I took turns with my two sisters taking care of her all summer. Her last words were, "Lin, be careful driving home. I love you." I was relieved to leave. She died before I made it home. I had no time to ask her back.

Yes—what you went through is terrifying—what you went through with your dad is heart-breaking. You must wake at night "lying in the bunk at Buchenwald." I'm so sorry you suffered so much.

I read somewhere that you write so no one will forget. I will not let my students forget either. Thank you for writing this book.

Melissa wrote a letter to Farley Mowat, not because it was a class assignment but because she had legitimate questions she wanted answered.

Mr. Farley Mowat
c/o McClelland and Stewart Inc.
481 University Ave, Suite 900
Toronto M5G 2E9 Ontario, Canada

Dear Mr. Mowat,

I have recently read your book *A Whale for the Killing* and found that I had many questions when I was finished. Was the whale really pregnant? What happened to you after the whale was gone? Did people forget about the whale or do most people still remember her? What did they do with the whale after she died? How did they get her out of the pond?

The biggest question that I have would be: In the time that you were trying to help the whale, did you ever doubt yourself? Did you ever think about giving up? turning away?

I think you did a very brave thing in saving the whale. I'm proud of you and I hope you are proud of yourself.

Thank you so much for writing that book. In it you have answered many of my questions, and I hope you will answer these. Thank you for reading my letter.

Sincerely,
Melissa

Dear Melissa,

Many thanks for your excellent letter. I am delighted that *A Whale for the Killing* has been informative for you. No, I couldn't give up on the whale, even though it would have been easier. I think we can all stop abuse of other forms of life on this planet, each in our own way. Man is only one of many forms of life and has no right to take precedence over the others. I hope you will keep your interest in animals alive as you grow older.

Farley Mowat

Abby, who like Sandy had strong opinions and beliefs about important world issues, seldom shared her writing with anyone but her mom or me. "I rarely show my writing to other students," said Abby, "because it is not written to satisfy teenage fancies." However, she had no hesitation at sharing her opinions with the appropriate audience. While Melissa wrote to Mowat, Abby wrote to the music teacher:

Dear Mr. Butler,

I am writing this letter to hopefully help you understand my reasons for not wanting to participate in the performance of the eighth grade school play, *The Mikado*.

My instant impression of this play was that it was sexist. The female characters are portrayed as being either ugly and mean, like poor Katisha, or silly and yummy, like the three sisters , Peep-Bo, Pitti-Sing, and Yum-Yum. The male characters are equally unappealing in that they are either brutal and controlling or dippy-do idiots. This type of stereotypical imagery has hurt women a lot, as well as men.

I have read over this play and I believe it is very degrading to the Japanese people in that there is a very narrow, simplistic portrayal of their culture. In my opinion, it does not promote a well-rounded understanding of them at all. They're not as silly and superficial as this play presents.

I am very willing to participate in a play with a socially redeeming message to the audience and a healthy balance of male and female characters. This play is extremely outdated and I find it embarrassing that we are still being entertained by this sort of silly, old-fashioned, obsolete, and, in my mind, damaging melodrama.

I think it is a violation of the first amendment of the Constitution that I have to say or sing words I do not like.

Sincerely,
Abby

Andy, who read science fiction and fantasy voraciously (Margaret Weis and Tracy Hickman, Douglas Adams, Piers Anthony), tried on the styles of the authors he was reading as he wrote descriptive pieces like this lead to "Night Walk."

We donned our jackets, chose our walking sticks and embarked on our journey into the College Woods . . . On our way we decorated our staves . . . my staff was decorated with a lush green fern, two bright red cardinal feathers and three blue jay feathers, along with a shiney plate of mica, and a pure white rock. I wrote ancient runes and glyphs with crushed berry juice. My most prized possession was a collection of porcupine quills that dangled from leather thongs attached to the crest of my staff . . .

He attempted numerous fantasy stories, often drawing out his characters first (Figure 1.1).

While many students were more comfortable writing real-world non-fiction, Andy focused on science fiction and fantasy. When I asked him how his writing was going one day in a roving conference, he told me

Figure 1.1
Andy often drew his
characters before
writing a story

casually that he was "writing for a magazine." He had practiced enough: on his own he had sent a query letter to the editor of *DUNGEON Adventures* proposing an "idea for a series of three modules" in which he discussed plot, characters, and possibilities. The letter was complex, well organized, and thoughtful. Inspired by his own compelling purposes, Andy joined the real world of writers.

Gillian reached beyond the classroom too. After tucking a painful story about her father into her working folder, she turned to one about a fishing trip with her grandfather. The last paragraphs read:

> I can see the tip of the fish's head over the side of the boat. It breaks through the water like a groundhog poking his nose through fresh, spring soil. Opa bends down and with one swift movement, scoops the fish into the net. My first fish—a little six-inch sunfish.
>
> I use the rest of my strength to paddle home. I need to tell Grandma about the "whale of a fish" I caught, before Opa comes up the path, carrying the truth.

Scholastic Scope bought and published Gillian's story (Nye 1987). She, too, was on her way to knowing that she is a writer. In time, she will work again on the piece about her father.

Because my students try fiction and nonfiction, I do too. I need to know what makes different kinds of writing work, so I can help students see the differences. But I hated the science fiction I wrote. It didn't make any sense: I didn't want anyone to read it. If my students are going to try science fiction, they will have to seek responses from each other. I will have to find other ways of helping them, perhaps by reading more science fiction or by finding someone who understands the genre.

My difficulties with science fiction taught me that choice is so important. What works for one student may not work for another. I ask them to try all kinds of writing, but the choice of what pieces, what genres, work best is always the decision of the student.

I tried realistic fiction, to exaggerate the truth. My lead to "Reunion" begins:

> She didn't give much thought to him as she entered the crowded country club hall. But as she plunged into an ocean of bobbing heads and waves of laughter, it was his face she recognized. Her mind, like a video-tape on rewind, darted backwards and replayed the scene twenty years earlier when she had last seen him . . .
>
> . . . She surveyed the still life as she would a Norman Rockwell painting. Pink plastic curlers rolled tightly in her hair. A crying baby wriggling in her arms. A toddler clinging precariously to a fistfull of jeans. And his face, this same face, as she opened the front door. A surprise visit. He didn't tell her he was on his way to Vietnam. He didn't tell her he came to say goodbye. He didn't tell her he was scared. She didn't ask him where he had been for three years. She didn't tell him how worried she had been . . . how much she had missed him. She didn't need to tell him she hadn't waited this time. She only remembered his words, "Sorry I . . . I missed your husband. You take care," and he was gone.

And I tried nonfiction. My lead to "First Love" begins:

> Bryan, my seventeen-year-old son, slams open the back door and swaggers down the hallway. He wraps his arm around my shoulder and

announces triumphantly, "Mom, wait'll you see the bargain I got. Why the wheels alone are worth two hundred dollars!"

Immersed in reading and writing in all genres for their own genuine purposes, my students take on serious issues. Often I don't know where their ideas come from. I suspect that some come from sharing each other's drafts and from the books they read, and that others come from the literature I share with them, the topics I'm writing about, or the occasional exercises I ask them to try. But most of their ideas are their own. Because they know they will write and read every day, writing topics and books they want to read are constantly on their minds. They read as writers and write as readers. Sarah says, "The more you read, the better you can find words to describe what you're talking about."

"I get a lot of stuff from the books I read," says Andy. "I just kind of record it in my brain—like describing words and vocabulary—and *ideas*, definitely. It might be something the author says . . . just that little bit of information might inspire me or just trigger something in my mind to try to expand on."

"The more you read," says Jessie, "the more you get different styles of writing, and you can bring them together as your own . . . and things you don't like you can try to avoid in your own writing."

Jeremy ties it all together. As a writer, "if you want to please the reader, then you gotta read what the reader is reading."

I shared my experiences and writing while on a fellowship at the Kennedy Center for the Performing Arts in Washington, D.C. An excerpt from "Why Should We Remember?: Reflections from the Vietnam Veterans Memorial" reads:

I watch as a little boy drags his fingers across The Wall, as carelessly as Tom Sawyer thumped a stick across a white picket fence. He stares at the names.
"These are all people killed in Vietnam?" he asks his dad.
"Yes," answers his father.
"Even the ones on that side?" he asks, sweeping his hand to the right.
"Yes," his father says again.
He cups his hands behind his head. Scans left, then right.
"Why?" he asks. "Why?"

. . . Families huddle in its arms. Veterans pace out their pain. Photographers, the dogtags of their profession chained around their necks—Nikon, Minolta, Pentax—adjust their lenses, move in move out, on the names on the memorabilia on the faces of grief.
Yankee-capped, tank-topped, leather-clad; cowboy-booted, barefoot; thick-bearded, clean-shaven; grey-haired, crew-cut—they all come—to eulogize, apologize, commemorate, communicate, or simply to remember.
. . . Sink deep into its vortex, where the names nearly bury you, and you can almost hear it whisper, "What a waste . . . What a waste . . . "

Lizzie listened to my words. She thumbed carefully through the book *The Wall* (Lopes 1987). Over a vacation, she visited Washington, D.C. When she came back, she wrote:

A young child pulls at his father's pant leg. The father stares mindlessly at a name carved in black on a mournful wall. His face remains emotionless.

The child, realizing his father's lack of response, changes the pull to a cling of fright. He stands, arms around his father's leg.

The father gently pulls out of his pocket a small scarlet case made of velvet. He stares at it aimlessly in his hand. "You mean more to me than this ever could," the man says to the name on the wall. He opens up the case and lays it underneath the name. The dim light of the sun shimmers on the purple heart in its new resting place.

A long period of silence accompanied by a lone tear passes by. The father turns to leave and the boy follows. They walk, side by side in silence, along the thousands of names carved in black and the endless numbers of people honoring, remembering, and hurting.

While Lizzie chose a global issue through the eyes of one man, Holly chose a personal issue that touched too close to home.

Petals on my Slippers

I threw my leg up on the barre and made the final adjustments on my multi-colored leg warmers. I tried to relax, but with no luck. Sinking into a deep plié, I tried to pretend it was just another Monday class, but I couldn't. With every developé I imagined the drunken bastard was right in front of me, and if I could just get my leg high enough I would be able to kick him. My frappés and grand battements were stronger than ever. With every combination I wished that I could put him in front of me, right in the line of my foot, and watch him die before my eyes, as he had done when he ran her over while driving under the influence. Was it that difficult to get a friend to drive you home?

I fell out of my double pirouette. No, it wasn't his life he sacrificed, so why should he care? She was only in tenth grade. She couldn't have meant much to anyone, could she?

Every time I thought of it, my adrenaline pumped stronger. I couldn't feel the pain in my toes; I could only picture him under them. Yes—she had meant a lot to some people. Her friends had cried. Tears had been shed for her while hate for him burned deep inside. Life was made to be lived—not cut short by misused alcohol.

Flowers were thrown on her grave. I thought I saw petals on my slippers.

Lizzie told me, "In elementary school we were encouraged to write about happy things. This year has been my first experience writing about dramatic, sad things—things that affect people, things that make people think about really uncomfortable things we need to focus on." Holly agreed.

Sometimes they can't let go of topics and go back again and again to issues that are important to them. A trip to a local nursing home stayed with Karen for a long time. She drafted this poem in her head for a long while, then wrote it down with few revisions:

The Nursing Home

They reach out,
   wrinkled,
      dying hands,
  in desperation

To touch,
　　swiftly passing people.
I see
　　their pain
But
　　am scared
　　　　by vacant eyes
　　and walk by.
I hear stories
　　　　　of lives
　　lived long ago,
But
　do not want
　　　to get involved.
So
　　I listen,
　　　out-of-place,
to tales issuing
　　　from gaping mouths
And leave,
　　with guilt
　　　and relief.

For Sandy, a similar thing happened. She read frequently about world issues and showed her concern in her writing. Sandy wrote:

To Seabrook

We
students
not registered voters
not tax payers
without any reason to believe
we can make a difference
for some reason decided
that although we don't have influence
will never say never
the future is ours
but the toxic waste
in the fine print
is yours
This may simply be a list of personless names to you
but it's more than that
it's our way of saying
You can't do this
Even if you disregard this
as we expect you will
Even if you put Seabrook into operation
as we expect you will
Even if you give us your toxic waste or your
"minimal" chance of annihilation
You can't forget
an unimportant minority called
the future
that cries out to be saved

In a lighter vein Scotti used language playfully when he borrowed the style of Marc Gallant's *More Fun with Dick and Jane* (1986):

Poison Rocks

"Ami," said Mrs. Norton. "What are you listening to?"
"I'm listening to Poison," said Ami.
"What is Poison?" asked Mrs. Norton.
"Poison is a rock group that is rad," said Ami.
"That type of music will make you go deaf," said Mrs. Norton.
"What?" said Ami.
"I said it will make you go deaf," said Mrs. Norton. [See Figure 1.2.]
"What?" said Ami.
"It has already happened," said Mrs. Norton.
"What?" said Ami.

In our classroom we are constantly playing with language: we immerse ourselves in language when we read, write, speak, and listen. We read children's literature, adolescent fiction, mysteries, nonfiction, poetry, newspapers, magazines, plays, fantasy—anything and everything that is good literature, literature that I love and literature the students love. I cannot immerse my students in literature I don't like, anymore than any of us can write effectively on topics in which we have little interest. We are

Figure 1.2 *"That type of music will make you go deaf."*

writing letters, poetry, essays, personal narratives, short stories, sports articles, even novels—anything and everything that is important to us. We talk about what makes good writing and about how we solve our reading and writing problems. We laugh, cry, and enjoy. In our classroom we are all learners, and we are all teachers.

My curriculum is controlled by the students—not by some publisher in a distant city who thinks all students should learn the same things, not by some administrator who doesn't trust the teacher as a literate human being.

Yet teachers need to trust themselves as learners too. At the NCTE Spring Conference in Boston (1988) I noticed a publisher's display table surrounded by a throng of teachers. Thinking only a well-known author could draw such a crowd, I stood back, hoping to catch a glimpse of Chris Van Allsburg, Lois Lowry, or Robert Cormier. But there was no author at this display. Instead, I discovered that teachers were shoving and pushing to get to this publisher's answer to our educational woes—teachers' manuals for *Cliff Notes*. Publishers will continue to control our classrooms with this nonsense as long as we continue to buy it.

Another publisher's representative must have seen the look of disappointment on my face. It's too bad he misread it. He pointed to a nearby table and said, "Here, we have manuals for *Velveteen Rabbit* and *Charlotte's Web* filled with comprehension questions, tests . . . "

"Why?" I interrupted. "Whatever happened to real reading and honest writing?"

He turned to a paying customer.

Just as I settle back into my book, *The Education of Little Tree* (1987) by Forrest Carter, on this October Friday morning, a teacher cranes his head around the frame of my door. He steps over sprawled bodies, bends down, and whispers to me, "Boy, you sure planned hard for this lesson."

I am somewhat annoyed by this interruption, but take the time to glance around the room. There is hardly a sound. Marissa and Missy are propped against one wall on a pillow. Marissa turns a page of *The Great Gilly Hopkins* (Paterson 1978). Missy's face is intent, almost worried, as she reads *Go Ask Alice* (Anonymous, 1978). Julie cradles *Taking Terri Mueller* (Mazer 1983) in her lap. George is halfway through *Never Cry Wolf* (Mowat 1981).

Mandy and Jen are conferencing quietly about a piece of writing about Mandy's grandmother. Val is revising a mystery piece she "just got an idea for." Brandon, who has just read *It Was a Dark and Stormy Night* (Rice 1987), is writing out some of his own best "worst" leads for the Bulwer-Lytton "bad" writing contest.

There is a commotion in the hallway as students pass. Across from us, the class is very loud while they wait for the teacher. Pat, one of my eighth graders, stands and walks slowly to the door, stepping over classmates, eyes remaining fixed on *One Child* (Hayden 1980). He reaches out, feels for the knob, and pulls the door shut, never missing a word as he and Sheila, the main character, sink back down into his chair.

My colleague is right. I have prepared very hard for this lesson.

Works Cited

Anonymous. 1978. *Go Ask Alice*. New York: Avon Books.

Atwell, Nancie. 1987. *In the Middle*. Portsmouth, N.H.: Boynton/Cook.

Carter, Forrest. 1987. *The Education of Little Tree*. Albuquerque, N.M.: University of New Mexico Press.

Coatsworth, Elizabeth. 1958. *The Cat Who Went to Heaven*. New York: Macmillan.

Fitzgerald, John. 1972. *Great Brain* series. New York: Dell

Frank, Anne. 1972. *The Diary of a Young Girl*. New York: Washington Square Press.

Gallant, Marc Gregory. 1986. *More Fun with Dick and Jane*. New York: Penguin Books.

Giacobbe, Mary Ellen. 1985. Reading Writing Connection. Seminar held at the University of New Hampshire, Durham, N.H.

Graves, Donald. 1983. *Writing: Teachers and Children at Work*. Portsmouth, N.H.: Heinemann.

Hayden, Torey. 1980. *One Child*. New York: Avon Books.

Innocenti, Roberto. 1985. *Rose Blanche*. Mankato, Minn.: Creative Education Inc.

Lopes, Sal. 1987. *The Wall: Images and Offerings from the Vietnam Veterans Memorial*. New York: Collins.

Maruki, Toshi. 1980. *Hiroshima No Pika*. New York: Lothrop, Lee and Shephard Books.

Maslow, Abraham M. 1982. *Toward a Psychology of Being*. New York: Van Nostrand Reinhold.

Mazer, Norma Fox. 1983. *Taking Terri Mueller*. New York: Morrow.

Mowat, Farley. 1979. *A Whale for the Killing*. New York: Bantam.

———. 1981. *Never Cry Wolf*. New York: Bantam.

Nye, Gillian. 1987. "Opa." *Scholastic Scope* (May 18):7.

Paterson, Katherine. 1978. *The Great Gilly Hopkins*. New York: Avon.

Peck, Robert Newton. 1975. *Wild Cat*. New York: Avon.

Rice, Scott. 1987. *It Was a Dark and Stormy Night*. New York: Viking Penguin.

Rodriguez, Richard. 1983. *Hunger Of Memory*. New York: Bantam.

Saint-Exupery, Antoine de. 1971. *The Little Prince*. San Diego, Calif.: Harcourt Brace Jovanovich.

Staudinger, Hans. 1981. *The Inner Nazi*. Baton Rouge, La.: Louisiana State University Press.

Steinbeck, John. 1984. *Of Mice and Men*. New York: Bantam.

White, E. B. 1980. *Charlotte's Web*. New York: Harper & Row.

Wiesel, Elie. 1986. *Night*. New York: Bantam.

Williams, Margery. 1975. *The Velveteen Rabbit*. New York: Avon.

Organizing the Room, Materials, Expectations

In her book *In the Middle* (1987), Nancie Atwell tells of a conversation with Don Graves:

" 'You know what makes you such a good writing teacher?'

"Oh God. I thought. Here it comes: validation, from one of the world's greatest writing teachers. In a split second I flipped through the best possibilities. Was he going to remark on my intelligence? My commitment? My sensitivity?

" 'What?' I asked.

"And he answered, 'You're so damned organized.' " (53–54)

Organization. It's one of the reasons Atwell's room ran so smoothly. I don't function well in chaos. Few people do. Check out the students' backpacks, notebooks, or lockers sometime, and you can tell fairly accurately which kids are succeeding and which ones are not.

Areas in the room, materials and supplies for reading and writing, time and expectations need to be well organized and predictable. I need to be organized for myself, as well as for my students.

Areas in the Room

My room is of average size. I have five hexagonal tables (although I would prefer smaller, round tables) where four to five students sit together and face each other. No matter how I arrange the tables, the students are elbow-to-elbow. If they're capable of choosing their own seats and working well together, I let that happen. If not, I move them around.

There is a reading area—a space defined by a 9 × 12 rug that I bought for $20 at a carpet warehouse. All the books, reading reference materials, and book sign-out boxes are in this area. The publication area, in another corner of the room, includes all writing reference materials and our one computer.

I have a small round table I use for one-on-one conferences. The table and stools are placed so I can see all students in all corners of the room as I talk to individual students.

23

Most of my time, however, is spent moving from student to student, discussing their writing or reading. If I sit down, it's at one of their tables, or on the floor talking or reading with them.

Materials and Supplies for Reading and Writing
Books

In order to keep track of the books, I have all the titles listed on the database of the classroom computer. I use subcategories for each book: author, title, genre, theme, number of copies, and price. With all this information listed, I can call up and run off lists of recommended titles based on any category. I also have a master list from which to order replacement copies for those lost or "misplaced."

In the references at the end of each chapter, I have included the books I use based on theme, or genre, in that chapter. In addition to those books, I have a classroom library of paperbacks which I have listed in Appendix B: Best-liked Books. I have more than six hundred books. I have listed only those that have been most popular over the years.

Building a classroom library

Once I started building a library of my own, I found the reading increased dramatically in my room. The students no longer had an excuse for not reading. All the books were right there. I started building the library by asking the students at the end of the year to list three books I should absolutely not be without for the next year. With my budget for textbooks, I purchased their suggestions. I do the same thing each year.

I ask the students in September to look over the books in my library. "What's missing? What book do you want me to buy?" I ask them. I want the new students to know right away I trust their choices. The kids can't wait for me to return with "their" books.

If I had a self-contained classroom, I would take the students with me to the bookstore and let them pick out their choices.

I support the book clubs—TAB, Troll, and Trumpet—pointing out each month the books I recommend or students recommend. With the bonus points I buy more books.

I ask parents for any books they no longer want. I suggest to parents that on their student's birthday, they give a copy of their child's favorite book as a gift to the classroom—paperback or hardback—with an inscription from the parents.

Maintaining a library

Part of my supply budget is spent on clear contact paper. I cover each paperback with contact, including a strip down the inside cover and end pages. The books usually last for several years this way. I print my name across the binding in permanent black marker, so there is less chance of losing them or getting them mixed up with the school library. I put a library card inside each book with the author's name, title, and price of the book. The student borrowers list the date of borrowing, their name and class section on the card, and file the cards alphabetically by author in their class index box, one for each section.

Each class section has a librarian (a volunteer chosen monthly), who is responsible for requesting the return of overdue books, making sure students don't take out too many books, and collecting money for lost books (the price of the book plus fifty cents for the inconvenience and the contact cover). The class librarian sends notices home at the end of each trimester if he or she has been unsuccessful with the student returning a book (see Appendix C: Book-borrowing Rules and Appendix D: Late- or Missing-Book Notice).

Writing Materials

I keep shelves with all kinds of paper available to the students—different textures, colors, and sizes. I keep a box labeled "Rough Draft Paper" filled with discarded mimeo-sheets from the aides' room. Kids write on the backs. I try each year to keep a holder filled with pencils and pens, flagged with my name on masking tape. Within days it is *always* empty. Writing utensils disappear like socks in my dryer.

I buy old calendars and specialty notebooks on sale, and tear them apart. One year students used parchment pages from a Writers Notebook to write letters to Soviet students. I cut pictures from calendars into strips and show them how to use the strips for bookmarks, writing down favorite lines from each book on the back of the bookmark. They cover the bookmarks of favorite quotes with clear contact paper and use them throughout the year.

Consumable Materials

The school supplies a manila file folder to each student to use as a portfolio—for their finished pieces. These folders are stored in open boxes by section and always kept at school.

Students purchase

- writing-reading logs
- *Merlyn's Pen* magazine
- *Writing* magazine
- *Punctuation Pocket* or *Topics Folder*
- *Writers Inc*

Logs: I either purchase a whole supply of notebooks at discount prices and sell them to the students at cost, or I have a local store set up a display for recommended sizes, usually about 7″ × 9″, because they fit nicely in the students' laps as they're reading paperbacks. Students still have the option of using any size or style that is most comfortable for them to write in. I tell them about Jean Little, who says: "Getting a journal is like buying shoes. You have to find the one that fits. And you are the only person who can tell if it pinches" (1986, p. 74).

Merlyn's Pen (P.O. Box 1058, E. Greenwich, RI 02818, 800-247-2027) and *Writing* magazine (Field Publications, 4343 Equity Drive, P.O. Box

16600, Columbus, Ohio 43272–2452, 800-999-7100) provide a nice balance between student and professional writing. They both provide a rich diversity of types of writing (including the visual arts), interviews with authors, and skills in context. Neither magazine waters down or adapts the language—it is real writing by real authors for authentic audiences.

Punctuation Pocket and *Topics Folder* are put out by the same company that publishes *Writers Inc.* The students choose the one folder (for holding their works in progress) that they feel would be most beneficial to them. Do they need the most help finding topics to write about or using punctuation as they edit? The choice is theirs. These folders are also stored in open boxes by class section so the students have easy access to them each day.

Writers Inc is a comprehensive reference book. I prefer a handbook for usage and skills rather than a textbook filled with meaningless grammar exercises. This book is so comprehensive I have numerous parents request additional copies for their offices or homes. *Writers Inc* and the folders are available from Write Source Publishing House, Box J, Burlington, WI 53105.

References

On a round table in the center of the room, or in magazine racks mounted on the wall, I keep several copies of the *Punctuation Pocket*, the *Topics Folder, Writers Inc,* books like *Writing Down the Bones, In the Middle,* and *Clearing the Way;* dictionaries, thesauri; and the NCTE reference books: *Your Reading, Adventuring with Books, Books for You.*

Jane Kearns, a teacher and writing specialist for the Manchester, New Hampshire, school district, showed me how to keep what I call a "Writers' Reference Box." In the box are file folders for a wide variety of genres of writing: short stories, essays, editorials, résumés, interviews, poetry, and so on. In each folder I put copies of students' and professionals' writing written in that genre, so students and I have easy access to good models. I also include articles related to the genre. *Writing* magazine is a particularly good resource for genre articles and interviews with authors. On the inside cover of each folder, I list some of the characteristics of that particular genre of writing.

I invite students to bring in articles or pieces they find that they think should be included in this reference box. I have added file folders that I use for minilessons: point of view, leads, endings, voice, setting, description, mood, and so on.

Until I started separating the writing by genre and keeping it filed and accessible, I usually lost the fine examples or forgot where they were. I use the examples for minilessons. The students can use them without depending on me if they have a question.

Near the class library I have a large loose-leaf notebook called "Authors." Each time we find an intriguing article about, or by, any author, we put a copy in a plastic binder insert and file it alphabetically in the notebook. It has become a rich resource for students wanting to know more about their favorite authors, or for me when I want to discuss certain authors.

Publication Materials and Opportunities

This area of the room is set up so students can independently publish their writing for the audience of their choice—themselves, a relative, friend, the classroom, or a national publication. It is an extension of the area where writing materials are kept. This is where the computer is located.

Materials might simply be special kinds of paper—parchment paper for students to write a final draft of a poem in calligraphy, materials and instructions for binding a book (½-yard pieces of cloth, discarded wallpaper books, cardboard, needles, heavy thread, poster board, scissors, glue sticks, rubber cement, paper cutter, nails, hammer, clamps, etc.), and information needed for sending writing to professional publications.

Binding books

Two kinds of book binding have worked especially well for my students, whether they choose to bind an individual story or an anthology: center-binding, as described by Mary Ellen Giacobbe (cited in Graves, 1983, 59–60), or outer-binding (called Japanese binding or blanket stitch).

I prepare several copies of the book-binding instructions and have them available at the publishing center for the students to use. I have samples of books using these bindings also there for them to look at and supplies of materials needed as well (see Appendix E: Center-Binding, and Appendix F: Outer-Binding).

Sundance Publishers (Newtown Rd., P.O. Box 1326, Littleton, MA 01460, 508-486-9201) sells blank books with hard covers in two sizes. I offer the information to the students, with samples available for them to see.

Professional publications

I encourage the students to send their writing and artwork to local and national publications. Among national publications, the ones I have found to be the most supportive and accessible to adolescent writers are *Merlyn's Pen*, *Writing* magazine, and Scholastic's *Scope*, *Voice*, and *Literary Cavalcade*.

I keep the latest issues of these magazines along with a handout that tells how to submit a manuscript for each. These handouts (see Appendix G: *Merlyn's Pen* Submissions) are kept in open manila envelopes stapled to the bulletin board and labeled according to publication.

I have a sign that reads: SASE—Don't forget a Self-Addressed Stamped Envelope, no matter who you are sending manuscripts or letters to. I also have the *Market Guide for Young Writers* edited by Kathy Henderson and published by Shoe Tree Press. It's a comprehensive listing of publishing possibilities and contests. It includes interviews with young writers and professional editors, sections on how to get started and how to prepare a manuscript, and further references for young writers.

The publishing center is also the focal point for any writing contests; NCTE and Scholastic are two good contests. *In the Middle* (Atwell 1987) is also a wonderful resource for publication possibilities.

I try to make sure that the potential publisher is not only legitimate, but offers encouragement, not discouragement. Jim Stahl, editor at *Merlyn's Pen*, understands the delicacies of rejection. His rejection letters offer enough encouragement to keep the writer trying. He is one of the few editors who also takes time to nurture growth by offering positive comments and suggestions with an offer to resubmit a piece when he sees potential.

We keep a large loose-leaf notebook called "Professionally Published Writing." It is filled with all the writing and artwork that my students and I have published since I began teaching. It includes letters to the editor in the local newspaper, short pieces for the parent newsletter, pieces published in *Merlyn's Pen*, and the winning pieces from our classroom for any contests, including those from the Scholastic Writing Contest. I want students to see their peers as writers. I want them to see me as a writer.

This collection of writing shows the growth and change I've undergone as a writer and teacher. My first piece of writing is a short summary for the parent newsletter of what we were doing in our social studies/reading class ten years ago. My latest piece is an article for *Educational Leadership*. This miniportfolio (of one classroom) also shows how much the students' writing has changed, or perhaps, how my expectations of what students can do has changed. The first professionally published student piece is Karen's ending to a short story in a student magazine. The latest student pieces are Jay's poem "School Daze," published in *Merlyn's Pen*, pictures of three "Art of Literature" projects (exhibited at the University of New Hampshire), and Gillian's poem "Grandpa" published in the August 1990 issue of *Seventeen*.

Literary magazine

The eighth grade publishes an annual literary magazine. It is a collection of the best pieces from each eighth grader: poetry, plays, short stories, personal narratives, drawings, sometimes even a best sentence.

The school budgets for the magazine, so that each eighth grader receives a free copy. (Soliciting ads would be one way of paying for the magazine, should the school be unable to fund the publication.) The students submit ideas for covers and names. Student volunteers help lay out the magazine and letter titles and page numbers. Each piece submitted by a student is supposed to be typed on the school computer and saved on a class disk. As the final editor, I can call up the pieces to make sure they are mechanically correct. I want writing that goes beyond my classroom to be the best it can be. Every writer has an editor who saves him or her from the embarrassment of mechanical errors. Student writing that goes beyond the classroom should be handled the same way. I am the final editor for my students' writing. The printouts are then used for layout.

My goal is to publish the writing of every student, not just the best. No matter what the students say or do, it is important to every one of them to have a piece in this class anthology. A few years ago I spent months reminding one student to get his piece submitted. I finally gave up. "Well, it's his own fault," I concluded and stopped nagging. On the last day of school I watched him flip through the pages quickly trying to find his piece, then start back again slowly page by page when he couldn't find it. Tears welled up in his eyes. He was devastated it wasn't there.

I know all about teaching "responsibility," but it wasn't worth the look on that boy's face when he realized he wasn't part of that magazine. It doesn't matter who disagrees with me; if I have to type that one last piece to make sure every student is published, I will.

I encourage publication—going public with our words. Our students have important things to say and unique ways of saying them. Making the piece correct mechanically seems to be a by-product of knowing their words are meaningful. The drive to write for an audience betters the product, but more importantly, what a writer has to say touches another person's heart or mind. It feels good to know that our words affect someone, causing confirmation or change of feeling or thought.

Greg wanted to give his mother a poem he wrote for Christmas. He went over his rough drafts again and again to make sure every word was spelled correctly and every mark of punctuation was just right before he started the final piece. He used parchment paper and calligraphy pens. His dad framed it.

Craig was furious when an especially thorough letter (see chapter 7 for Craig's letter) in response to a *USA Today* spread of "Why Johnny Can't Write" was ignored, while my cover letter was published. No one valued his opinion because he was a student, not the teacher. He had important things to say to an audience, about audience, and this editor only confirmed what he was trying to say—students' opinions are often ignored.

Publication must be a choice for each student. In my classroom, publication means taking a piece to finished product, making it the best it can be for the writer first, and then deciding if the piece goes to a wider audience. Whom is this for? What would you like to do with it? Do you want to simply leave this in your portfolio? Would you like to read this to the class? Do you want to post this for the class, take it home, send it to someone? Do you want to put this in the parent newsletter, the school newspaper, or submit it to the literary magazine? Do you want to try *Merlyn's Pen*, a writing contest, or *Seventeen*? The options are up to the students. I am there to provide the opportunities.

Publication ideas I want to try

I would like to try publishing a calendar, using the students' best writing and artwork throughout the months. The calendar might be a district project, listing all school activities for the year.

I would also like to try "Postcard Poetry." My idea would be similar to what NCTE puts out each year: Postcard Poems. Postcards could be packaged by tens in an envelope, with a different student's poetry from any year on each card, or packaged with the same poetry, or just one student's work, on each card. They could be sold to parents and other students to help raise money for books in the classroom, to bring writers into the schools, or simply to meet the costs of publishing the postcards professionally.

Both projects give students an audience beyond the classroom. They also show an audience how articulate, sophisticated, and literate middle school students can be.

I keep a manila file folder called "Ideas to Try" on the bulletin board. Students know they can add to it. Last year Monica added an entire list of books, articles, and movies she read or viewed about the Holocaust to my recommendations.

Each year students surprise me with their own ideas. This year Lindsay got an idea from the book *One Child* (Hayden 1980, 55–56) and set up a Kobold's Box, a place for students to send anonymous messages to each other when they notice something nice that the person did. She typed the page from Hayden's book that described what the box was about and posted it above the box. She put an envelope of blank paper near the box and a pencil to make note writing to each other easier. She also added a note suggesting that students remember the box was for "nice" notes. (When Lindsay heard some students fretting about never getting notes from the Kobold's Box, she made sure the "Kobold" noticed. Notes appear for everyone, even me.)

Time, Structure, and Expectations

Time

My five language arts classes are approximately fifty minutes long. There are no other reading, writing, or English classes. Because of the time constraints I have to make choices—about curriculum, about goals, about objectives, about what is best done with the time we have. Although I have a hard time delineating "writing" time or "reading" time, I spend more of my time on writing in class and reading out of class.

Students are expected to read for at least a half hour each night. I do devote one whole day to in-class reading, however. I think it is important for students to see I value reading and take the time to do it. By taking time to read, I also know more about books and authors than ever before.

We take at least thirty minutes of each Monday for silent writing. I begin my writing with the students during this time. But my obligations are to help them. I have to spend my own time on writing outside the classroom, so I can move around helping them. I want them to see me take some time to begin the writing with them. I share my struggles and my successes throughout the year.

Structure of the Reading-Writing Workshop

Monday through Thursday

5–10 minutes	read aloud and/or minilesson
30–35 minutes	Monday: silent writing
	Tuesday through Wednesday: individual conferences
	Thursday: response/small-group sharing
5–10 minutes	whole-class sharing/minilesson/reading aloud

Friday

10 minutes	reading aloud
35 minutes	silent reading
5 minutes	whole class sharing

Expectations

Within this framework, I have certain expectations of my students:

Writing

- to write from three to five pages of rough draft writing per week (maintenance of working folder)
- to take *two pieces* of writing to final draft every six weeks (maintenance of portfolio)
- to read writing to at least three peers and at least once in teacher conference before final drafting

Reading

- to read for a half hour five nights per week
- to keep a list of books read

Reader's-Writer's log

- to respond to/reflect on reading and writing and observations in log with three to five pages of writing per week
- to find at least three words per week, which are new or unknown (from reading, writing, listening, etc.)
- to write those words in log with date found, citing reference and sentence or context in which word was found, and dictionary definition in own words
- to maintain a spelling list in log of words consistently misspelled
- to maintain a section on class notes in response to minilessons on skills and learning strategies

Self-evaluation

- to set goals every twelve weeks with regard to what the student wants to be able to do better as a reader, writer, speaker, and listener
- to evaluate him- or herself as a learner every twelve weeks based on goals set and achieved

Discussion of Expectations

Writing

To become better at anything, it takes practice. I want my students to become better writers; therefore, they must do a lot of writing. From the quantity of writing they produce (three to five pages per week, on one topic or a variety of topics), they select the most effective drafts to craft into their best pieces. They know they have two finished pieces due every six weeks. The pieces might be one-page poems, fifteen-page short stories, three-page essays, or letters to an editor. The finished pieces might come from specific genres or themes I ask the students to try or from ideas of their own. It's the student's choice.

Once I started requiring a number of pages of writing a week from which the students select their most effective pieces, students began looking at their writing more reflectively. Evaluation begins with these decisions. This selection process allows writers to begin to recognize the characteristics of effective writing.

I seldom read all these rough-draft pages. I can't, and there's no reason to. I do a lot of rough-draft writing in my search for what I want to say and the best way to say it. I don't want anyone to read some of that writing. It's awful. Most writers work the same way. Kids need to be given the same opportunities—to write terribly as they search for the good writing. I simply count that these rough-draft pages are done or not done.

The pieces going best are the ones taken to peer responders and to conferences with me. Over the years I've set up response groups in a variety of ways. Based on student feedback, I now ask only that students read their writing to at least three peers for feedback before they read it to me. Students asked for that freedom for two reasons: depending on the genre, they want to go to the students who write that genre best ("I prefer to read this poem to Sandy, she's a poet") and because relationships between adolescents often fluctuate so rapidly, only they know who will give positive feedback ("Scott just broke up with Amy and she is *not* going to give him good response to his writing"). So long as this works, I'll stick with the students' choices. If it doesn't, then I'll set up response groups.

Reading

It is impossible to fit all the writing and reading into a fifty-minute period. For that reason I have to make choices. I can best help students in class with their writing. Therefore, my standing homework assignment is to read for a half hour each night, to keep a list of books read, and to respond in a writer's/reader's log with three to five pages of writing per week. The kind of reading students choose to do and the ways they respond to that reading are discussed in detail in chapter 3.

I want students to know I value reading enough to take class time to do it. In addition to reading aloud frequently during the week, we all read silently for the entire period on Friday. I read during all Friday classes, and I do *not* feel guilty. I know more about books now than I ever did. We see each other committed to books. We learn to read *by* reading, not by reading *about* reading.

Within this framework of expectations, I use different themes and ideas for getting at various kinds of writing and reading. The students have a lot of choice as to what they read and write. But at times we read the same literature. At times I ask them to try certain kinds of writing. They are always the final decision makers, however, about what works or doesn't work for them—what gets evaluated, what doesn't.

Keeping Parents Informed

Right up front, parents need to know what they can expect for, and of, their children in my classroom. The first day of school I send a letter home that briefly describes my philosophy and curriculum, what is expected of their children overall and nightly, and how they can help us achieve those goals (see Appendix H: Letter to Parents).

Trimester I

- 6 weeks: introduction to reading-writing workshop; free choice: writing topic/format, book; read aloud: reading-writing-learning process
- 6 weeks: relationships: self/parents/grandparents

Trimester II

- 6 weeks: reading-writing workshop
- 6 weeks: relationships/cultural diversity

Trimester III

- 6 weeks: relationships/cultural diversity and reading-writing workshop
- 6 weeks: reading-writing project

The following chapters are laid out in a manner similar to my year's outline:

- chapter 3 addresses immersion in a reading-writing workshop
- chapter 4 discusses a thematic approach through generational relationships
- chapter 5 maintains a workshop setting through the whole-class reading of one book (cultural diversity)
- chapter 6 ties all the writing and reading together

Organization cannot be relegated to the back of the writing-reading teacher's mind, any more than it can be placed as the last chapter in a book. Organization has to be right up front. I want to be "so damned organized" I can concentrate on the students: Who are they? What do they know? How have they come to know that? How can I help them become better readers, writers, and learners?

Works Cited

Abrahamson, Richard F., and Betty Carter. (Eds.). *Books for You:—A Booklist for Senior High Students*. Urbana, Ill.: NCTE.

Atwell, Nancie. 1987. *In the Middle*. Portsmouth, N.H.: Boynton/Cook.

Davis, James E., and Hazel K. Davis. (Eds.). 1988. *Your Reading: A Booklist for Junior High and Middle School Students*. Urbana, Ill.: NCTE.

Goldberg, Natalie. 1986. *Writing Down the Bones*. Boston, Mass.: Shambhala.

Graves, Donald. 1983. *Writing: Teachers and Children at Work*. Portsmouth, N.H.: Heinemann.

Hayden, Torey. 1980. *One Child*. New York: Avon Books.

Henderson, Kathy. 1988. *Market Guide for Young Writers*. Belvidere, N.J.: Shoe Tree Press.

Jett-Simpson, Mary. (Ed.). 1989. *Adventuring with Books: A Booklist for Pre-K–Grade 6*. Urbana, Ill.: NCTE.

Little, Jean. 1986. *Hey World, Here I Am!*. New York: Harper and Row.

Romano, Tom. 1987. *Clearing the Way*. Portsmouth, N.H.: Heinemann.

Writing and Reading

"Lemme get this straight," Shawn asked in chapter 1. "We get to *read* during reading and *write* during writing?"

A few years ago that wasn't such a strange question. The majority of students now, however, come from elementary classrooms where they've been immersed in reading and writing since kindergarten. They come into our middle school expecting to read and write. Most know how to find books and topics. Most know how to respond to each others' writing and reading.

Even when students come from other districts, they catch on very quickly. Kyle's first year at Oyster River was in eighth grade. After explaining what I hoped for each of us on that first day, I heard an audible sigh of relief and Kyle's voice, "At last, a creative teacher."

Kyle had it wrong. I wasn't so creative, he was. All I did was offer him the opportunity to show all of us what he could do and how he did it. The same choices were there for all of us. Kyle accepted the invitation. By the end of the year he had written two 250-page novels. If I hadn't watched him compose most of the writing on the computer in my room, I wouldn't have believed he knew what he knew. The following is an excerpt from his novel *Speak Softly and Carry a Big Stick: Theodore Franklin Roosevelt*

Chapter 1: A Brief History of U.S./U.S.S.R. Relations, 1989–1993

The delicate political situation that is the latter half of the Twentieth Century began with an eight year skirmish between the countries of Iran and Iraq. In its seventh year the United States of America intervened by placing a large naval force in the Persian Gulf and escorting Kuwaite supertankers. The tension escalated when President Ronald Reagan invoked the War Powers Act. Petty skirmishes followed and the breaking point was reached when the U.S. blew five of Iran's minelayers out of the water along with their ten speedboat escorts. The Iranians retaliated hastily and left the USS *Enterprise*, *Bridgeton*, and *Missouri* merely hole-filled derelicts lying 200 meters underwater with their vicious Silkworm missile attacks. The country of Iran finally got what it had coming for the past eight years. The nuclear missile carrier Arkansas unleashed its full load of hydrogen heathens upon the windswept sands of Iran. The world's first nuclear strike left the world in a state of shock and Iran a very large glazed crater.

After the deadly radioactive haze cleared, the world's leaders finally found their voices. The Soviets wanted reprisals, allies questioned the wiseness of their decisions so long ago to become U.S. allies, sides were taken, and, most of all, the White House was in turmoil. The new president was not to be envied. She had to sweep up the ashes of Reagan's impulsive, half-brained decision. Reagan certainly went out with a two million degree bang . . .

Chapter 2: The Roosevelt Project

In late 1979 a technology leap-frogging project was initiated by the Delta Foundation, the Navy's think tank that even the Navy didn't know about. They were instructed to design the ultimate battleship: specifically a trimaran attack sub that could hold its own in both a surface and subsurface role. The result was the Roosevelt Project. The prototype was the USS Theodore Franklin Roosevelt and it made the Soviets' subs look like tramp steamers. It fell short in the weapons department but nothing could hit it because it was so fast.

When the National Strike Force went looking for a flagship they stumbled upon the Roosevelt Project and realized its potential . . . They gutted the prototype and added lasers, Electron Beam Weapons (affectionately called ELBOWS by their designers), and Mark 67 torpedoes. They replaced the now ancient conventional drive system with a prototype MHD (MagneticHydro-Dynamics). The MHD drive uses magnetic fields to propel streams of water scooped from an inlet at up to 100 knots down a tunnel to be expelled from the rear of the sub to produce speeds of up to 80 knots with barely any noise! A Microfusion reactor was installed as a powerplant and the modifications were nearing completion. The electronics package included the world's first artificially intelligent personality and EW equipment. The hull was coated with a material that made it, under proper conditions, immune to sonar and radar waves . . .

Chapter 3: A Brewing Storm

The White House 1993 A.D.

Even at age 47 President Amanda Reilly's beauty hadn't faded. Not that many Senators cared to notice. Her political prowess and style made them think of her as "just another one of the guys" and as a formidable opponent. She had her auburn hair braided and her lanky six foot stature was bent over a folder on her desk in the Oval Office. She was studying the Delta Section of the folder on the Roosevelt Project that Al Randlin, the chief engineer for the Theodore Roosevelt and Sandra Lakmince, Delta Section's Special Projects Director, had handed her.

Her brow furrowed even deeper as she scanned the contents of the file. Suddenly she turned very red and her eyes widened. She slammed down the folder and began to smolder at the two engineers. "Uh-oh, what do you want to bet she read the part about our mutual friend's position as the commanding officer? Christ, we knew that she hated his guts, why couldn't we have left that small fact out?" Randlin whispered to Lakmince . . .

Kyle read Tom Clancy, Richard Ludlum, and Isaac Asimov voraciously. He tried on their styles in his fiction. The technical descriptions of aircraft and ships were so detailed in his writing that I felt like I was at the controls. "How do you know all that?" I asked.

"I've been building military models for years," Kyle said. "I just pretend they're the real thing in my descriptions."

A sly smile crept across Kyle's face. "Many people claim that I indulge in plagiarism—shades of *Hunt for Red October* . . . but I must admit, once I get an idea, from anywhere, a joke, TV, a story, my pen starts to itch and whoosh—I have a whole chapter. . . . I'm a very visual person though. Once I get an idea, I sit back, close my eyes, and see a whole scene before I write it. The more I read, the more I can visualize what I want to say. The more I write about what I read, the more I understand."

Prior to eighth grade Kyle was a closet reader and writer. He did all his own reading and writing at home. He now has a computer program at home he calls "technobabble." Whenever he reads something in a book that he loves or that sparks an idea, he goes right to the computer and jots it down in "technobabble." He isn't sure when he'll use the information—he just knows he wants to hold it for future reference.

No teacher had ever asked Kyle what he knew as a reader or writer or had invited him to use that knowledge in school. He was bursting to read and write. Some students don't accept the invitation as readily. They need to find out who they are as learners at this time. They need to find out who I am as a learner, and how I will respond and react to what they know. They need to know me as a person. They need to trust me. We need to know and trust each other. It takes time.

It took a lot longer to get Steve to write what he knew about—dirt bikes. I seldom convinced him to read a book. He did read *Hatchet* by Gary Paulsen, but most of the time he had a dirt-biking magazine hidden under the book. It was the only reading I could get him to respond to in his log:

> The CR 500 is kick but the whole way it has a superer strong motor, awsome clutch, good brakes, and the suspintion is as smooth as a babeys butt . . . the honda 250 X is awsome it is a medeam bore 4 wheeler 4 stroke so that means less werk in the garge and saves gas . . . I think the YZ sucks it doesent have a strong motor, good suspintion, and the only way to get the hole shot is every body bike blew up. all yz really bite Like the yz 80 hasent been changed in 4 years thats bad! . . . How not to become a winter icecycle sense its starting to get really cold I thought it might be smart to read this artical sense I plan on rideing in the snow Im wearing goggles ski mask cover alls boots coat levis sweater 3 pairs of socks and two pairs of gloves why so much. When you start rideing at a tempetcher at 32 degrees and start rideing the wind chill factor is below zero . . .

Kyle and Steve have unique voices. Kyle needs *time* to read and write. Steve needs to be convinced that someone cares what he knows before he will invest in reading and writing. Once he takes responsibility for his learning, I'll be able to teach him the mechanics he needs so badly. At the beginning of the year I immerse all of us in reading and writing experiences, so we can begin to find our own voices and so the students can trust me as a learner also.

Writing

I don't give my kids topics, but there are times, especially at the beginning of a new school year, when they need maps of the territory—their own territory.

Getting students started writing isn't getting students writing. It is getting them thinking about themselves: what they know and what they want to know.

The writing exercises I take my students through are merely that— exercises to get them thinking about themselves, their territories. From their responses I learn about each one as an individual. It gives them direction for the year and it gives me direction for working with them. The exercises give them strategies for gathering and sharing information.

I use the following writing exercises at the beginning of the year for all the reasons I discuss within each category. What the students produce in response to these exercises counts toward their rough-draft requirements. Sometimes they continue working on the pieces and they become final drafts. It's always the student's decision. I ask students to

- fill out an anecdotal card
- complete a writing-reading survey
  - design a visual representation of the most significant information on the survey
- compose a résumé and cover letter
- conduct an in-class interview
- design a visual representation of their lives, based on a brainstormed list of positives and negatives
- in their reader's-writer's log, discuss selves as readers, writers, speakers, and listeners; answering the following questions:
  - what do they already do well in each of those categories?
  - what are three goals (things they'd like to be able to do better) in each of those categories?
  - what are three topics they want to write about?
  - what are three books they really want to read?

Anecdotal Card

I hand out a 5 × 8 index card to each student. They write their name, mailing address, phone number, and class section in the top right-hand corner. On the card I have them answer the following questions:

- What makes learning *hard* for you? or What makes learning *easy* for you?
- Write one sentence about yourself that you also think is true for everyone in this room (e.g., I am a human being).
- Write one sentence about yourself that you strongly suspect is true *only* for you (e.g., I spent a week on an island in the Philippines with no electricity and no running water).

We share responses. Talk. Laugh. Get to know each other.

I file these cards on my desk. Every year I try to focus on one thing that I don't feel I do very well. I have a difficult time keeping anecdotal records on my students. With anywhere from 100 to 125 students per year, and fifty-minute classes, it is difficult to keep such records.

Evaluation continues to be my focus. Because I believe it is important to keep anecdotal records as part of my evaluation system, this will be one of my areas of concern this coming year. Each day I will clip two students' cards from each class to my clipboard. I will make a point to notice something about what they are doing and jot that down on the back of their cards. By using these same cards I can also remind myself of what makes learning difficult or easy for each of these students. If I make an effort to do this, I should be able to make at least ten notations on each student and not lose the notes. It should also help the "quiet" kids from slipping totally through the cracks.

Reading-Writing Survey

I give this to the students the first week of school (see Appendix I). I want their answers fresh and individual, so I usually give it to them before we've had time to discuss reading, writing, or my expectations. I collect their answers. I give them the same survey at the end of the year. I then give them back the one they responded to in September and ask them to comment on any observations they make about themselves as readers and writers. These surveys and their observations become part of their port-folios at the end of the year.

Visual representation

Before the students hand in their surveys I ask them to look back at what they wrote about writing and reading. If they had to pick the most significant point to tell someone about writing or reading, what would that one thing be? I ask them to write that one thing and then represent it visually. I post these visual representations and some of their quotations all around the room, with all the professional quotations I especially like, such as Annie Dillard's: "[The writer] is careful of what he reads, for that is what he will write. He is careful of what he learns, because that is what he will know" (1989, 68). I want kids' words and thoughts published that first week of school.

Students wrote:

"Reading is like getting advice from someone who is a professional at writing."—Melissa

"Writing has power if you give power to it. It can change what people think about things. It teaches you about yourself."—Abigail

"If a person reads a lot and likes to read, then he/she is actually studying writing. What you read helps you develop your own writing style."—Nathan

Lance made a cardboard figure to represent his statement (Figure 3.1). Kari wrote a "minibook," because she had too many "important points to make" (Figure 3.2).

Résumé

I am frequently asked to conduct workshops for teachers in other school districts. Several years ago I thought, why am I doing this alone? The students are the ones who know what works best for them. I began taking my students (all volunteers)—ten to fourteen at a time. The students work with the teachers, taking them through writing exercises, sharing their reading and writing, and answering any questions the teachers have.

Figure 3.1
Lance's cardboard
figure

YOU HAVE TO GO OUT ON A LIMB TO WRITE WELL

Lance

Before we go on the first workshop, I ask the students to submit a résumé, to show me who they are as readers and writers and convince me they have something to teach teachers about learning. I share a résumé with them, showing them what's important to include. They also write a cover letter describing in narrative form why they would like to be considered for the "job."

These résumés and cover letters are an opportunity for me to teach business letters, and they are the students first persuasive pieces. This is writing for a purpose. It's a format they can use in the future. It also lets the students reflect on and summarize what they've succeeded at. It lets me see how they feel about themselves and what's important to them. The workshops let them articulate what they know. For the students these are purposeful "orals."

Figure 3.2 Kari's mini-book

I used to ask only for a letter, until Andy taught me to ask for more. He included this résumé with his cover letter:

NAME
Address
Telephone Number

OBJECTIVE:
To teach teachers how to teach better and to share problems and creative ideas about teaching.

SKILLS:
Writes creative stories
Good at poetry
Has a wide vocabulary
Reads well

EDUCATIONAL BACKGROUND:
My education began in Mt. Airy, Maryland where I went to Mrs. Libby's Nursery School for a year, then moved on to Green Valley Kindergarton.

Next we moved to Durham, New Hampshire where I went to Mast Way Elementary School and worked my way up and out through grades 1–5 having Mrs. McClure for first and second grade, then I had Mrs. Roberts for third, I continued on with Mrs. McKinzie, then last but not least I had Mrs. Kolbjornsen for a fifth grade teacher. During this time I was doing very well in English getting good grades and remarks and having the major role in a school play.

I then started my sixth grade year having Mrs. Bechtell for an English teacher; that year I got an A+ for a final English grade. Then I moved up to having Mrs. Rief for an English teacher in seventh and eighth grade.

BIOGRAPHICAL INFORMATION:
    Date of Birth—March 25, 1973
    Place of Birth—Vallejo, California
    I was born in Vallejo, California then our family moved to Mt. Airy, Maryland and resided there for five years. Then we moved to Durham, New Hampshire where we presently live.

REFERENCES:
Mrs. Bechtell   ORMS English Teacher, 868-2820
parent's name

Brian's cover letter included the following sentence: "I feel I would be good for this job because I've had some good teachers and some bad teachers and could help people out." He also included this section in his résumé:

PUBLICATIONS
1986 (7th grade)   "Life of a Woodchuck"
1984 (5th grade)   "The All-Star Team"
                        "My Cat"
                        "Mt. Rushmore"
                        "Why I Hate Airline Food"
1983 (4th grade)   "My Trip to the Northwest"

I use Marie's cover letter and résumé (Appendix J) as a complete example to show the class, with excerpts from other résumés. Before the students actually write their own cover letters and résumés and after they've seen several student examples, we brainstorm and list all the possible categories of information they might include. Last year some of the students even included a copy of their best piece of writing from previous years. They choose the categories with the most meaning for them.

We talk about the format for a business letter and what is essential to include. We discuss the presentation of information. I teach all the skills that go into this kind of writing and the mechanics they need. They draft ideas, share, conference, redraft, and edit—the same way they would for any piece of writing.

We've brainstormed the following categories:

Objective

Skills (What are you good at?)

Educational Background (schools, teachers, experiences)

Related Experiences (visual or performing arts)

Publications

1. Choose a partner—the person in the room you know the least about. Teacher chooses a partner also, if there is an odd number of students. (When there isn't an odd number, I sit by a pair that intrigues me for one reason or another, and jot down questions, observations or overheard comments, so I have something to write about.)

2. Hand out 3 5 × 8 index cards to each person. Fold two of the index cards in half. Number the cards:

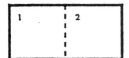 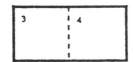 

3. Section 1 (3 minutes): Write down 5 questions you want to ask your partner to get to know them better. (This can be left totally open, or you can specify: to get to know them as a writer, as a reader, as a learner, etc.)

4. Section 2 (5–7 minutes): Ask your partner your questions. Record his or her answer in Section 2. (If you discover you are getting just "yes" or "no" answers, reformulate your questions so your partner is doing the talking.)

5. Section 3 (3–5 minutes): Look over the answers to your questions. What appeared on the card that you didn't expect? What surprised you? intrigued you? do you want to know more about?
List 5 more questions in Section 3 that FOCUS on that one thing.

6. Section 4 (5–7 minutes): Ask your questions and record your partner's answers in Section 4. This time, try also for direct quotes, significant words which show the voice, the uniqueness of the person. Note body language also. (How does the person respond nonverbally to your questions?)

7. Section 5 (11 minutes): Develop a draft of a piece of writing which would help a reader get to know the person you are interviewing, OR about anything you heard your partner say that you would like to talk more about. Follow the writing wherever it takes you. Try for direct quotes.
One of the things to keep in mind about a short piece of writing is the LEAD, the one or two sentences that pull the reader in and focus the piece. Listen to all the different ways other students and teachers began their writing:

> *Mark stood in line for eight hours waiting to board the aircraft carrier John F. Kennedy. Is he obsessed with aircraft and ships? Yes . . .*

> *"I know this sounds like a strange question," I said to Margaret, "but I wonder how you would relate your playing in a wind ensemble to your teaching."*

> *Circled in red, Brad's journal entry for August 13th reads BAD DAY, in capital letters.*

> *Sandy leaned forward in her chair, held Sheila with her eyes, and asked, "Do you ever take time to do absolutely nothing—time just for you?"*

> *"It has its advantages, but it's, well, kinda lonely," Becky said, explaining what it's like to be an only child.*

> *"Experiencing of pain," she said, calmly, seemingly matter-of-factly. Yet the inner tone of those words had more of an impact on me. She seemed to be speaking louder inside than what was said outside.*

Figure 3.3 Interviewing exercises

continued

Yaygers, double-back fly-aways, free-hip circle hand-stands, yachenkoffs, round-offs, back tucks, sukaharas. Sarah makes me feel like I'm at the Olympics.

Amy desires motivation. "I run faster," she says, "if people cheer for me and watch me run."

"I always thought I had to have all the answers," said Roger. "Now I know it's more important to have the questions."

"Look at this, Miss Chisholm," Karen said, calling me over to the gravestone. "Look at what it says on this stone. 'She done good.' That's not proper grammar."

"You know, Karen. I hope someday someone can say that about me."

Have you ever felt the urge to get even? Amanda did.

8. Read your rough draft to your partner. (2–3 minutes) Partner responds to what she likes or hears that sticks with her—words, phrases, information—, asks any questions, and confirms the information or interpretation.

9. Revise your draft (1–3 minutes): add, delete, reorganize, clarify, etc.

10. Ask if anyone would like to share his or her piece of writing. (I share mine, especially if no one else volunteers.)

Figure 3.3 Continued: Interviewing exercise

Biographical Data

Work Experiences

Significant Books or Authors

Professional Presentations

References

In-Class Interview

One of the most important strategies we can give kids for gathering information is how to conduct an interview. This is only one of the reasons I take my students through an interviewing exercise (Figure 3.3). After the exercise, which takes from one to two class periods, I ask the students to talk about everything we've done. We discover we have

... experienced the entire writing process as a writer, reader, speaker, and listener.

Although I give the students a broad topic because I ask them to get into partners, they have choice of ultimate topics by the questions they ask. When I ask them to look at the information that surprised them or they want to know more about from section 2, they are making the topic their own. They choose the focus and the format for presentation and even decide whether they want to go public, by sharing what they have written. I'm always amazed at the quality of what they produce under the time constraints. I think the limited topic helps them do that.

We all tend to write more than we need to—the "bed-to-bed stories" Don Graves (1983) talks about. By having the writers choose a focus for their second set of five questions, I am showing them how to discard superfluous information. I am teaching them how to evaluate. They are the ones choosing. This is similar to the same questions I might ask in a conference: "You wrote about this and this and this. If you had to choose the one thing that surprised you the most or was most important to you, what might it be?"

By asking them to come up with five questions on that one area that surprised them or they want to know more about, I am teaching them how to develop a topic.

Throughout the entire process they are drafting, revising, reading, speaking, and listening—to themselves and to a partner. Throughout every phase they are evaluating—making judgments about what they want to know, what works, what is important, what makes sense. Watch their eyes. Watch the thinking.

. . . been introduced to the kind of response I reinforce throughout the year.

Over the years I've discovered the most helpful response to me as a writer has come from adapting Peter Elbow's technique in *Writing Without Teachers* (1982). Writers need to hear what the listeners heard that stuck with them, any questions they have, and a suggestion or two. It has worked for me and seems to work for the majority of students.

I intentionally have the students find the person in the room they know the least about. If they can feel comfortable working and sharing their writing with that person, they most likely will feel comfortable and safe with anyone. By having the students write about their partners, you insure positive response. The partners are not going to make derogatory remarks about the writing if it is about them. The listeners give positive responses because it is about the listeners.

The writers have the opportunity to revise their texts. After receiving responses, they can add, delete, reorganize information in a way that makes the most sense to them. The writing always remains in the hands of the writers. I want the writers maintaining full responsibility for that writing.

. . . been introduced to skills we will need and I will reinforce throughout the year.

I think one of the most important skills we can teach our kids is how to gather information from primary sources—real people. Whether they are interviewing someone or gathering information from first-hand-account books, they need to know how to formulate questions that elicit information. Yes or no answers give the writers little to work with. But I want them to discover that. Even in this exercise, once they know they have to write from the information they gather, they discover they have to get the partner talking. We talk about how they did that. What happened when they asked certain questions? How did they restructure what they asked?

They researched a topic by brainstorming questions; reflected on the information by looking for surprise, finding specifics, connecting ideas; rehearsed their own voices as they heard about themselves and read their own writing; revealed what they each knew; and restructured a meaningful text for themselves and others.

When they conduct research to make their writing the best it can be throughout the year, I will remind them of this exercise. How did you do that? I will ask. What can you do now?

. . . *learned skills in context.*

In one to two short class periods with this exercise, I introduce the students to interviewing techniques, leads, and dialogue. I model the interviewing by participating in the process and discussing everything we did. I share and discuss a variety of leads and explain how dialogue, the voice of the speaker, is embedded in the lead or in the text—if they choose to do that.

This is plenty for now. Am I worried about spelling, punctuation, and so on? Not yet. I want the students to focus first on the content and the process of making meaning. This writing is for them. When it's time to go public, then we talk about mechanics.

. . . *chosen publication for our own reasons.*

Do all students read their pieces aloud? Not at all. Do I make them read them? Never. Only when they are ready. I want kids to know they can trust me to believe they have responsibility for determining when their writing is worthy of publication.

In this exercise, some writers feel they haven't done justice to the topic. Some aren't committed to the writing. Others simply couldn't get their partners to talk. It works for some, not for others. Every time I apologize to Don Murray for my inadequacies as a writer he tells me that you have to do a lot of bad writing to get the good writing. I share that with the students. I may read my piece because it is bad. I want them to know writing is hard work, and sometimes it doesn't work—especially for me.

This is simply an exercise. If the writing is going well, the students may decide to continue with their pieces. Otherwise, the writing becomes part of their five pages of rough-draft writing for the week. I may see it only to count it as a page of writing. It is not graded. Whatever happens to the writing, it is the writer's decision.

This exercise also has a number of fringe benefits for writers and readers:

- It establishes a social environment in which there is a lot of talk, laughter, and sharing. Students get to know one person in a way they normally wouldn't.
- They hear a variety of voices—kids they might not normally talk to. They hear their own voices when other students read the pieces about them. They hear their own voices as writers when they reread what they've written.
- It feels good to know someone needs and uses our words.
- It establishes a positive, nonthreatening atmosphere in which kids feel good about each other and are willing to share because they know they don't have to.
- They become more confident as they start talking about everything they know and can do. Even in such a short time span they hear effective pieces of writing and know they can do it.

- It reinforces the value in reading, writing, speaking, and listening when the teacher is a participant in the same process and talks about what worked or didn't work along with them.
- It gives them ideas for future writing: topics come from their own questions, something a partner said, or something they heard someone else write about.
- Most importantly, we learn about each other—as people.

I adapted this exercise from a writing-response exercise of Donald Murray's. He has partners write on index cards about an event, person, or place important to them. One partner role-plays the teacher, getting the writer to talk about what surprised her. With each card, the writing gets more specific. The writer is always writing about him- or herself. I use this exercise also with my students.

I found, however, that my adaptation works better with adolescents because they feel less intimidated writing about someone else. (They know they have the option to write about themselves or wherever the writing is taking them as they draft.) They respond with positive comments when the writing is about them. They feel good about themselves when someone needs and wants their information. It teaches them information-gathering and interviewing techniques they can use for any writing. It still teaches them they have things to say and unique ways of saying them.

Karen's interview contained just snippets of information, even after she focused her questions. But look at how she tied everything together:

Have you ever felt the urge to get even? Amanda did.

During the summer of 1984 Amanda went to Cape Cod. One day in a restaurant there, she dumped a bottle of cream over a girl's head. "She was a little brat who we hated," was the explanation Amanda gave.

Maybe the reaction was a little hot-tempered, but Amanda is a Taurus, born May 8, 1972 (the year of the rat). She can spend plenty of time analyzing her behavior when she becomes a psychologist, or defending herself, when she becomes a lawyer.

Of course, if Amanda gets her one wish for "a million bucks," she won't have to work at all. She can just travel. She has already lived in New Hampshire, Florida and Georgia.

I wrote this piece in a workshop in Massachusetts:

"I know this sounds like a strange question," I said to Margaret, "but I wonder how you would relate your playing in a wind ensemble to your teaching?"

Not to worry. Margaret's face lit up, she broke into a smile and her hands began to weave and play like a master conductor.

"In our wind ensemble we're doing a piece now by a composer where we have to concentrate and shut out every other instrument or part around us, in order to master our part. Yet, when it comes time to putting it all together, it's essential to the melody and composition that we all work together."

Margaret sat back, almost surprised that that's how this musical composition has to work. She's equally surprised by the similarities in her classroom. The kids are very private about their writing until they get a sense that they've succeeded, mastered "their part—their composition." Like the musicians, they can't wait to share their success. "It surprises me how excited they get by their writing. It surprises me how much they know and how much they want to share. They want to read everything."

Knowing Margaret's most important objective in working with learning disabled kids is giving them self-confidence and the balance between safety and responsibility, I'm not surprised.

Amanda, one of my students, wrote this piece about Sandra, a teacher she interviewed in a workshop we conducted in Brookline, Massachusetts:

"I don't really have time to write anymore, but I think I might enjoy it if I went back. I used to write for T.V." A cloudy look came over Sandra's face as the silence grew. Then, like the snap of fingers, her expression cleared and her eyes were bright. "I want to make ethnographic films, and now I realize writing will be a major part of that."

She seemed a little surprised at the thought that just occurred to her, even though it seemed to have been there all along. "I guess I'll have to take the risk and do some writing!" A smile came over her and with it I could already smell the popcorn and hear the applause at the end of her first screening.

Positive-Negative Graphs

"What are twenty-one of the *best* things that ever happened to you?" I ask. For ten minutes we all list.

"What are seventeen of the *worst* things that ever happened to you?" For another ten minutes we list.

"Look back at your list and star the three most positive and the three most negative things," I say. "Is anyone willing to share what you starred?"

Hands shoot up. If not, I share first. After at least half the class has shared their positives and negatives, I give the students another seven minutes to add to their lists. The discussion always gives them ideas, similar things that happened to them, but they forgot. (I always ask for odd numbers because it surprises the kids. I also ask for a lot, so they'll work hard and get some.)

"Look through your list and star all those things that you feel are most significant to you for one reason or another. You might have a lot more positives than negatives, or vice versa."

The students then chart their most significant positives and negatives on a graph. I give them graph paper to do a rough draft of their ideas. I share Chris's and Sara's graphs on an overhead (Figures 3.4 and 3.5).

I point out the use of the time line. Chris used it as age. Sara represented it with years. We discuss scale and legend and how the connecting lines move across the page in relation to time. The positive and negative scales run down the left side of the chart, +5 being the highest, −5 being the lowest. These numbers too are charted by scale.

I put one sample of charts from previous students on each round table so students can see what other students did as they drafted. We talk about being as specific as possible, focusing on the most important information in the limited space. We talk about aesthetics. How can you make your graph appealing to a reader? How can you visually represent what you are saying with words?

The students share their rough drafts with each other, pointing out what they liked, asking any questions they have, and offering one or two

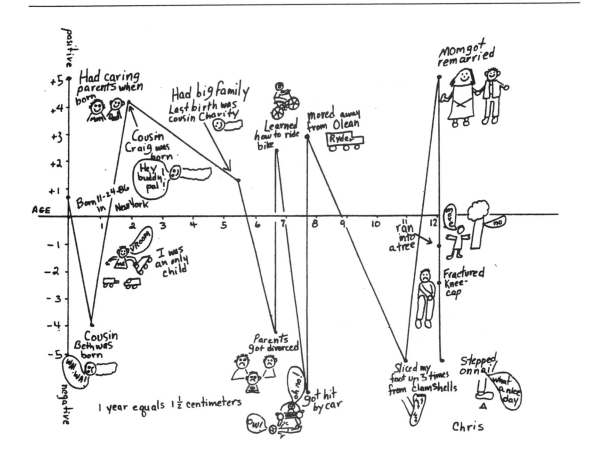

Figure 3.4 Chris's "significant positives and negatives graph"

suggestions. Final drafts are only published if the students choose to have them hung on the walls. Otherwise, the charts remain in their working folders.

I've found that the better the students know me and trust me, the more honest they are. Do they *have* to include very personal negatives or positives? Not at all. Do they *have* to share their lists or final charts with the whole class? Not unless they choose to. They might choose not to share them with me. The charts and lists may end up in their working folder as rough drafts. Most kids like doing this, however, and the pieces end up as finals.

As teachers of adolescents, we can't avoid our kids' self-esteem. From these charts I have an instant visual clue as to how each student feels about himself or herself. I know what their lives have been like. I can talk to them as people.

I've found, too, that these charts have become invaluable resources for topics for writing. When I look at numerous portfolios of finished pieces, I can usually trace a few pieces for each student back to something that was listed on the chart.

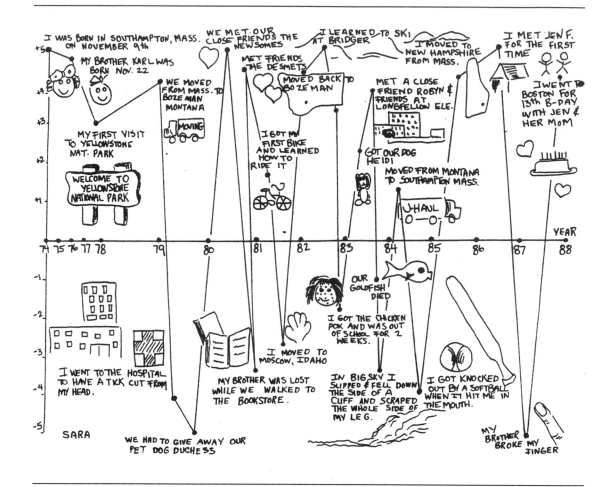

Figure 3.5 Sara's "significant positives and negatives graph"

After the kids finish their charts I start them thinking even more closely about writing from them. I ask them to fold three standard size pieces of paper in half lengthwise. I have them choose six topics from their charts that either surprised them, that they want to know more about, or that they'd like to talk more about. They list a different topic at the top of each column. I then give them three minutes to brainstorm and list every word, phrase, sentence, question, or idea that comes to mind when they think of that topic. The longer the list, the more they know they have to write. We do that for each of the six topics. Sometimes they write from this exercise, sometimes they don't.

Throughout the year, when the kids are stuck I can refer them to their chart or look at it with them.

"Sara, I noticed from your chart that you've moved a lot. What's that been like for you?"

"Chris, you indicated that the worst thing that ever happened to you was getting hit by a car. Tell me more about that."

50   "Sara, what makes your friendship with Jen so special?"

These charts get the students thinking about themselves and what they know. It starts them out writing from their own point of view. I want that. I want to hear their voices right away so they can teach me who they are and what they think. Thinking about themselves leads into my thematic approach with the eighth graders all year: *relationships*—with peers, parents, grandparents, the elderly, those people like them, and those people different from them.

I ask them to tell me who they are, what they think, and how they fit into this world. I want them to ask the same questions of themselves and those around them. I am seeking diversity. I ask them to do the same.

Reading

"When I'm reading a book I really can't stand, the words on the paper don't mean anything."

Chris, a student

"If a person reads a lot and likes to read, then he/she is actually studying writing."

Nathan, a student

I want to be as immersed in reading as we are in writing, so Chris can find meaning in the words, so Nathan can study writing. We read in many different ways: aloud, individually, and together.

Reading Aloud

Day One: Right away, I'm reading aloud to the class—before I even introduce myself. It might be *Super Dooper Jezebel* by Tony Ross (1988), *The Teacher from the Black Lagoon* by Michael Thaler (1989), a poem from *Brickyard Summer* by Paul Janeczko (1989), or an excerpt from *They Shoot Canoes, Don't They?* by Patrick McManus (1982).

Whatever I read, it will be something I love and will feel right at the time. It will seldom be the same piece year after year, because I am never the same reader year after year. What will be the same is my attempt to read something out loud every day to every class.

I want the students to *hear* good writing. I want the students to see all the options open to them for ways of writing. I want them to practice listening, to feel something, to connect their worlds with other worlds, to get ideas, to think, to laugh, cry, ask questions, criticize, analyze, react, and to enjoy. I want them to see me as a reader. I want my students to read. Reading out loud does all this, and more. It is just one way of reading in my classroom.

For the first few months of the year I am primarily introducing kids to authors, genres, topics, words—to the process of reading and writing from the perspective of writers. I read a variety of excerpts from novels or whole pieces, if they are poetry, essays, short stories, or children's literature. We talk and share whatever the reading brings to mind. Sometimes I ask the students to write about this reading. Other times I just want them listening and enjoying and trusting me. That is why I *never* begin the year with a book we all read together. I want the students to trust me as a reader first by giving them choice and recommending books and authors based on what they like to read. (At the end of this chapter I have listed the books I read aloud to introduce students to the reading and writing process and to authors.)

After the first few months, I turn the reading aloud over to the students. Each student is responsible for sharing an author for the first five minutes of each class for a week. They read favorite excerpts Monday through Thursday and then on Friday share the most extraordinary information they discovered about the author's life as it relates to his or her writing. Students may work in pairs or alone.

Abby and Julie read T. S. Eliot's poetry for three days, played a tape of several songs from the show *Cats* the last day, and shared their discoveries the last day. Andy memorized and acted out several Shel Silverstein poems, put them on overheads, and asked students to discuss the elements of humor. Sarah presented Jane Yolen. She read *Owl Moon* (1987) and excerpts from *The Devil's Arithmetic* (1990), *The Girl Who Cried Flowers* (1981), and *Dream Weaver* (1989). Lindsay read excerpts from Susan Fromberg Schaeffer's *Buffalo Afternoon* (1990), a novel about Vietnam, all four days. On Friday she read a letter she received from Schaeffer describing what prompted her to write the novel and how she went about doing it.

Each of the students prepares a poster to display his or her author of the week and hands in an author/book-share sheet the week before the presentation (see Appendix K).

Reading books and excerpts allows kids to get comfortable in front of a whole class. It allows them to hear a diversity of good writing. It allows them to research a topic and to evaluate the information by choosing what they want to present and why. It lets them use voice to get *into* character roles by reading from various points of view. It introduces all of us to other authors and writing styles. It lets kids show how articulate they are, when they speak from what they know. It gives kids another way of sharing what they know about reading without always having to write it. Reading aloud gets kids reading for themselves and invites kids into other books.

Individualized Reading

In addition to sharing a lot of books out loud, the students begin reading on their own that first night of school. They choose books from my library, the school library, or from home.

I show them the book I'm reading and explain why I chose it. I ask them to share all the ways they choose a book. We read. I expect us to read for at least a half hour each night. We spend at least one full class period each week reading the book of our choice. I do not feel guilty about taking a full teaching day to read. I know more about books than I ever knew. It is most importantly my time for showing the students that I value reading. All the reading also helps me find examples for minilessons about writing and reading: leads, endings, character description, voice, metaphor, and so on.

After a few days I introduce the concept of a reader's-writer's log. This reader's-writer's log is not new to me, but it is to my classroom. Over the last ten years in my classroom, students' response to reading has progressed through the following stages:

- answering questions at the end of the literature anthology
- no response
- book projects
- oral sharing
- daily letters to me
- reader's logs, small-group and whole-class discussions
- reader's-writer's logs, small-group and whole-class discussions

It took me several years to get comfortable with writing. I had to figure out how to give kids choice, time, positive response, and an organized structure, and still find time to write and share my own writing. I couldn't let go of everything at once, so I held on to the literature anthology several days a week. I had the kids answer the questions the publisher asked.

The more I learned about writing, the less right those questions felt. I let go of the questions. I didn't know what to replace them with, so I did nothing. We just read the stories.

That didn't feel right either. I asked the other English teachers, Paula Ickeringill and Michele McInnes, what they were doing. We came up with recommended lists of books based on units or themes, and book projects to go along with the lists: a story wheel, a bookmark, a poster to "sell" the book, a book cover, and the like. We worked hard on the lists and the projects. They looked great on display. But I was beginning to realize all these ideas were ours, all the lists were ours . . . and all this work wasn't teaching me what the students knew about literature and authors. I began to suspect they weren't even reading. How would I know? I never saw them read or asked them about their reading. None of us was reading in class.

I signed up for a course at UNH: Reading-Writing Connections. The instructor, Mary Ellen Giacobbe, taught me to take a hard look at reading for myself. Giacobbe asked us to bring five favorite books we had read recently or were currently reading to the first class. Five, I thought. This woman obviously doesn't teach English. I don't have time to read. Just thinking that scared me. I was the reading teacher—language arts included everything. I had concentrated so hard on writing, that I couldn't find the time for reading.

Giacobbe took us through the same process as readers that I knew as a writer. Once I realized readers also need time, choice, response, an organizational structure, and models who value reading, I had to change what I did.

I brought books into the classroom. I built a library of several hundred paperbacks in my room and we all read silently at least one day a week. Just getting the library together, helping kids choose books, and convincing them they could *read* during reading time took all my energy. I looked up one day to the overwhelming silence of the reading period and realized I needed to find a way to find out what these kids knew and felt about the books they were reading. We needed to talk. Talking felt good for awhile, but an uneasy feeling kept gnawing at me. I really didn't know what these kids knew as they read. What were they thinking? What were they feeling? How were they interpreting and connecting with what they read?

At the time, my students were doing little in the way of reacting or responding to reading as writers. We talked in small groups and class discussions, but I wanted to find a way to get them really thinking about their reading.

I read Nancie Atwell's chapter (1987, 149–98) about letters students wrote to her about their reading. The kids' responses sounded wonderful—exactly the way I wanted my students to react. I jumped right in. I had my students—all 127 of them—write to me every night. I wrote back . . . for *two* nights. I worked on three hours sleep those two nights. Who *was* this woman? How many students did she have? Either she's crazy or she's superwoman, I thought. I can't do this.

I reread the chapter I had *misread* and apologized for cursing this woman's idea. The students didn't write every day. She didn't respond every night. She had half as many kids. Her reading class was separate from her writing class. I tried it again. By the end of two weeks I had letters everywhere: some were answered, some unanswered; some kids had written several; others had written none. This wasn't working exactly the way Atwell did it.

I still liked the way her students responded to their reading and the way she responded to them. They connected books to their lives, to their writing, and to each other. I wanted that to happen, but needed to organize it in a different way, a way that fit my classroom and me.

I looked at my personal journal. I suggested students use a similar small notebook and call it a reading log. I gave the students a sheet with instructions and several open-ended questions, in case they were stuck for what to write, and a blank reading list.

At first many students wrote summaries of the books they were reading. They seemed to be trying to prove to me they had read the books. No one else had asked them what they thought or felt as they read. I gave them examples of responses students in previous years wrote. I shared my response.

We shared the best responses, in partners, and whole-class discussions. Kids noted, "I didn't know we could write like that." They still say that. I still make sure they hear all those wonderful responses.

Reading logs worked for more than five years. I changed because I noticed these last two years that I had a hard time keeping my comments to reading. I wanted to write about more than just books. While I tried to keep a reading log at school, I kept a reader's-writer's log at home. The notebook was a collection of every thought I had about reading or writing; snippets of conversation I overheard in restaurants, at school, at any gathering; poems and pieces of writing cut from magazines or newspapers; interviews from television; song lyrics I liked; observations as I travelled; drafts of pieces of writing; conversations with kids in the classroom; unique things they said or did—this was real writing, for me. I started sneaking it to school in place of the reading log.

By calling the notebook the students wrote in a "reading log," I think I limited their response. It's difficult for me to separate reading and writing times in my classroom now. How can I limit their response just to reading, when they have so much to say about writing? The change is part of the whole evolutionary process of learning—not just in my classroom, but in our entire school district. The students come to the middle school with a much deeper understanding of their own processes as learners. They *are* readers and writers, and they expect to continue reading and writing. I am constantly raising my expectations of what these kids can do, because the range and depth of their learning gets deeper every year. They come from classrooms with teachers who challenge these students through real literature and writing for real reasons. My challenge is to keep up with their learning, so I can push them further. That's the fun of teaching, however. It is not stagnant. It is not quiet. It is rewarding.

Because I only have the students for fifty minutes a day, I have to make choices about what best helps them in school and what they can best accomplish on their own at home. I still ask them to read for a half hour every night and write in their log three to five times per week.

I show on an overhead examples of the different ways students and I have responded to our reading and writing (see Figures 3.6 and 3.7 and additional entries in Appendix A.). I don't like to give the students too many examples, as it seems either to overwhelm them or lead them in someone else's direction.

I give the students two handouts: one reiterates the requirements for reading and questions/suggestions for ways they might respond if they are stuck (Appendix L); the other is a Reading List (Appendix M). The students staple these handouts on the inside front cover of their logs. The students' lists become record-keeping systems for who they are as readers. Based on their lists and their responses I can nudge them into more challenging books, different genres, and/or different authors.

Responding to reading-writing logs

Readers want response to what they write. They want that response to be a dialogue—a confirmation of what they know in a conversational tone. I try to do that. Many times I ask too many questions, give too many suggestions, or push too hard in my direction. My aim is to convince them these logs are for them—a place to collect first-draft thoughts, observations, quotes, anything that's meaningful to them in their reading-writing lives. I try to direct my response to that goal.

One of Lizzie's entries spoke about her reaction to the book *The Wall* and a visit to the Vietnam Veterans Memorial—what she read in the letters and how her mother reacted to the names. Part of her response read:

> I noticed the large directory books that stated where certain names could be found. My mom said she knew someone who went to school with her who died in Vietnam. She found his name. For a long time we just stood there. My mom began to cry and I didn't know why because at that moment I didn't know who the person was or how my mom knew him. I didn't want to leave the wall and I wasn't sure why . . . When I think of how many people died and how awful the survivers feel it really hurts me. I hope we never have another war like Vietnam or any war ever again.

I wrote back:

Figure 3.6
Scotti's log response

11-6

I just took out The Outsiders by S.E. Hinton from Mrs. Rief's library. I took it out for I saw the movie and loved it. It was one of the best movies I've ever seen. It was one of the best for It drew me into it like a book would. It was like watching a book if you understand what I mean.

11-7
½ hour
page 35 to 81
The Outsiders

I came across a poem that I had heard in the movie. I just have to write it down in my log.

> Nature's first green is gold,
> Her hardest hue to hold.
> Her leaf's a flower ;
> But only so an hour.
> Then leaf subsides to leaf,
> So Eden sank to grief,
> So dawn goes down to day.
> Nothing gold can stay.
>
> -Robert Frost-

I wrote it down for I wanted a copy, and cause I can't quite understand what it means. I have a vague idea...
That when your born your pure like gold. But as you grow older you become less pure and the gold goes away. Or it means that a mother tries to keep her children safe and pure, like gold, from all the rotten things in the world. But she can't hold on to her pure children forever for things don't stay pure forever.

Lizzie, you really touched me with your reaction to *The Wall*. I started lumping up and crying as I read it. I also have the book you talked about. I'll bring it in if you'd like to see it again. My task in Washington in '88 was to find a piece of artwork which affected me. After seeing the Wall and other peoples' reactions, there was no question that this was the most powerful piece of sculpture I'd seen. Your experience confirms it all. Thanks for sharing it.

D. Murray- good poetry = makes me see the ordinary as extraordinary
8/89   = makes me see, feel, think
      bad poetry = pretentious, fake, using long, complicated words
      when Ss are curious about poetry, give them an anthology
            to find one or two they like
      from D.M.'s writing:
      "What happens if someone's shadow is on you + he is killed?"
      "He smiled his way across the yearbook."
      "... the wedding cake in the middle of the road..."
      - responder should read with writer's eye, so writer learns
            to read for self - in the matter of reading and writing,
            the ear is the final test - it's the sound that helps
            you respect the text - talks to you...
      - are there different ways of reading one text ???

"Teacher research is more a wondering to pursue, than a hypothesis
      to test." Nancie Atwell

9/11/89: finally we danced
          circling the cobblestones
          drunk with silence
          you stopped me turned me held me
          held my face to the summernight
          why? I asked because
          I want to hold the moment
          you said as and your shadow fell
          across me and stayed
          after you left
(D.M.'s line -- made me write this poem- it wouldn't go away...)

9/12: When Craig was 12 he ruptured his spleen while skiing.
       Waiting for him in the operating room I was alone - a
       policeman arrived with a boy he'd arrested in a car
       accident - the boy was injured - I was convinced the
       policeman was Satan - there to take Craig. I knew his
       injury was serious. The examining doctor had laid
       a hand on Craig's midsection and said, "Linda, I
       don't want to panic you, but this is serious. He's
       ruptured his spleen." They plunged a needle into
       Craig the size of a shish-kebob skewer. It sounded
       like a suction cup inhaling as it pierced his skin. I fainted...

Figure 3.7 My log response   continued   **57**

2/27: Boy- no one can say these kids don't keep writing after 8th gr.
- Kyle brings me a 1st chap. to his book instead of a whole book- he tells me he's trying to concentrate on characters, instead of getting too involved w/ the plot
- Monica (still 8th) makes an "appt." to work on her writing during recess- she asks me how I like her briefcase she's now carrying to keep her writing "neat + organized"
- Gillian (10th gr.) calls w/ good news - out of 24,000 entries in Scholastic Writing Contest, her short story "Skeeter" receives an Honorable Mention as one of 60 - now if my math is right - that's ¼ of 1% or .25% - and she's disappointed - she knew it was a good piece (I did too!) - and hoped it would receive a 1st thru 3rd - she didn't have to sign a release for it - so she's sending it off to _Seventeen_ right away - M.P. next...

3/5: I am really angry with myself. I can't seem to keep up with a reading log and response. What makes me so angry is that as I read books I want to jot thoughts down or I forget them. On the flights to and from Winnepeg I read and finished Annie Dillard's _An American Childhood_. So many things in it reminded me that it's okay to have obsessions - she spent months collecting stamps, months collecting rocks, months studying amoeba under a microscope. She uses language so beautifully. Sure- I underlined lots of passages, but those words never stick in my mind the way words I write down do. I finished _Prince of Tides_ - I read _Fallen Angels_ - only put in one entry. I learned so much from that book - fear/regret/useless destruction. The book started out slowly - simply, almost paralleling the main character's naïveté about war. As his attitudes and fears intensified and became more complex, so did the writing. Myers seemed to choose more complex words - more complex sentence structures - and the character's thinking became more complex . I should write the letter to Myers I wanted to write when I finished reading _Fallen Angels_. He captured what this undeclared war was about: not knowing...

Figure 3.7 Continued My log response

For weeks Graham's entries were so generic ("This book is so cool, I just can't find the words to describe it . . . The dialogue is so complex and rich that it would make a great movie.") that I knew little about the books or him as a reader. I latched on to his "request" to me:

YO, RIEF: I notice that I have 1 week of extra entries and pages. In this great world of ours, something, like grades, matter emensly to the forefathers of my generation (Pop). It is with this thought in mind that I ask you to strongly emphazize this extra (and extradinary) initiative on my part in coming up with evaluations (grades) for this first third of the nine moons. Happy pilgrims day and may squanto be with you,
Love, Graham

I replied:

Dear Graham,
Although I *immensely* enjoyed your entry enlightening me as to the importance of grades—especially with respect to your "forefathers"—I feel the rest of your entries concentrated too much on the cranberry and olives and not enough on the turkey. Squanto wasn't with you! . . . You are a wonderful reader but you really have to reflect more on what you are thinking as you read and why? As a writer yourself, what do authors do to keep you involved in a book? How do you relate to their experiences? What do you learn from these authors? . . . You might like Tom Clancy's book *The Hunt for Red October*, especially if you liked *The Caine Mutiny* so much.
"Love," Mrs. Rief

To Holly, I wrote:

Holly, I just bought *The Lottery Rose*—read 30 pages. What a sad/wonderful book! The poor little guy. It is so incomprehensible to me that there are parents who can be *so* cruel. Thanks for recommending the book. . . . I just read Torey Hayden's *The Sunflower Forest*. I think you'd like it. It's fiction— the story of a family tragedy as a result of the mother's experiences with the Nazis during WWII. . . . I wish everyone saw how to get the most from their reading, as you do. How could I get more students to use their logs better?

Responding to 125 logs is difficult; yet the depth and diversity of writing make them a lot more valuable and more fun to read than the traditional book reports. I don't want to be the only person who reads the logs, however. For the first few months I collect the logs every two weeks (one class each day of the week) and respond to them myself. After three or four months I set up a rotating schedule where the students get response from a peer of their choice one week, me the next week, a parent (or older sibling) the third week, and me the fourth week. (See Appendix A for more examples of peer and parent responses.)

Lizzie's mom wrote to her:

Dear Lizzie,
I have not read *Hound of the Baskervilles* but your entries have inspired me and tempted me to read the mystery for myself. From what you have written, I am curious to read more. I am enthralled with the relationship between Watson and Holmes as you have described it. . . . I especially like your description of the Moors with the Hell Hound lurking out there somewhere. I think I would react as you have to the suspense of wondering where and

when the hound may strike. . . . I think you uncovered the author's ploy for creating suspense when you said the author "pre-occupied your mind and therefore surprised you" as you discovered that Holmes was living on the Moors. . . . The way your words flow, your sensitivity, your interpretations all impress and delight me. Tonight I can feel the integration of your reading with your writing.
Love, Mom

Charlie's mom tried a unique way of responding to his log:

Better Late Than Never
It's late and I am tired
And my mind is somewhat mired
By a day long spent
Trying to make a dent
In housework's minutia dire.

It's late and so utterly quiet
Not like the earlier riot
Of the radio rocking and singing
And the unceasable telephone ringing
And the two year old's "Why not?"

It's late and as I and the cat
Sit here where you earlier sat
I'm pondering the you
That just suddenly grew
And my joy at witnessing that.

Sarah (not wanting to share her log with her parents, for the first time) shared her log with her sister Karen, who wrote:

I'm incredibly impressed. I've never gotten the idea that Sarah thinks so deeply, she never shares it with me. . . . She's been reading good books—I usually read them too. She's really good at setting out her thoughts so other people can understand them. *Rich Man, Poor Man* is a lot of plot twists and is probably easy to respond to when the twists have straightened out. . . . I want to read her log more often and maybe I can get her to *talk* to me about her reading 'cause what she says is extremely interesting.

Sarah replied:

I'm really happy that Karen liked my responses so much. That kind of worried me when she said, 'I've never gotten the idea that Sarah thinks so deeply—at least she never shares it with me!' I hope that I don't appear shallow to everybody who doesn't read my log, what a depressing thought!

Sometimes I ask the students to look through their logs and star their most effective entry, letting them know that's the only one I will read and respond to for a particular two-week segment. It helps cut down on some of the time, and still lets me speak to the student and his or her writing and reading. It also puts the onus of evaluation back on the student.

Other times I ask the students to stand back and take an outsider's look at their own logs. Sara wrote:

I have trouble thinking of things to write. Often the only thing that helps me is by asking myself how does this book make me feel. Or what does this book make me think of or what memories does it bring back. By asking these questions I can usually think of a response and they keep me away from summarizing the book.

Kate wrote:

I love this log! I never thought I could get attached to something not living (besides books). Maybe it's that you can get attached to anything that you care a lot about. . . . Even when I'm done with this log I will still carry it around with me. I mean, it's like when you have a baby you love it and she or he grows on you, sorta like this log.

Reading Together

I am always trying to get at reading in three different ways: reading aloud, individual choice, and reading with the students. Writing is similar: writing to the students (as feedback and response), individual writing, and writing with the students (pieces I ask them to try, I try).

Once I have the students trusting me as a reader and writer and they know I trust them to make choices for themselves, I share literature we read together. In the next two chapters I describe those times. I base my selections on a general theme of relationships, generational and cultural. Everything we do is still based on a workshop approach, with the similar requirements existing throughout the year. My expectations are the same. Now, however, I am inviting the students to try certain kinds of writing and reading. They are still the final decision makers about whether to take the writing to finished draft and determining how to respond to the literature. They determine what works, what doesn't work.

Works Cited

Atwell, Nancie. 1987. *In the Middle*. Portsmouth, N.H.: Boynton/Cook.

Clancy, Tom. 1984. *The Hunt for Red October*. Annapolis, Md.: United States Naval Institute.

Doyle, Sir Arthur Conan. 1964. *The Hound of the Baskervilles*. New York: Scholastic.

Elbow, Peter. 1982. *Writing Without Teachers*. New York: Oxford University Press.

Hayden, Torey. 1984. *The Sunflower Forest*. New York: Avon.

Hunt, Irene. 1976. *The Lottery Rose*. New York: Scribner's.

Janeczko, Paul. 1989. *Brickyard Summer*. New York: Orchard Press.

McManus, Patrick. 1982. *They Shoot Canoes, Don't They?* New York: Henry Holt.

Ross, Tony. 1988. *Super Dooper Jezebel*. New York: Farrar Straus Giroux.

Schaeffer, Susan Fromberg. 1990. *Buffalo Afternoon*. New York: Ballantine Books.

Shaw, Irwin. 1989. *Rich Man, Poor Man*. New York: Dell.

Thaler, Michael. 1989. *The Teacher from the Black Lagoon*. New York: Scholastic.

Wouk, Herman. 1954. *The Caine Mutiny*. New York: Doubleday.

Yolen, Jane. 1990. *The Devil's Arithmetic*. New York: Puffin Books.

———. 1989. *Dream Weaver*. New York: Philomel.

———. 1987. *Owl Moon*. New York: Philomel.

———. 1981. *The Girl Who Cried Flowers*. New York: Schocken Books.

Further References: Books Focused on the Reading-Writing Process

Children's Literature

Bunting, Eve. 1989. *The Wednesday Surprise*. New York: Clarion.
On Wednesday nights when Grandma stays with Anna everyone thinks she is teaching Anna to read.

De Paola, Tomie. 1989. *The Art Lesson*. New York: Putnam's.
Having learned to be creative at drawing pictures at home, Tommy is dismayed when he goes to school and finds the art lesson there much regimented.

Hoban, Russell. 1989. *Monsters*. New York: Scholastic.
John's obsession with drawing monsters takes him to a doctor, where a startling discovery is made about the reality of John's drawings.

Hyman, Trina Schart. 1981. *Self-Portrait: Trina Schart Hyman*. Reading, Mass.: Addison-Wesley.

Lionni, Leo. 1967. *Frederick*. New York: Knopf.
The fable of a field mouse, suggesting it is our ability to recreate memories through words that sustains the human spirit.

Reading Is Fundamental. 1986. *Once Upon a Time*. New York: Putnam's.
Through memories, stories, and anecdotes, well-known children's authors and illustrators celebrate the magic of reading.

Rylant, Cynthia. 1988. *All I See*. New York: Orchard.
A young boy paints with an artist friend. They each paint what they see.

Wiesner, David. 1988. *Free Fall*. New York: William Morrow.
Inspired by books and things surrounding his bed, a young boy dreams of daring adventures.

Young-Adult Fiction

Hamley, Dennis. 1988. *Hare's Choice*. New York: Delacorte.
A hare killed by a car is found by two children who take it to school where they and their classmates write a story about it—giving it a choice to make in the afterworld. (Really about the way the artistic process alters the writer as well as how the characters created take on a life of their own.)

Jarrell, Randall. 1963. *The Bat-Poet*. New York: Macmillan.
A bat who can't sleep days makes up poems about the woodland creatures he now perceives for the first time. (Especially insightful into how various responses help/hinder the writer.)

Snyder, Zilpha Keatley. 1990. *Libby on Wednesday*. New York: Delacorte.
Having been put ahead in an accelerated eighth-grade program by her bizarre and creative family, eleven-year-old Libby hates her "socialization" process, until she makes some highly original friends in a writing workshop.

Poetry

Janeczko, Paul, ed. 1990. *The Place My Words Are Looking For*. New York: Bradbury.

Thirty-nine U.S. poets share their poems, inspirations, thoughts, anecdotes, and memories.

Janeczko, Paul, ed. 1983. *Poetspeak.* New York: Bradbury.
A collection of 148 poems on a variety of topics by 62 modern poets who provide commentary on their individual work.

Little, Jean. 1986. *Hey World, Here I Am!.* New York: Harper and Row.
A collection of poems and brief vignettes from the perspective of a girl named Kate Bloomfield, reflecting her views on friendship, school, family life, and the world.

Collections About and from Authors

Gallo, Donald R., ed. 1990. *Speaking for Ourselves.* Urbana, Ill.: NCTE.
Autobiographical sketches by notable authors of books for young adults.

Gilbar, Steven. 1989. *The Open Door.* Boston, Mass.: David R. Godine.
When writers first learned to read.

Zinsser, William. 1990. *Worlds of Childhood: The Art and Craft of Writing for Children.* Boston, Mass.: Houghton Mifflin.

————. 1987. *Inventing the Truth: The Art and Craft of Memoir.* Boston, Mass.: Houghton Mifflin.

Biography and Autobiography

Campbell, Patricia J. 1985. *Presenting Robert Cormier.* Boston: G. K. Hall.

Carter, Forrest. 1976. *The Education of Little Tree.* Albuquerque, N.M.: University of New Mexico Press.

Dahl, Roald. 1984. *Boy: Tales of Childhood.* New York: Viking Penguin.

Daly, Jay. 1987. *Presenting S. E. Hinton.* New York: Dell.

Dillard, Annie. 1989. *The Writing Life.* New York: Harper and Row.

————. 1987. *An American Childhood.* New York: Harper and Row.

Duncan, Lois. 1982. *Chapters: My Growth as a Writer.* Boston, Mass.: Little, Brown.

Fox, Mem. 1990. *Mem's the Word.* New York: Viking Penguin.

Grotta, Daniel. 1978. *The Biography of J. R. R. Tolkien.* Philad. Pa.: Running Press.

Holtze, Sally Holmes. 1987. *Presenting Norma Fox Mazer.* New York: Dell.

Lee, Betsy. 1981. *Judy Blume's Story.* New York: Scholastic.

Nilsen, Alleen Pace. 1987. *Presenting M. E. Kerr.* New York: Dell.

Paterson, Katherine. 1988. *Gates of Excellence: On Reading and Writing Books for Children.* New York: E. P. Dutton.

Paterson, Katherine. 1989. *The Spying Heart: More Thoughts on Reading and Writing Books for Children.* New York: E. P. Dutton.

Novel Excerpts

Reading excerpts from novels opens the book covers for kids. I don't find it unusual that the pieces I read most often to them come from the books the students list as best liked. They hear a chunk of wonderful writing. They want more. I only read language I love. I may hate *what* it says, but love the way the author said it. Especially powerful excerpts related to teaching and learning can be found in

Angelou, Maya. 1970. *I Know Why the Caged Bird Sings.* New York: Random House.

Conroy, Pat. 1986. *The Prince of Tides*. Boston: Houghton Mifflin.

Cormier, Robert. 1974. *The Chocolate War*. New York: Dell.

Dahl, Roald. 1988. *Matilda*. New York: Puffin.

Hayden, Torey. 1980. *One Child*. New York: Avon.

Hinton, S. E. 1967. *The Outsiders*. New York: Dell.

Lowry, Lois. 1979. *Anastasia Krupnik*. New York: Bantam.

Marsden, John. 1987. *So Much to Tell You*. New York: Ballantine.

Sebestynen, Ouida. 1988. *The Girl in the Box*. Boston: Little, Brown.

Wright, Richard. 1966. *Black Boy*. New York: Harper and Row.

Writing and Reading
for Life

"One mother can take care of ten children," she said, "but ten children can't take care of one mother."

Rose Blue, *Grandma Didn't Wave Back*

The Elderly

The words hit me hard. I couldn't get them out of my head. I kept thinking about my grandmother in the nursing home. With four children and nineteen grandchildren, why was there no one to care for her? She had been in the home for a month. For the first time I went to visit her.

The smell of urine gagged me as I entered the state-run facility. It took my eyes several minutes to get used to the dark corridor. The single attendant/nurse shuffled by me without a word. On my own I found Gram, her name mounted with seven others next to the third door on the left. Dark green shades shut out any sunlight. In the shadows I found figures curled up in beds or hunched over in chairs.

Gram sat slumped in a wheelchair. I knelt down beside her. Her thin cotton robe was soaked in urine, from her waist to her knees. Apparently the catheter had slipped out. There was no one to replace it. I took her hand. Dried food circled her lips, caked her fingers, and solidified her gold wedding band to her ring finger. "Help me, Lin," she whispered. "Help me."

Sobbing, I called my father. Within an hour he had an ambulance at the door. We took my grandmother to my parents' house. She lived there for three years, then with my sister for seven, before too many body parts gave out and she had to be put in a nursing home—a reputable one, where she was well cared for.

I tell this story to my students every year. I'm not sure if it was thinking about my grandmother or the story I read about Oscar in the local paper that started me reading and writing about grandparents and the elderly with my students.

Oscar had been living in a private nursing home run by a couple who had just been arrested when iron shackles were discovered in the basement. The shackles were used when patients "misbehaved": spilled food, yelled too much, or couldn't control their bladders. Like everyone else, I was appalled.

When Joan Savage, the home economics teacher in our school, said she was taking her students to visit several nursing homes, I asked to join her. Clean, happy, and well fed, Oscar was now at this home. Brandon, one of the students, picked up crayons, and began coloring Scooby-Doo with Oscar. Brandon knew where Oscar had been. He was touched to think he had survived such punishment.

Joan still takes the students every year to a nursing home. It leads to a lot of reading, writing, and thinking about generations in my classroom. Before they go to the nursing home I try to take the students through the following exercise. If scheduling doesn't permit, I do it after they have all visited the home.

I hand out five different colored index cards (4 × 6) to each student. For the first fifteen minutes of each class period for five days in a row, or during two full class periods, I read two or three of the following pieces per section out loud to the students and ask them to respond to specific questions on designated index cards, with words, phrases, and whole sentences that come to mind. I tell them to think about the oldest person they know, or have met, or can think of, as they answer the questions.

Day One. I read

"Aunt Carrie and Uncle Frank" by Ted Kooser (1976)

"Uncle Adler" by Ted Kooser (1976)

" Endings" by Lynn Kozma (in Martz 1987)

"Here, Take My Words" by Karen Brodine (in Martz 1987)

"Mrs. Thurstone" by Jean Little (1986)

On the pink index card I ask the students to describe "old": "What do you see, hear, smell, think of, feel, when you think of 'old?' "

Day Two. I read

Nana Upstairs and Nana Downstairs by Tomie de Paola (1973)

The Two of Them by Aliki (1979)

Tales of a Gambling Grandma by Dayal Kaur Khalsa (1986)

Babushka's Doll by Patricia Polacco (1990)

To Hell with Dying by Alice Walker (1988)

I tell the students to describe on the green index card what old people can do and what they like about old people.

Day Three. I read

How Does It Feel to Be Old? by Norma Farber (1979)

"The Very Old" by Ted Kooser (1976)

"Old Soldiers' Home" by Ted Kooser (1976)

On the blue index card the students describe what old people cannot do and what bothers them about old people.

Day Four. I read

You're Only Old Once by Dr. Seuss (1986)

"The Little Boy and the Old Man" by Shel Silverstein (1981)

The little boy and the old man

Said the little boy, "Sometimes I drop my spoon."
Said the little old man, "I do that too."
The little boy whispered, "I wet my pants."
"I do that too," laughed the little old man.
Said the little boy, "I often cry."
The old man nodded, "So do I."
"But worst of all," said the boy, " it seems
Grown-ups don't pay attention to me."
And he felt the warmth of a wrinkled old hand.
"I know what you mean," said the little old man.

and the following two articles from Ann Landers's columns:

His very special day

It was Grandfather's birthday. He was 79. He got up early, shaved, showered, combed his hair and put on his Sunday best so he would look nice when they came.

He skipped his daily walk to the town cafe where he had coffee with his cronies. He wanted to be home when they came.

He put his porch chair on the sidewalk so he could get a better view of the street when they drove up to help celebrate his birthday.

At noon he got tired but decided to forgo his nap so he could be there when they came. Most of the rest of the afternoon he spent near the telephone so he could answer it when they called.

He has five married children, 13 grandchildren and three great-grandchildren. One son and daughter live within 10 miles of his place. They hadn't visited him for a long time. But today was his birthday and they were sure to come.

At suppertime he left the cake untouched so they could cut it and have dessert with him. After supper he sat on the porch waiting.

At 8:30, he went to his room to prepare for bed. Before retiring he left a note on the door that read, "Be sure to wake me up when they come."

It was Grandfather's birthday. He was 79.

Dear Ann Landers: Yesterday, after a couple of hours shopping, my husband and I stopped at a small cafe for a sandwich. I didn't notice the teenage lad who was sitting at the next table until I heard him speak. He was reading the menu to his grandfather.

It was apparent that the old gentleman had had a stroke. One side of his face was distorted, and his eye was almost closed. The young man was very cheerful, ordered two tuna salad sandwiches and two glasses of milk, "One with a straw, please."

When their orders arrived the old man started to eat his sandwich. He managed fairly well but then began to drool. The lad took his napkin and ever so tenderly wiped Grandpa's mouth. This went on throughout the entire lunch. The young fellow kept up a lively conversation, laughing and talking with Grandpa, and mopping up the drools.

When they got up to leave, the boy gently helped his grandfather out of his chair, handed him his cane, took his arm and they both walked slowly to the cashier's stand.

I thought to myself, "We hear so much about how badly behaved teenagers are today. There is endless hand-wringing about their hair, their clothes, their music and their bad manners. How did this young man turn out to be so sensitive and caring, so thoughtful and kind? What made the difference? . . .

I ask the students to describe on the yellow index card how old people are treated by others.

Day Five. I read *Wilfred Gordon McDonald Partridge* by Mem Fox (1984). On the white index card, students describe in as much detail as possible a vivid memory they have of a grandparent, a great-grandparent, or any old person.

I have the students initial their index cards on the back so they can get them back. I collect their cards by color, separating them into five stacks. I have the students count off by fives. All the students who are number one get together in one corner and are given all the pink index cards. Each group divides the same way, all the twos get the green index cards, and so forth.

As a group, their task is twofold: to come up with a collaborative piece of writing that answers the question or statement they have, and to present that piece of writing orally in the most creative way possible (a play, a rap, a mime, a skit, etc.). Every student has to take part in the presentation. They are to use as much information as possible from the cards, but they may add to the information to create a polished, cohesive piece of writing.

It takes several class periods for the students to share and disseminate information from the cards, to come up with a plan, to write their ideas, and to present them. This exercise lets students work collaboratively and gives all of them ideas for individual pieces when they hear all the ideas presented. After the presentations they get their cards back and can use them as springboards for individual pieces of writing.

Kristen, Scott, Pam, Jeremy, and Brian wrote the following piece:

I hate it when . . .
I start telling my grandfather a private secret that I want him to keep and I turn around in the middle and he's falling asleep.
I hate it when . . .
my grandmother and I are having fun but she has to take a seat and starts complaining about the corns on her feet.
I hate it when . . .
I need to get somewhere fast and my grandfather offers to be the chauffeur.
I hate it when . . .
my grandmother and I are talking and even though we are sitting really near each other she starts shouting her words when she's the one who can't hear.

I hate it when . . .

I've been talking to my grandfather but notice he hasn't moved or even coughed, so I look behind his ear and I see his hearing aid is turned off.
I hate it when . . .

my friends come over to our house to visit me and my grandfather has left his dentures out for everyone to see.
I hate it when . . .

my grandmother wants to take me shopping because she's seen the "perfect outfit." So we get to the store and it's called "The Off-Price Outlet."
I say I hate all these the most
But that's a lie
Because what I hate the most
Is when they die.

For their oral presentation, the five students stood with their backs to the audience and said in unison, "I hate it when . . . " Then they each took turns one at a time facing the audience and sharing one of the vignettes. At the conclusion, they all faced the audience for their last four lines. All of it had been memorized.

Josh, Jamie, Nicole, and Lindsay designed a board game, complete with fold-out board, game pieces ("chocolate chip cookies" and "mothballs") made from a flour-dough mix, and the following directions:

You will soon be going on a class trip to a nursing home. Along the way you will discover things that people like and dislike about the elderly. Since you especially like your grandmother's chocolate chip cookies, these will be used to keep score of positive qualities. Because the smell of mothballs in your great-aunt Sylvia's house bothers you, we will use these to keep track of negative qualities.

The object of the game is NOT to get to the nursing home (FINISH) first, but to accumulate as many cookies (positive points) as possible. The game should go something like this:

- Player 1 rolls the die and travels forward that number of spaces.
- If Player 1 lands on a space instructing him/her to pick up a card, s/he chooses a card from the correct pile.
- There are two types of cards, Cookie Cards and Mothball Cards.

Cookie Cards tell about positive qualities

Ex. You help cook dinner at a rest home. Take 2 cookies.

When you were little, your grandmother read you stories. Take 2 cookies.

Mothball Cards tell about negative qualities

Ex. Your great-grandmother dies. Take 3 mothballs.

Your great-aunt Sylvia forgets your name. Take 2 mothballs.

S/he must pick up the number of cookies or mothballs indicated.

- If Player 1 lands on a blank space, s/he stays there and it is Player 2's turn.
- Player 2 rolls die and continues pattern, then Player 3, etc.
- As each player reaches the finish, they count up the number of points they have. Cookies are worth +1 point each, and mothballs are worth −1 point each.

Examples: 1 cookie + 1 mothball = 0 points
3 cookies + 1 mothball = 2 points
1 cookie + 3 mothballs = −2 points

- The player with the most points when everyone has finished is the winner.

Each year the students surprise me with presentations of similar quality: poems for several voices, ads and letters of response from the lonely, skits around the family breakfast table, one-act plays set in a nursing home, and so on.

The exercise immerses the students in reading, writing, and thinking about the elderly and about grandparents. I think adolescents and the elderly share similar identity crises. Perhaps that's why I've found adolescents tend to empathize with the problems of the elderly.

Teenagers are seeking independence and their own identity as they move away from the authority and control of their parents. They are trying constantly to figure out who they are and where they belong, as they develop a sense of self-confidence in themselves.

The elderly face a similar identity crisis as they're confronted with threats to their sense of self-esteem, their own adequacy. It doesn't feel good to lose that control over one's life and have to become dependent again on others.

Adolescents are self-centered. Letting them consider the plight of other generations lets them get outside themselves to reflect on others for a change. They are going through so many physical, social, emotional, and psychological changes, that we sometimes think adolescents can't get outside themselves to think about the plight of others. They can, whether it's in understanding what motivates characters in a story or in empathizing with real people.

Jeff thought going to the nursing home would be a real "drag." He went reluctantly, only because it was required. Yet, he drafted, revised, and wrote "Cecile," after his visit to the nursing home. It was one of the few pieces he did all year.

Cecile has cerebral palsy and has been at the nursing home for more than thirty years. She is in her early fifties.

Cecile

Cecile sat humped over in her wheelchair, sipping coffee through a straw. Her black hair hung limp and straggly. It took great effort to control her head.

At first I felt uncomfortable. Who was this person? I thought. What if I say something to hurt her feelings? But I was easily persuaded to sit next to her. I don't know why. She stared at me.

"What's that?" she asked, trying to keep her eyes on my earring. "Does your mother let you wear that?" Questions, why so many questions? I'm the one who is supposed to be asking the questions. Even though her head bobbed uncontrollably, her eyes focused on my one earring of dangling crosses. "That looks ridiculous," she said with a laugh.

If it had been anyone else, any one of my friends, I probably would have belted them by now. But not Cecile. Her gray-blue, almost playful eyes, looked innocent, magical. Cecile could have controlled the world with one look.

Cecile continued to stare at my earring. "Take it off," she demanded. "You're much too handsome to wear that thing."

A wise woman, I thought. But her words cut through to my heart and soul. "Why do you wear it? I want to know why you wear it?" she continued. "Why do guys wear earrings? Doesn't it make you feel like a sissy, a mamma's boy?"

I laughed. "No," I said.

For anyone else I would never have even thought of taking my earring off, but for Cecile I would. I reached up and, taking the earring out, placed it in my pocket.

"Now give it to me," she demanded. She might have been magical, but not that magical.

I felt good being able to make a lonely soul happy. I had accomplished my goal, making one person happy—so happy she would never forget me. I won't let her forget me. I will go back and laugh with her again.

I want students to see learning as connected to situations beyond our classroom walls. I want them to listen to, think about, and interact with people outside the classroom about real issues.

I shared my response to a vivid memory about a grandparent. "Gram said she had something special in her top drawer to show me. The smell of April Violets powder permeated the room as she took out a pink-flowered satin case. Inside rolled in tight coils, were two yard-long braids of jet black hair. It stunned me to think the braids had been cut from my grandmother's hair when she was a teen-ager. I couldn't imagine her as being someone other than my white-haired, roly-poly grandmother. She told me about the beach trips to Manomet in Grampa's Model-T. I couldn't imagine her in high-button boots. I could only imagine her in cotton aprons and standing at the stove stirring fudge frosting for vanilla cupcakes . . . "

I asked my students, "How often do you speak to your grandparents long enough to find out what memories they hold dear? What life was like for them as a teenager? What life is like for them now? How they feel about growing older?"

Silence.

"How would you find out?" I asked.

"Ask them!" the students replied.

We brainstormed and came up with all kinds of questions:

What kind of relationship did you have with your parents?

What was expected of you as a member of a family?

What were the latest fads when you were a teenager?

What was school like for you?

What concerned or worried you the most as a teenager—in your personal life? in the world around you?

In what ways is life better now than when you were a young person?

In what ways was life better when you were young?

What does *old* mean to you? What does *young* mean to you?

What worries you the most about growing old?

If you could give any advice to young people, what would it be?

We then talked about interviewing. Few opportunities exist in schools for students to gather information from primary sources. Students don't realize that people often give them more useful information for writing than do books or encyclopedias. Not only is the information more valuable, but the process involved in using the gathered material is invaluable to students as writers. Writers must think out their own arrangements of words, their own formats, and synthesize this information with their own perceptions.

At the beginning of the year (see chapter 3), I take the students through an interview with a classmate. When we talk about generations I simply review what we did. I remind the students, when you interviewed each other earlier in the year, what kinds of questions worked? They do remember. For example: Ask the right questions, those that get the most information, not just yes or no answers. Ask follow-up questions to preliminary questions—things you want to know more about, or are still wondering about. Use questions that ask how and why. Say to your interviewees, "Tell me more about that . . . "

I share good interviews related to the topic we are now discussing, so the students can see how follow-up questions elicit more information.

> In what ways was life better when you were a teenager than now?
> The world was a better place to live thirty years ago as far as I'm concerned.
> In what way?
> If you wanted something you had to work and save for it. But not anymore.
> How's that different from now?
> Now, you just hold a plastic card and you can have it. Kids ask and they get. They don't know the value of hard work and don't have any appreciation of what it takes to earn a living.
> Whose fault is that?
> I'm not saying it's the kids' fault. It's the fault of the parents and our society.
>
> —*Charlie, age seventy*

> What was school like when you were a teenager?
> Teachers made school humiliating.
> What do you mean by "humiliating"?
> Sometimes I got in trouble in school. I skipped school once and had to sit in the corridor all day with a dunce cap on my head.
> What else did they do?
> Once I got caught chewing gum and had to put it on the end of my nose all day. They also made bad grades public.
> How did they do that?
> By announcing to the whole class what you got or posting your bad grade as an example to others.
>
> —*Elaine, age sixty-two*

Like William Zinsser (1980) I advise my students not to use tape recorders. Recording is not writing. Equipment can malfunction, it takes hours to transcribe the words, and there is no longer any involvement with the person or in the writing process.

I also tell them to be extremely observant, with hand, heart, and eyes. Take down as many exact quotes as possible from the person being interviewed to make the story alive. Try to capture the feelings behind the words—note any sadness, delight, enthusiasm, confusion, and so on. Note what people look like—a physical description, especially of their faces—

and observe what they do as they talk or react to you—fidget, lean forward, wring their hands, whisper, yell, watch you, or look away.

Pieter, for example, wrote of Charlie, age seventy: "Charlie is a man of average height, with a kind and caring face that seems fatigued from years of hard work. His hands are worn and slightly wrinkled and his face is permanently darkened from long endless days of working under the hot summer sun."

Becky wrote of Elsie: "Elsie, a small, frail, white-haired lady in her mid-seventies, sits sleepily in an overstuffed chair, occasionally dozing off."

The students set up interviews with grandparents. We talk about using nursing home patients to interview in place of, in addition to, a grandparent. I agree with the students that sometimes the patients are not the most cooperative or easy to talk to, and questions should therefore be asked selectively. One good question is often enough, such as "What's the most vivid memory you have of growing up?" The nursing home is a particularly good place to be observant, however, with every sense.

When the students return from the nursing home and/or have conducted their interviews, we begin in earnest to read and write about generations. I cannot separate reading and writing. I choose books, poems, essays, and short stories as models for the writing I hope the students can accomplish.

I have three ways of getting at reading, all designed around writing.

Literature I read to the students

I read poems, short stories, or selected sections of books to the class. Often I read selections from books that I recommend they read on their own. Just reading the lead or an especially well-written passage invites readers in. I also show movies. Sometimes they simply listen, or watch. Other times we talk. At times I ask them to respond in their logs. After showing the film *Shopping Bag Lady* I wrote in my log:

> "Are you lost? Are you hungry?" Annie says to her new friend. She hugs him. She loves him. My god, that film just about tore my heart out. It makes me wonder, is a person better off in the streets than in a nursing home? Patients are portrayed as incoherent, motionless vegetables. I have so many questions. What's wrong with our society that we don't *care* about the elderly? Why can't even *one* person show some love for Annie? What happened to her family? What's wrong with us that a papier-mâché doll provides more love and attention than a human being? Why don't I go to visit my grandmother more? Why do I leave her alone in that nursing home? What kind of life does she have there? Why is it that I could find time to go to her funeral if she died, but I have no time just to go and talk to her? What if my kids don't want me? How can I feel such compassion for a fictional character in a film, but when my mother-in-law implies she's lonely, I think, she'll just have to get used to it.

Literature I read with the students

We read selected stories, poems, or essays as the various pieces of writing are being drafted. In addition to modeling good literature, I think students learn to read better if they see words as they listen to someone reading.

Students wrote responses to these readings in their logs. Steve wrote:

The poem "Meditation on His Ninety-First Year," by John Haag, puzzled me, for the man accepted death so calmly. How can anyone sit there and wait until it happens? Death scares me, but puzzles me. I wonder how I will accept death? The poem reminds me of my grandmother. Recently my grandmother went to the hospital. She had a stroke. It was hard for me because nobody would tell me her condition and I couldn't go see her because I was sick . . .

As soon as she got out of the hospital, I rode my bike straight to her house. She told me she had a stroke and thought she might not live. But she said she had too many things to do and lots left to accomplish. Summer was coming and she had to open the camp. She said she didn't want anyone to miss her when she was gone, so she said, "I decided. I'm *not* going!"

Literature the students read on their own

From a recommended list of books, students continue to read on their own at night and respond in their reader's-writer's log. I ask them to relate their immediate response and feelings to the books, to relate the stories to their own experiences, to talk about the main characters and how they changed, or to talk about what they might have done differently if they had written the book.

Lisa wrote:

In the story *Getting Nowhere* by Constance Greene, I liked the phrase, "The day dragged on like a turtle out for a walk." It described a boring day excellently. I disliked the phrase, "That's some piece!" I've never heard a reference to a female that's so crude. It's as if she was a new Porsche or something. If I had written this story, I would have put less emphasis on the stepmother and more on the stepson's problems facing reality.

While we are reading, we are writing. The writing comes from many sources: interviews with grandparents or the elderly, observations at the nursing home, notes and ideas jotted on the index cards in our initial exercise, and as memories stirred and jarred as we read or drafted particular types of writing.

The students still have five rough draft pages of writing due a week. This time they have to attempt specific kinds of writing I ask them to try. They still have the option of choosing those pieces that work to take to final draft. They still have to read nightly and respond in their log. They have to try several books from the recommended list.

We talk about good writing. The students think that good writing pulls readers in and keeps them reading. It causes some reaction in the readers and the readers can usually identify with the writing. Because middle-school students are still imitating styles, they need *models* of good writing of the type they are asked to do. That is why I cannot separate reading and writing.

I ask the students to attempt several kinds of writing: impression, a personality portrait (description), an experience, a persuasive piece, or a research paper. In addition, the students are writing responses to numerous pieces of literature as they write.

Impression

I want the students to see the impressions left on others by older people and grandparents, especially those in nursing homes. I read Karen's poem "The Nursing Home" (see chapter 1) and Lisa's poem:

> The nursing home
> was a series of small,
> cramped,
> cubicles,
> which granted one
> just barely enough room
> to exist within.
>   In one such cubicle
> sat an elder woman
> shying from
> any outer form
> of activity.
>   "I'm old now. I'm unable to do that anymore,"
> was her excuse
> for not playing any games,
> not taking walks,
> not staying to socialize when a room fills,
> and many other pleasurable pastimes.
>   So she sat
> silently,
> alone,
> in front of her window,
> looking out at a world
> of fluttering autumn leaves
> framed
> against a seemingly never-ending,
> bright-blue sky.

Together we read the short stories "A Visit of Charity" by Eudora Welty (1969) and "The Moustache" by Robert Cormier (1982). "The Moustache" is about a teenage boy who visits his grandmother at a nursing home. He feels guilty because he hasn't seen her in so long. At the nursing home, he discovers she is not just his grandmother, but a real person, with guilt feelings of her own.

After reading the stories, I ask the students to close their eyes: What words come to mind about the nursing home? What's the one dominant feeling or impression you had? What stands out still in your mind?

The students draft, revise, and write. Jill wrote "Children's Games":

> Wheelchairs, walkers, canes.
> Wrinkled, sagging, gray skin.
> Limp, splayed, lifeless limbs.
> Glazed, confused, wandering eyes.
> Quiet, mumbling voices.

> In the living room of a home for the elderly, Amy, a volunteer, sets up plastic bowling pins in the center of a circle of elderly residents. She hands Catherine a ball.

Silently, they watch each other bowl.

"Come on, Millie," Amy encourages.

"You can do it, Joe," she urges.

"Good job, Artice," she congratulates.

I look at Marion. She appears hostile and angry. Her head is cocked to one side. Her chair and body are pulled back from the group. She stares at us out of the corners of her eyes.

I glance at Earl. He looks bored, unstimulated. His hands are folded neatly on his lap. His legs are crossed one on top of the other. His eyes stare at the ceiling.

"Throw it to the right a little, Mary, and with a little more power if you can. We're getting such good exercise! This is so much fun! Isn't it, Carl?"

Isn't it, Carl?

Janet took her sparse notes from talking to a patient at the home and wrote from the woman's point of view:

A quiet, sunny day
walking through the breezeway, the canaries
caught my eye like earrings on a child
My books in hand
Like long hair shaking, my cat Shar followed me
his tail straight up and blue eyes intense
Walking through the garden, the lush green
wet my hands and the scent of the lilies
of the valley was thick in my head
The stately blue spruce stood in the
middle of the rock garden like a soldier
in the midst of family
I stopped to look,
then continued on my way to sit
under the tree
Shar came right up next to me
His bright cat eyes closed against the sun
and he rubbed his silky cheek against my knees
His long sea-black fur brushed softly
against my fingers
His paws reached to my chest and felt
like stamps on a soft envelope
His 20 pounds curled itself upon my chest
I breathed in and began to read
And now, under that stately tree,
there is a box
entwined in the arms of the roots
secured by the earth
In it is my sunny day
only it's black as night
I am three thousand miles away
but Shar's blue eyes still look at me
on a sunny day

Kristan described her reaction through "Alice":

Her head wobbled
like a jack-in-the-box still
bouncing on its spring.

Her translucent fingers
grasped a red checker, my checker.
I said nothing
as she jumped her own man.
"Do you have any kids?" I asked.
I waited for a response.
When none came
I repeated my question.
"Oh, yes." Her lips pursed together.
"A son. He put me in here."
"Do you have any grandchildren?"
"No," her eyes said. "Too busy.
He's not even married."
I moved a checker
then put it back.
" That one," she smiled, pointing to another checker.
"I'd like to see you win this time."
She had beaten me at four games.
"How long have you been here?" I asked.
"Fourteen years," she said. Her red lips pursed.
Had she done her own makeup? I wondered.
And her shaky hand
jumped four of my men.
"Do you like it here?"
I wished for no answer.
"It's okay, but it's not my home."
Fourteen years with no home.
A woman's voice filled the room.
"I'm just here to wait," it said.
"Wait to die."
Alice's head nodded intentionally.
I stood to leave.
"What's your name?"
"Kristan," I said.
"Kristan, you'll be with me when you
come next time, right?"
"Yes," I said. There would be no next time.

    I tried my hand at poetry, trying to capture the woman with whom I spoke.

Trying to remember

"My memory is one . . . loud . . . void . . . "
Margaret chuckles softly.

Nothing moves but her arm, mechanically,
like a water-wheel turning from her lap.

She clutches the edge
of a blue, terrycloth bib,
   smoothing the corners
     until they all touch
       like rabbits' noses.

Two silver wedding bands drift
on parchment sticks,

as Margaret's arm moves up, and down, back,
    and forth,
        in the sign of the cross.

"Where are you from?" I ask.

Faded blue eyes dart to my words.
Her hand cups her chin.
One finger slowly caresses
the side of her nose,
and her vacant stare
returns my question.
She folds her bib again, all corners matching,
then spreads the towel across her mouth
and nose like an oxygen mask.

"Margaret?" I whisper. "Are you sleeping?"

"I'm trying to remember . . . " she exhales,
    and slowly turns out another cross.

Dorn turned his impression into a piece of fiction entitled "Time-flyer."

I had just landed my Pitts special, my tired but well-loved red biplane. I was walking around the hangars, waiting for the air show to begin, when I noticed one of my old friends running toward me. He had been one of my partners on a recovery team for destroyed aircraft. He came up to me and patted me on the back.
"You still stuck in the pits?" he said jokingly.
"Sure, I think this airplane has been repaired about a dozen times in the last month, but I still love it," I said.
"You gotta see my new war bird!"
"What is it?" I asked.
"Do you think I'm going to tell you?" he said brightly.
I followed him down to the airfield. He ran ahead. I kept up only so I could see where he went, weaving in and out of the many parked airplanes.
Finally, I saw it across the field. It was a Mustang! My dream finally came through! The P51 Mustang was the dream of most pilots of that time. Here I was at an air show in Colorado on a spur of the moment whim, and there it was! I was glad I had come.
I reached the spot where my friend was standing, a grin on his face that looked as if he would split his lip if it grew any bigger. I was still looking at every part of that wonderful plane with my jaw open. My friend sat me in the cockpit. It was just as I had fantasized in my dreams. The headrest fit me perfectly, the paddles seemed born to my feet; the whole plane fit like a hand in a glove—I felt as if I had been born to it.
"Would you like to try it out?" he asked proudly.
"Would I? Would I?" I said, still in awe.
I had flown Corsairs and 38 Lightnings, and most of the others, but this was the unbelievable best. He explained briefly the special points of the cockpit controls, and showed me all the comforts of the incredible machine, but I didn't listen. I had read so much about the Mustang, I could have built it from scratch.
"Off you go!
My friend backed up, away from the plane. I could feel the plane's power as the real stuff of my dream moved down the runway to wait in line for our turn to take off. There was a home-built aircraft in front of me. I used to

admire them, but now, had eyes for nothing but the dream I was in. I was on top of the world. I turned on my radio and listened for my signal to take off. Finally, I heard the call: "V683R, you are cleared for take-off."

I raised the trim tabs, closed up the cockpit, and pushed on the left pedal. I revved up on the turn, pulled back on the throttle, and faced the plane down the runway. I reached down with my left hand, grasping the throttle again, after a moment of rest. I pushed it forward, increasing the speed, and taxied down the runway pushing forward. I felt the tail wheel rising up and the wing wheels spinning off the pavement. I pulled back harder—and was airborn! World War II's greatest flying airplane was back again. But I didn't like to look at the Mustang as an instrument of death, rather as a beautiful flying machine.

I looked back through the bubble canopy at the airfield. The weather was as clear as it had been for days, with only a few clouds up high. I touched my right foot to the pedal and moved the stick to the right and forward. I did a clean right bank turn and came back toward the field.

"Oh, my gosh," I said to myself. I almost forgot to raise the landing gear in my ecstasy. I pulled up on the lever near the floor. All three lights blinked green. All was well.

I pulled back on the stick. The nose rose in front of me. We were heading for 30,000 feet, almost the ceiling. I pulled on my oxygen mask, turned off my radio, and watched the clouds nipping at the wings. I didn't want anyone to break the spell. I looked at the altimeter. It read 20,000 feet and climbing. I lay back and thought, 25 will have to do.

I leveled off, preparing to break the sound barrier. I pushed forward on the stick, and wow! The clouds parted in front of me—as if in fear. There was an abrupt woosh! I eased back on the stick. Nothing happened. Confident, I tried again and finally came out of the long dive. I had broken the sound barrier in the P51! I was so happy I did a snap roll. I turned around, headed back to the airfield, and turned my radio back on.

"V683R, you are cleared for landing on the west runway."

I was home again.

The sunroom of the rest home is bright and warm, keeping the sleepy old gray-haired men in wheelchairs comfortably mellow as they nap in the late afternoon sun. The crisp white nurses begin to wheel the men back to their rooms for supper.

"A quiet afternoon today?" asks one nurse of the old man sitting in the corner he always prefers. He has a clear view of the sunset. He smiles quietly to himself, lost in his own world of dreams and memories.

Personality Portrait

Together we read the personality portrait of "Annie Lane" from the book *don't send me flowers when I'm dead* by Eva J. Salber (1983). I want the students to read a good example of writing that reveals a character in several ways through an interview.

Annie Lane—71—"Don't send me flowers when I'm dead. I want them now." Annie Lane, tall and sturdy, her fair complexion shielded from the sun by an old-fashioned sunbonnet, works her land and is proud of it. In very good health, except for deafness, she wears a large hearing aid pinned to her dress.

I was next to the baby in the family. I had to work hard. I was a widow woman's child. I was raised to work. We had to dig our living out of the ground. I quit school when I was in the seventh grade and got married . . . My husband was a farmer, too, so I went right on helping him with the farm. We both worked hard. A farmer never gets rich but we lived a happy life . . .

I was raised to work and I still enjoy it. I'll go on working as long as I'm able. When I'm not able, when I get to that, I hope I'll do a big day's work and lay down at night and go to sleep and not wake up.

I tell people, "Don't send me flowers when I'm dead. I want them now." It wouldn't do me two cents worth of good after I'm dead to put me in my grave and put a pile of flowers on me as big as this house. If you've got a flower you want me to have, give it to me while I'm living. (20–21)

Lisa wrote:

Marjorie H., age 68: "Old is older than me. It isn't how you feel, it's how you think."

Marjorie sits down, sliding a pillow behind her somewhat disfigured back. She lowers her cigarette, takes up her beer, and picks a white hair off her stylish knit sweater. She is back from her job at a dry cleaners, where she does specialty sewing jobs.

The atmosphere is a comfortable one. The room is clean, although small and cluttered. Her well-favored Welsh Terrier wanders about in search of extra attention from her guests. As Marjorie talks, her eyes reflect the light of knowledge, knowledge of things past and present.

I went to Newton schools, which, according to the census were some of the best. I didn't like them. They were large schools with forty in a class. My graduating class was one of seven hundred. They weren't intimate, and the social life was a small clique excluding most of us.

If I ever had a chance to live my life over though, I would choose to skip my teen years. I was constantly worried about little things like deadlines on homework. I can't remember ever being that happy during those years. If I could give any advice to a young person it would be, "Keep your cool and you'll live longer." I think kids today still worry too much.

I read *The Polar Express* by Chris Van Allsburg (1985) out loud to the students. Together we read "A Worn Path" by Eudora Welty (1969), as we listen to Welty read it aloud to us on tape, and "Goodbye Grandma" by Ray Bradbury (Dristle, 1982). We compare the young child in *Polar Express* and the old woman in "A Worn Path", and note how no one listens to either the young or the old.

We talk about how a writer reveals a character's personality by describing his or her physical appearance, what the person says or does, the person's thoughts or feelings, and the person's effect on other people. I ask the students to try to describe the person they interview—a grandparent, an elderly person, or the person at the nursing home—by trying one sentence for each of the four ways.

I go around the room, asking the students to read their best sentence. Becky tried to show her grandfather through something he said: " 'Hey babe, you wanna try out the back of my car?' he wiggled his bushy eyebrows." When Becky read hers everyone wanted to know what happened next. Her piece "Dinner with Grandfather" came from that one sentence.

Dinner with grandfather

"Hey babe, you wanna try out the back of my car?" he wiggled his bushy eyebrows. The waitress just smiled and walked away. We all stared at my grandfather, shocked.

I looked around the restaurant to see if anyone else had heard him. I could just imagine the waitress in the kitchen telling all her friends that a three hundred and fifty pound sixty-two-year-old man had just tried to pick her up. I could almost hear them laughing.

The waitress came back, barely suppressing a smile. "May I please take your order?"

"Anytime . . . I'll have garlic bread, spaghetti, Greek salad . . . with Italian dressing, and a jug of house Chianti for the table." Grandpa handed her the menu. After we had all ordered and the waitress had left, we, somehow, got into a discussion about politics. Big mistake.

" . . . And what about medicare? Reagan doesn't care about that!" My father's eyes practically blazed.

"You young people don't know what you're talking about." Grandpa leaned back in his chair at the head of the table and smoothed back his silver-grey hair. "Reagan's an old frump too. He'll fix everything. No sweat."

"And the national debt? He's doubled that all by himself!"

"Blame that on nuclear missiles." He looked at his son-in-law as if he couldn't believe that anyone could be that stupid.

"We have enough nuclear missiles to blow up the world nine times! We don't need anymore!" I blurted, unable to keep from breaking in.

Grandpa gazed in my direction, turned up his oversized nose and kept talking to my dad, as if I wasn't even there. I could have slapped him.

Dinner arrived just in time. My uncle, who had had his fair share of Chianti, and my grandfather, were the life of the party. "What's hard and long when it goes in and soft and sticky when it comes out?" My uncle's eyes had a mischievous glint in them. Mom's lower jaw practically rested on the table top.

"BUBBLE GUM!"

My grandfather laughed so hard that a forkful of spaghetti fell into his lap. "Ooooops!" He picked it up with his fingers and put it on his bread plate.

I was so embarrassed. I was sure everybody was looking at us. "Becky, you didn't have that sunburn when we got here, did you?" My grandfather's booming laugh reached my ears. I closed my eyes and imagined myself someplace else.

"Hand me the bread." Aunt Sheryl pointed to the basket. Grandpa reached out with his beefy hand and knocked over his wine glass. Luckily it only had a little wine in it. Still, it landed in his lap.

The night wore on. By the time we left my grandfather looked like a smorgasbord. "Well, that was great! You can tell. I wore my dinner home! Spaghetti, coffee, rose, Chianti, spumonti . . . " He pointed to each spot as he talked. My mother and I exchanged looks.

"Well . . . you kids go on home. Your mother and I have some personal business to attend to!" he yelled out the window of his car, wiggling his eyebrows. My grandmother sat on the other side of the car with her hand over her face.

Other students drafted, revised, and wrote. Graham wrote "Papa's Jacket."

Papa is a tall man, his hair still black and his eyes bright blue. He lives in California with Grandma, in a house on the water . . .

. . . One day, when we are visiting Papa and Grandma for Christmas, the subject of Papa's time in the Air Force comes up. Papa was a pilot in World War II, a captain of his squadron. He's proud of being in the war and being a pilot.

Suddenly a thought strikes him and he walks out of the room. He returns with a faded and worn brown leather jacket.

"Graham, this seems to be a little small for me. I wonder if it would fit you?"

I glance around the room. Everyone is smiling and looking pleased. Only my brother's face looks different. He has a hurt expression spread over his features. Papa notices this and pulls from behind his back a pilot's hat.

"This might fit your head," he says, looking at Ian.

The hurt expression fades from his face and is replaced by one of mixed happiness and confusion. He seems unsure of whether he has been cheated, he getting a hat, me a jacket. The hat falls over his eyes and ears and he looks slightly comical.

I slowly put on the jacket. The smooth inside feels soft on my bare arms. The smell of leather enters my nostrils and the feel of the big coat on my shoulders makes me feel proud. I look at the faded right arm and think with pleasure of the battles Papa and this jacket have been through.

I am much older now, and as I look at the leather jacket hanging proudly on the back of my chair, I am compelled to tell of how it was given to me by the tall and loving man I call Papa.

Stacey captured "Big Grandpa" in one moment:

As we sit in my grandmother's kitchen, I study my big Grandpa. His large, darkened eyes dart back and forth from the spoon in his hand to his half-eaten bowl of mini-wheats.

I watch his unsteady, wrinkled hand grasp the spoon, his pointer finger and thumb are wrapped around the neck of the spoon. His ring finger is at the middle of the spoon, trying, but unsuccessfully, to steady the spoon.

After Big Grandpa steadies the spoon as best he can, he moves it downward towards the bowl. As I watch, the spoon moves around his bowl as if to pick out a specific mini-wheat. Finally, he selects a mini-wheat and with great trouble, he cuts the cereal in half and pushes it to the edge of the bowl to make the picking up of it easier.

Big Grandpa slides the mini-wheat onto the spoon. It hangs over one edge. His shaky hands start the long upward journey from the bowl to his mouth.

Big Grandpa's hands, aged and crooked from arthritis, get a better grip on the spoon and lift it. The vibration of the spoon caused by his shakiness makes the cereal drop back into the bowl. Milk splatters and lands on the table. Big Grandpa takes a handkerchief from his pocket and wipes up the milk. I take the handkerchief from his hand and wipe up the places he has missed. Big Grandpa pats my hand and gives me a gentle smile.

He picks up his spoon once again and pushes the same piece towards the edge of the bowl and begins the routine over again.

This time Big Grandpa doesn't get past the top of the bowl before it drops off. He sighs very loudly and again starts everything over.

This final time Big Grandpa bends his head down very close to the bowl to make sure he would get the cereal into his mouth. His spoon is shaking a little bit more this time; I have my eyes transfixed on the slow up and down motions of the spoon.

As the spoon gets just inches from his face, Big Grandpa opens his mouth wide . . . the spoon shakes wildly as he puts it into his mouth. He continues to hold the spoon even though he doesn't need it. Big Grandpa chews very slowly, as if savoring every bite. After he swallows, Big Grandpa looks at me as if to say, "I did it." Then he turns away, puts his spoon in his bowl, bends his head down, and starts the slow process again.

Toby, who received remedial reading services, struggled with language. He worked hard as he developed "He Is My Great-Grandfather."

He lives in Wisconsin
I live in New Hampshire
He is old
I am young
He is lonely
I am too

He is the one who put the worm on my fishing pole.
I am the one who threw the line into the water.
He is the one who helped me bring in my first fish.
I am the one who thought it was a great blue whale
when in reality it was a four and a half inch sunfish.

He is the one I love and he is the one I always will.

He is my great grandfather.
I am his great grandson.

Experience

I want students to hear how other writers tell of their experiences with
grandparents, to realize that what students might think of as trivial is what
is really important in the experience.

While writing an experience piece, I read aloud poems from *Waltzing
on Water* by Norma Fox Mazer (1989) or *Strings: A Gathering of Family
Poems* edited by Paul Janeczko (1984). I might also read one of the following children's stories: *Annie and the Old One* by Miska Miles (1971),
Knots on a Counting Rope by Bill Martin, Jr. and John Archambault (1987),
or *Waiting for May* by Thyrza Davey (1984).

Together we read "Another April" by Jesse Stuart (Harris, 1983). I
reread this one section in the story "Goody-bye Grandma" by Ray Bradbury, because I love the way the author describes the experience. Great-
grandma, at age ninety, has lain down in her bed knowing it is time to die.
Her family tries to tell her it is not the time.

"Grandma! Great-Grandma!"
The rumor of what she was doing dropped down the stairwell, hit, and
spread ripples through the rooms, out doors and windows, and along the
street of elms to the edge of the green ravine.
"Here now, here!"
The family surrounded her bed.
"Just let me lie," she whispered.
Her ailment could not be seen in any microscope; it was a mild but ever-
deepening tiredness, a dim weighting of her sparrow body; sleepy, sleepier,
sleepiest.
As for her children and her children's children—it seemed impossible that
with such a simple act, the most leisurely act in the world, she could cause
such apprehension.
"Great-grandma, now listen—what you're doing is no better than breaking a
lease. This house will fall down without you. You must give us at least a
year's notice!" . . . "Grandma, who'll shingle the roof next spring?"
Every April for as far back as there were calendars, you thought you heard
woodpeckers tapping the housetop. But no, it was Great-grandma somehow
transported, single, pounding nails, replacing shingles, high in the sky!
"Douglas," she whispered, "don't ever let anyone do the shingles unless it's
fun for them."
"Yes'm."

"Look around come April, and say, 'Who'd like to fix the roof?' And whichever face lights up is the face you want, Douglas. Because up there on that roof you can see the whole town going toward the edge of the earth . . . " (92–93)

We each write about an experience in an attempt to show the kind of relationship we have with that person. Several of the finished pieces can be found in the student portfolios in Appendix A: Andy's "Empty Stairs," Gillian's "Opa," and John's "Mr. King." (See also Nahanni's "Unable to Forward," Fig. 8.2.)

Scott "captured" an unknown side of his great-grandmother in China.

In the lighted alleyways of Peking, China, children run back and forth, hidden underneath the cover of homemade dragon suits. The same breeze that swings the bright red and green paper lanterns back and forth, lifts the dragons' skirts, to reveal little legs playing. The gentle wind sweeps the sound of laughter away.

The alley is crammed with small shops. Tourists dicker with stubborn vendors over cheap pots and pans, old jewelry, and books of little use to anyone. The sellers end up with the best deal.

A small, stout lady stands out in the crowd. Her white hair and fair skin identify her as from another culture. She stands before a preacher. She hands him something. Little does anyone know that she's a smuggler! Of what? Bibles!

This 80 year old American lady knows she can be killed. The Bible is illegal in China. Death is the penalty. The preacher quickly disappears. He will teach his followers Christianity.

Later, this same lady is seen waddling down the Great Wall of China, sightseeing under the noses of those who could arrest her . . .

Who was this lady? My great-grandmother. In 1982 she went to China with a group of other ladies she knew. Little did we know what she was doing over there. We didn't find out until she was out of China. The preacher who received the Bible was jailed and killed. We'll never let my great-grandmother go out of the country again!

Jill revealed the hurt and anger of a sad relationship in her poem.

Sitting in the family room, I think
about my grandparents,
and write:

They send no birthday cards.
We send no money.

They speak no praise.
We speak no admiration.

They show no concern.
We show no respect.

They hear no achievements.
We hear no complaints.

They see no beauty.
We see no compassion.

They feel no love.
We care
No more.

Figure 4.1
Shannon's illustration
for "Grandmother's
Whales"

Mark and Shannon collaborated on "Grandmother's Whales." Mark wrote, Shannon illustrated (Figure 4.1).

My grandfather lifts me up on to his knee and points out across the ocean. "See the whales out there, Mark?" he inquires.

"Where?" I demand.

My grandfather waits, then exclaims, "There! See the white line on the horizon?"

"No!" I say impatiently, and run to my mother. My grandmother catches me halfway across the deck, lifts me into her lap, gives me a kiss on my head, and tries to help me see the whales.

"Concentrate on the horizon," she says calmly.

My little eyes comb the horizon, then comb it again. I concentrate as hard as I can.

"There! See it?" my grandmother exclaims all of a sudden.

All my eyes see is the blue sky where it meets the water. I see no white spout of spray sent sailing into the sky from the blow-hole of a surfacing sperm whale.

"No," I mumble.

"Keep trying," she encourages me. "You can see them if you really want to."

I am bound and determined to see the whales. I squint, scanning the horizon like a cat watching a bug crawl across the floor. Then I see it! A white line on the horizon like a geyser at Yellowstone Park, spouting water into the air.

"There it is! I see it!" Looking up at my grandmother, I ask, "Did you see it?"

85

She looks down at me and says, "No!" and begins to laugh. I look at my mother, then at my grandmother, and laugh with them.

Persuasion

It's difficult to write well persuasively if you don't feel strongly about the topic. While many of the students do care about the elderly, especially after such immersion in reading and writing about them, many do not care enough to take a convincing stand on how to meet their needs. That's okay. We talk about the problems and possible solutions. I know most of the students care when I look at all the writing they have done.

For those students who want to write a position paper, I keep a file folder filled with articles, studies, and editorials.

I give the students two options:

Write a position paper that clearly states your opinion about who bears responsibility for care of the elderly. Society? Or families? Include in the piece:

What do you believe? (your position)

Why do you hold that position? (reasons cited)

What is wrong with the other position?

What should a person do?

Consider all points made in class discussions. Use facts and statistics from the readings and articles.

or

Write a research paper teaching us about something you read or heard about over the last few weeks and want to know more about such as, Alzheimer's Disease, making root beer, and so on. Use information from the people you interview, books, magazine articles, etc.

Jennifer wrote a comprehensive position paper.

Where should the elderly go?

There is no clear-cut solution as to who should bear responsibility for the care of the elderly, who are unable to care for themselves. Everyone has their opinion—the one that caters to their needs. I agree with society caring for the elderly, as in nursing homes, and I also agree with the families providing the care, but under different circumstances.

I feel that if the elderly person is capable mentally, they should be able to choose their future residence. If, however, the elder is not capable of making the decision, then the family should decide.

The reasons I support society in providing the care for the elderly are as follows: There is quick access to medical care in the nursing homes, and the elderly will be with people approximately their own age who know what they're going through and might have common interests. There will be experienced doctors and plenty of help in the nursing homes also. There are two kinds of nursing homes, the one where the government pays and then there is the one where the resident pays; if the occupant runs out of money they also run out of a place to stay. The nursing home idea is a good idea for people who are too busy to care for an elder at home or just plain can't afford to, because in some homes the government will provide the cash.

The reasons I support the families in providing the care for the elderly are because of the following: The elder might want to live out the rest of his or her life happily with their family among them. Large families will usually always have someone around to care for the aged. Well-off families can afford to provide the money to give the elderly a happy environment at home. Round the clock nursing care and "elder sitters" can be hired for the relative in the family. In the family setting it is as if the child were paying back the love that their parents gave to them.

The problems with the society position are that old people get depressed in the nursing home environment; they may also feel unloved and unwanted. The people in the home (doctors and such) rarely care about the person—just their health. In homes the patient might not get specialized attention as they would in a family setting with one doctor to a person. Doctors have taken studies that have proven that the emotional and physical health of elders in family settings is better than in the nursing homes.

The problems with the position of keeping the elderly at home are people have to quit their jobs and make babysitting an elder a fulltime career. This in turn may lead to resenting the elder and nervous exhaustion of the person who provides the care. Elder abuse might then be an outlet for revenge on the elder; in fact, 13% of the families who care for elders have admitted to doing this. The environment at home will thus not be a happy one for either party. It would also cost a lot to take care of an elder; they turn into a baby again and need that care. The person has to really want to live with their parents who are totally dependent on them to do this.

What should we do with these elderly people who need a place to be themselves? It is a tough question, but I definitely feel that if you have the time, money, and love, the family is the only way to go. If you have only one or none of these things it is not fair to anyone to have to suffer in an unhappy situation; place the elder in a nursing home, be sure they understand why, and visit them often, after all, elders are people too.

In Conclusion

In concluding our reading and writing about grandparents and the elderly, I read a children's book *The Wish Wind* by Peter Eyvindson (1987). It's the story of a little boy wishing always for the next season—until he receives his wish and arrives too quickly into the "autumn" of his years. I give the students the lyrics and we listen to the tape of "The Living Years" by Mike and the Mechanics. Lastly, I give them copies of, and read, Robert Frost's (Latham, 1969) three poems, "A Leaf-Treader," "Stopping by Woods on a Snowy Evening," and "Nothing Gold Can Stay," and Dylan Thomas's poem "Do Not Go Gentle into that Good Night" (Nims, 1983). The students respond to all of these in their logs first, and then we talk about what they think. We talk about all the images and symbols of the aging process and youth in these poems, how they relate what they read to themselves, and what they now know.

Alison wrote in her log: "If this is really what old is all about, I don't want to grow old. I'm not scared of being old. I'm scared of growing out of being young. Why can't we live like the monarch butterflies? They are ugly when they are young. Then when they spread their wings they soar and the world stands and watches this lovely butterfly."

Matt wrote two contrasting poems (see Figures 4.2 and 4.3).

Figure 4.2
"When I Grow Old" by
Matt

When I grow old I'll keep occupied.
I'll jog everyday, and I'll keep my stride.
I'll never play bingo after sixty-five,
While others die out, I'll be coming alive.
I won't sulk in chairs saying "Bottoms up!"
Alcohol will never, ever fill my cup.
I won't be caught
        with grandfather clocks—
        or rocking chairs—
        or holes in my socks.
I won't be seen
        in thirty foot cars—
        or smoking a pipe—
        not even cigars.
I will not move
        to Tampa Bay—
        or Miami Beach—
        where they sit all day.
When I grow old I want to be known,
As the crazy old man living down the road.

                Matt

Figure 4.3
"Old" by Matt

I will never, ever grow old.
I'll always stay so young and bold.
I'll never get
        dry wrinkled skin -
        or creaky bones -
        or double chin.
My hair will never turn white nor gray.
I will not sit in a chair all day.
I won't become so tired and slow,
Even when it nears my time to go.

It makes me shiver as if I'm cold,
Every time I realize
I WILL grow old.

                        Matt

Patty concluded:

Quiet as the hint of dawn
Quiet as the frightened fawn
Quieter than spiders' webs
Drawn with silver thread and sparkling dew
Quiet as a rabbit hidden in the sage
Quiet as all quiet things
  Slowly
    Silently
      Comes age.

Adolescents

While age comes "slowly" and "silently," in the minds of the young adolescence does not. It never has.

Beverly Cleary writes:

Our class was changing. A quiet boy who sat in front of me had so much trouble with arithmetic that he began to cry during an important test. The tears of a boy thirteen years old distressed me so much that, for the second and last time, I cheated in school. I slipped him some answers.

A bitter, scowling boy across the aisle from me spent his days drawing, in elaborate pencil, guns and battleships. He made me uneasy, and perhaps made Mrs. Drake, our eighth-grade teacher, uneasy, too, for she left him alone . . .

The boys who were so awful in the sixth grade and terrible in the seventh grade became really *horrible* in the eighth grade. They belched; they farted; they dropped garter snakes through basement windows into the girls' lavatory. In the days before zippers, a boy could, with one swipe of his hand, unbutton the fly of another boy's corduroy knickers—always in front of girls who, of course, nearly *died* of embarrassment while the red-faced victim turned his back to button up . . .

The horrible boys, whose favorite epithet was "horse collars!" shouted "Hubba-hubba!" at any girl whose developing breasts were beginning to push out her blouse.

Some girls changed, too, and were considered "fast" because they took to wearing lipstick and passing around two books, *The Sheik* and *Honey Lou: The Love Wrecker* . . .

Claudine peeked into *The Sheik* and reported, "Gee, kid, there was this sheik who kidnapped this girl and carried her off to his tent in the desert. He laid her on a bed, and when she woke up in the morning, he was gone, and then she discovered a dent on the pillow next to her, and she knew he had slept in the same bed with her. Wow!" Our innocent imaginations were incapable of filling in the crux of this scene. A dent in the pillow was shocking enough. (1988, 160–61)

Working with teenagers is not easy. It takes patience, humor, and love. Yes, love of kids who burp and fart their way through eighth grade. Who tell you "Life sucks!" and everything they do is "Boring!" Who literally roll to the floor in hysterical laughter when you separate the prefix and the suffix from the word "prediction" and ask them for the root and what it means. Who wear short, skin-tight skirts and leg-laced sandals, but carry teddy bears in their arms. Who use a paper clip to tattoo Jim Morrison's picture on their arm during quiet study, while defending the merits of Tigger's personality in *Winnie-the-Pooh*. Who send obscene notes that would make a football player blush, written in pink magic marker, blasting each other for stealing or not stealing a boyfriend, and sign the note "Love, _____ P.S. Please write back."

In *A Tale of Two Cities* Dickens was referring to a time, but he could just as easily have been referring to adolescence. It is the worst of times, the best of times. Charity James knew:

> At times it seems as if adolescence were seen as a social problem rather than a vital period of growth and establishment of the personality, a period which a society should provide for with affection, good sense and patience. (1974, 83)

> Some teachers know that the most important contribution they can make to young peoples' lives is to support them in the work of self-discovery and self-making, a search which makes it possible to be in this sense truthful in relation to others and oneself. (1974, 147)

And I hope I continue to know. When I stop liking these kids as people, I will get out of the classroom. Listening to them, accepting them for who they are as they go through this worst of times and best of times, laughing with them, respecting them, and helping them find what's good in their lives is what I am about. For many of the kids (broken homes, abusive relationships, temptations from drugs and alcohol, sexual awakenings, war, pollution, the pressure to succeed . . .), life is not easy. For many of them I agree, "You're right, life sucks! So what are you going to do about it?"

In an exercise similar to what I do with the elderly, I ask the students to think about themselves as adolescents. I hand out index cards and ask the students to number them from 1 to 5. For a week I spend the beginning of each period reading poems, excerpts from novels, pieces my students have written, or short things I have written. (Note: Many of the paperbacks listed in the appendix are works of adolescent fiction and contain excerpts I use when I'm going through the exercise described in this chapter. I think we need to become familiar with those books in particular, to find out which ones feel most comfortable with our kids, in our classrooms. I use a lot of former students' writing, and writing of my own, throughout the year, but especially when we're talking about adolescence.) Each piece read bears some semblance to the question I want them to answer on the card:

Card 1: Describe adolescent or teenager. What specific words, phrases, sentences come to mind?

Card 2: What do you like about being a teenager?

Card 3: What do you dislike about being a teenager?

Card 4: What troubles, frustrates, puzzles, or worries you?

Card 5: What's a vivid memory you have of something you did, feelings you had, relationships, the way you were treated, the way you treated someone else—as a teen?

I collect the cards at the end of the week. On Monday the kids count off by fives, I hand a set of cards out to each group and ask them to respond with a collaborative written and oral presentation. They are to find a way of answering the question they have, based on the information on the cards.

They are so honest, even about themselves.

Gabe, Dena, Greg, and Jay created a TV commercial:

Cycle teen commercial

(Gabe and Dena in "kitchen")

Dena: Kids, come down for breakfast!
(Kids enter, tired and bored.)

Gabe: We've got a new cereal this morning! Cycle Teen!
(Kids eat and start to come alive.)

Jay: Mikey likes it!

Dena: Yes, look at the children perk up!
(Kids exaggerated smiles!)

Gabe: Mikey, is there sugar in there?
(Mike checks box.)

Mikey: No.

Dena: Don't lie to your father!

Gabe: Let me see.
(Mike hands box to Gabe, who checks for ingredients.)

Jay: What *is* in this dad?

Gabe: Let's see . . . Filled with the best parts of growing up. More fun, freedom, respect, responsibility, growth hormones, and get this! It lasts from when you are 13 until you are 19!

Dena: And no sugar? So what makes the kids like it so much?
(Mikey grabs the box and looks at the back.)

Mikey: It's what comes with it! Friends, money, shopping, a phone, driving lessons, loud music . . .

Dena: I knew there was a gimmick!

Everybody: Cycle Teen. The right thing to do, and the only way to do it.

The students get their cards back for ideas they might have for individual pieces of writing. Often the writing is pretty serious, like Amanda's poem, Stacey's poem, or Abby's essay, because they take life seriously, whether they show it or not.

Things don't come easy

Things come so easy
when you are little
and your sweet smile melts another's heart
the way the sun melts
a chocolate bar in your pocket

Things don't come so easy
When your smile is full of braces
and you remind your mother
that time is passing
and you are now what she once was

Things don't come at all
when your pockets are empty of chocolate bars
and the sun has set

Amanda

And I'm not ready

My mother talks to me
as she's cooking
The roast she prepares
doesn't have its familiar smell
Outside the snow is cold
each flake falling like a lonely leaf
Mom drops her head in her hands
and I can see her reflection
off the glassy finish of the table
She looks
sad
tired
She says I will have to be more mature
to act older
Me?
The little girl who wears ponytails in her hair?
The little girl who somersaults down the hall?
The little girl who plays with dolls?
To act older?
No matter how much I try to prevent
her leaving
It's going to happen anyway

and it does

she leaves

It's like an intense wind
fiercely blowing
pushing at me
pushing the child
right out of me
and I'm not ready.

Stacey

I'm not afraid to wear the peace sign anymore

A tremendous feeling of accomplishment swept over me as I stepped up
onto the charter bus behind my mom. I slumped into my seat, happy and
exhausted. I peered out the window and watched as thousands of other
people clambered over and down the hill with signs reading, "Abolish
Apartheid," "Peace and Justice in South Africa," and "No Contra Aid" still
propped up against weary shoulders. I watched as people dropped into the
wet grass and slipped out of ponchos, sweaters, backpacks, hats, mittens,
and boots. Everyone was exhausted.

Groups of marchers huddled in pockets around guitars, humming beautiful anti-war songs, like, "Where Have All the Flowers Gone." I love that
song. I lay back in my seat and recaptured the day's events.

I could hardly believe that it was only this morning that I had pulled into
this enormous parking lot in Washington, D.C., now jammed full of

hundreds of buses, huddled like enormous herds of sheep, from all over the United States. It seemed like a lifetime ago; in a way, it was.

All the people who marched today were so different—all original, all intricately patterned like snowflakes. For the first time in my life I was not embarrassed for being different, for not trying to fit into a mold, for standing up for my beliefs.

My mom is a feminist. I've grown up in a family that is well aware of how women in this culture are treated, and how they treat each other. In school, girls who are not pretty, or who don't wear the right clothes bear the brunt of everyones' jokes. I not only recognize this treatment, I'm usually on the receiving end. This awareness of being different has separated me from most kids my age for a long time. I worried the separation was forever. But I'm not embarrassed anymore. I don't think of the peace sign as a cool trend. I write it on my shoes because I believe in it.

The feelings I experienced during this march gave me confidence in myself. There were no put-downs, no insulting remarks about what you looked like or what you wore. No one cared about anyone's past. No one was ignored or left out. This march was a group effort. It didn't matter if you had your ear pierced in the left or the right ear, if you had long hair or short hair, if you wore corduroys or jeans. Everyone was respected as a human being, for what each had to offer on the inside, not for what each looked like on the outside.

When there was a problem, everyone cooperated. When a large sign flopped over, a woman rushed over to help the other marchers carry the limp sign. Even as strangers, they worked and joked together for a common cause.

Even though I was only one of over one hundred and twenty thousand people at the demonstration, we were all there supporting each other. That's all that mattered. Despite the rain and wind, the atmosphere was intense and alive, sustained by a sense of humor and the constant sound of laughter. No one whined or complained about the weather.

The biggest lesson of my life? This is a cruel world. A lot of people don't take the time to get to know someone on the inside; they judge them on the outside. Stereotypes are too often set in peoples' minds. If you are too different, you are not acceptable. If only people would think.

I'm not afraid to wear the peace sign on my sneakers anymore.

Abby

Holly wrote a less serious piece:

"Anyone who doesn't have their newspaper, get out—GET OUT!" I slowly stand up. Sara joins me. Nahanni. Then Chris and Danny. Michelle and Greg pull themselves to their feet and trot out—they've been sent there before. The rest of us stand rooted to the spot, too dumbfounded to move.

It finally hits us. This is going to be—fun! We make our way out. Staggering across the corridor we remain silent until we hear her preaching again.

Then—out comes the small, but much loved rubber frog from Chris's pocket. He throws it up, and performs his famous "catch the froggie in the mouth" act. We take turns, and soon become experts. She thinks she is spiting us by sending Dan way down the hall, but it doesn't rattle us. It just makes the show more entertaining for us and the class as the froggie flies by the door aimed for Dan's mouth. We practice our shots, wiping it on our shirts after each catch. Soon the froggie begins to fly freely from mouth to mouth. Heck—who cares—it's not as gross as making out.

What do teachers think they're doing when they send people to the hall? Do they actually know what goes on?

And I tried my hand at fiction, after an intern told me what was going on under the table as I tried unsuccessfully to hold a writing conference with Heather. I wrote:

Sitting at the writing table in her tenth grade English class, Heather attempted to work out the theme running through Hawthorne's *The Scarlett Letter*. The white-lined composition paper stared blankly back at her. It made no sense to Heather that the woman always bore the brunt of abuse from an affair. At least, that's what she thought had happened. And that sleaze minister, she thought, standing piously by in silence.

Heather was having trouble concentrating. James was sitting across from her. For three years they'd sat next to each other. They had the same last initials. Teachers weren't creative enough to do anything but alphabetize their seating plans. The arrangement was painful to Heather. She'd been in love with James since seventh grade. To him, she was invisible. Until Monday. He not only spoke to her, he looked at her like he'd never seen her before, like she was new at school. He'd even asked her out after the basketball game this Saturday. Heather was having trouble concentrating on Hester Prynne.

"Why, Heather, we haven't worked out our ideas yet, have we?" clicked Miss Meserve through dentures. "So what do we think the theme is, Heather?" her teacher inquired, tapping her gnarled, arthritic knuckles on the blank paper.

It wasn't the smell of mothballs that choked Heather's words. Just as she opened her mouth James ran his foot up Heather's leg. She coughed, and shot a deadly look across the table. James had kicked off his Nike. As Miss Meserve continued to question Heather, he continued writing industriously, his head bent into his composition. From under the table he ran his foot up and down Heather's leg, up under her denim skirt, up and over her knee cap, massaging the inside of her leg with his toes.

Heather shifted positions and stumbled through a, "Well, I, uh, that is, we, I mean, I, gosh, I'm having trouble, I, uh . . . " James's hand and head remained bent into his writing, while his foot continued its upward climb.

"Well, Heather," humphed Miss Meserve, "I'll come back when you have something more substantial to say. Keep up the good work, James."

"I will," James smiled. "I will."

As long as we continue to know our students will "keep up the good work," we will be able to work with them effectively as readers and writers. Whether they are writing about grandparents or themselves, they are reading and writing for life.

Works Cited

Elderly

Aliki. 1979. *The Two of Them*. New York: Mulberry.

Blue, Rose. 1972. *Grandma Didn't Wave Back*. New York: Watts.

Cormier, Robert. 1982. *Eight Plus One*. New York: Bantam.

Davey, Thyrza. 1984. *Waiting for May*. New York: Doubleday.

de Paola, Tomie. 1973. *Nana Upstairs and Nana Downstairs*. New York: Putnam's.

Dristle, Gail A. & Luann W. Glick. 1982. *Literature*. Evanston, IL: McDougal, Littell.

Eyvindson, Peter. 1987. *The Wish Wind*. Winnipeg: Pemmican Publications.

Farber, Norma. 1979. *How Does It Feel to Be Old?* New York: Dutton.

Fox, Mem. 1984. *Wilfred Gordon McDonald Partridge*. New York: Viking Penguin.

Greene, Constance. 1977. *Getting Nowhere*. New York: Viking.

Harris, Raymond. (Ed.). 1983. *Best Short Stories: Middle Level*. Providence, R.I.: Jamestown.

Janeczko, Paul. (Ed.). 1984. *Strings: A Gathering of Family Poems*. New York: Bradbury Press.

Khalsa, Dayal Kaur. 1986. *Tales of a Gambling Grandma*. Montreal: Tundra Books.

Kooser, Ted. 1976. *Not Coming to be Barked at*. Milwaukee, Wis.: Pentagram Press.

Latham, Edward Connery, (Ed.). 1969. *The Poetry of Robert Frost*. New York: Holt, Rinehart and Winston.

Little, Jean. 1986. *Hey World, Here I Am!* New York: Harper and Row.

Martin, Bill, Jr., and John Archambault. 1987. *Knots on a Counting Rope*. New York: Henry Holt and Company.

Martz, Sandra, (Ed.). 1987. *When I Am an Old Woman I Shall Wear Purple*. Watsonville, Calif.: Papier-Maché Press.

Mazer, Norma Fox. 1989. *Waltzing on Water: Poetry by Women*. New York: Dell.

Miles, Miska. 1971. *Annie and the Old One*. Boston: Little, Brown.

Nims, John Frederick. 1983. *Western Wind: An Introduction to Poetry*. New York: Random House.

Polacco, Patricia. 1990. *Babushka's Doll*. New York: Simon and Schuster.

Salber, Eva J. 1983. *don't send me flowers when I'm dead*. Durham, N.C.: Duke University Press.

Seuss, Dr. 1986. *You're Only Old Once*. New York: Random House.

Silverstein, Shel. 1981. *A Light in the Attic*. New York: Harper and Row.

Van Allsburg, Chris. 1985. *The Polar Express*. Boston: Houghton Mifflin.

Walker, Alice. 1988. *To Hell with Dying*. New York: Harcourt Brace Jovanovich.

Welty, Eudora. 1969. *A Curtain of Green and Other Stories*. New York: Harcourt Brace Jovanovich.

Zinsser, William. 1980. *On Writing Well*. New York: Harper and Row.

Adolescents

Cleary, Beverly. 1988. *A Girl from Yamhill*. New York: William Morrow.

Dickens, Charles. 1983. *A Tale of Two Cities*. Mahwah, N.J.: Watermill Press.

Hawthorne, Nathaniel. 1983. *The Scarlet Letter*. Mahwah, N.J.: Watermill Press.

James, Charity. 1974. *Beyond Customs: An Educator's Journey*. New York: Agathon Press.

Milne, A. A. 1954. *Winnie-the-Pooh*. New York: Dell.

Further References
Exploring Generations—Novels and Essays

Alcott, *Little Women*

Auel, *Clan of the Cave Bear*

Babbitt, *Tuck Everlasting*

Blos, *A Gathering of Days*

Bosse, *The 79 Squares*

Bradbury, *Dandelion Wine*

Brancato, *Sweet Bells Jangled Out of Tune*

Branfield, *The Fox in Winter*

Bridgers, *All Together Now; Notes for Another Life; Permanent Connections; Home Before Dark*

Burch, *Two That Were Tough*

Burns, *Cold Sassy Tree*

Butterworth, *Leroy and the Old Man*

Byars, *After the Goat Man; Cracker Jackson; House of Wings*

Cleaver, *Queen of Hearts; Where the Lilies Bloom*

Coleman, *Weekend Sisters*

Cooper, *The Dark is Rising; Silver in the Tree; The Grey King; Over Sea, Under Stone*

Cormier, *Take Me Where the Good Times Are*

Culin, *Cages of Glass, Flowers of Time*

Edgerton, *Walking Across Egypt*

Evernden, *The Dream Keeper*

Fast, *The Immigrants*

Fox, *One-Eyed Cat*

Freedman, *Mrs. Mike*

Fritz, *Homesick*

Gardiner, *Stone Fox*

Garcia, *Spirit on the Wall*

Gilbreth, *Cheaper by the Dozen*

Girion, *A Tangle of Roots*

Graber, R., *Doc*

Green, *Grandmother Orphan*

Greene, *Unmaking of Rabbit*

Greenfield, *Childtimes: A Three Generation Memoir*

Gurganus, *Oldest Living Confederate Widow Tells All*

Hemingway, *The Old Man and the Sea*

Hilton, *Goodbye, Mr. Chips*

Hoover, *The Shepherd Moon*

Hunt, *Up a Road Slowly*

Irwin, *What About Grandma?*

Johnston, *Carlisle's Hope*

Kerr, *Gentlehands*

Klein, *Mom, the Wolfman and Me*

Konigsburg, *Throwing Shadows*

Lasky, *The Night Journey*

L'Engle, *Ring of Endless Light; The Summer of the Great Grandmother*

Little, *Mama's Going to Buy You a Mockingbird*

Lowry, *Autumn Street; Anastasia Krupnik; Anastasia Again*

MacLachlan, *Sarah, Plain and Tall; The Facts and Fictions of Minna Pratt*

Magorian, *Good Night Mr. Tom*

Mahy, *Memory*

Majerus, *Grandpa and Frank*

Mazer, *A Figure of Speech; After the Rain*

McCullough, *The Thornbirds*

Myers, *Won't Know Till I Get There*

Nixon, *A Place Apart; Caught in the Act*

Painter, *Gifts of Age: Portraits and Essays of 32 Remarkable Women*

Paterson, *Jacob Have I Loved; Come Sing, Jimmy Jo*

Paulsen, *Tracker*

Peck, *Father Figure; A Day No Pigs Would Die*

Riley, *Crazy Quilt*

Sarton, *As We Are Now; A Reckoning*

Schulman, ed., *Autumn Light: Illuminations of Age*

Sebestyen, *Words by Heart; Far From Home*

Shanks, *Old is What You Get: Dialogues on Aging by the Young and Old*

Siegal, *Upon the Head of the Goat*

Smith, R. K., *The War with Grandpa*

Smith, *A Tree Grows in Brooklyn*

Smucker, *Amish Adventure*

Snyder, *Egypt Game*

Sobol, *Grandpa: A Young Man Grown Old*

Tate, *The Secret of Gumbo Grove*

Taylor, *The Cay; Let the Circle Be Unbroken*

Tolan, *Grandpa and Me*

Voigt, *Dicey's Song; Homecoming; Tree by Leaf; Sons from Afar*

Wersba, *The Dream Watcher*

Wilkinson, *Killing Frost*

Williams, *The Glass Menagerie*

Wojciechowska, *Shadow of a Bull*

Wrightson, *A Little Fear*

Yep, *Child of the Owl*

Zindel, *The Pigman; The Pigman's Legacy*

Exploring Old Age—Children's Picture Books

Ackerman, *Song and Dance Man* (Knopf)

Andrews, *The Auction* (Groundwood)

Baylor, *The Best Town in the World* (Aladdin)

Beattie, *Spectacles* (Ariel)

Blos, *Old Henry* (Morrow)

Bonners, *The Wooden Doll* (Lothrop, Lee & Shepard)

Capote, *A Christmas Memory* (Knopf) *I Remember Grandpa* (Peachtree)

Cole, *The Trouble with Gran* (Putnam)

Cooney, *Hattie and the Wild Waves* (Viking) *Miss Rumphius* (Puffin)

Denslow, *At Taylor's Place* (Bradbury)

de Paola, *Now One Foot. Now the Other* (Putnam)

Egger, *Marianne's Grandmother* (Dutton)

Eyvindson, *old enough* (Pemmican)

Flournoy, *The Patchwork Quilt* (Dial)

Fox, *Night Noises* (HBJ)

Hall, *Ox-Cart Man* (Viking) *The Man Who Lived Alone* (Godine)

Haseley, *The Old Banjo* (Aladdin)

Hest, *The Crack-of-Dawn Walkers* (Puffin) *The Ring and the Window Seat* (Scholastic)

Johnson, *When I Am Old with You* (Orchard)

Kesselman, *Emma* (Harper and Row)

Kurelek, *They Sought a New World* (Tundra)

Levinson, *Watch the Stars Come Out* (Dutton)

Lindgren, *My Nightingale is Singing* (Viking Kestrel)

Lyon, *Come a Tide* (Orchard)

MacLachlan, *Through Grandpa's Eyes* (Harper)

Mathis, *The Hundred Penny Box* (Puffin)

Miller, *My Grandmother's Cookie Jar* (Price Stern Sloan)

Munsch, *Love You Forever* (Firefly)

Nobisso, *Grandma's Scrapbook* (Green Tiger) *Grandpa Loved* (Green Tiger)

Precek, *Penny in the Road* (Macmillan)

Polacco: *The Keeping Quilt* (Simon and Schuster) *Thunder Cake* (Philomel)

Rylant, *When I Was Young in the Mountains* (Dutton)

Sakai, *Sachiko Means Happiness* (Children's Book Press)

Schein, *Forget-me-not* (Annick)

Silverstein, *The Giving Tree* (Harper and Row)

Steig, *Caleb and Kate* (Farrar Straus Giroux)

Stolz, *Storm in the Night* (Harper and Row)

Turkle, *Do Not Open* (Dutton)

Turner, *Dakota Dugout* (Macmillan)

Van Allsburg, *The Wreck of the Zephyr* (Houghton Mifflin)

Wagner, *John Brown, Rose and the Midnight Cat* (Viking Kestrel)

Wheatley and Rawlins, *My Place* (Australia in Print)

Wild, *The Very Best of Friends* (HBJ)

Willard, *The Mountains of Quilts* (HBJ) *The High Rise Glorious Skittle Skat Roarious Sky Pie Angel Food Cake* (HBJ)

Woolf, *Nurse Lugton's Curtain* (HBJ)

Zolotow, *I Know a Lady* (Puffin)

Exploring Old Age—Films, Videos, and Songs

From the series *Learning to be Human*
  The String Bean
  Shopping Bag Lady
  Death of a Gandy Dancer
A Christmas Memory
Driving Miss Daisy
On Golden Pond
Trip to Bountiful
Stone Fox
Makem and Clancy, "Dutchman"
Dan Fogelberg, "Windows and Walls"
Mike and the Mechanics, "The Living Years"

Adoff, Arnold. 1986. *Sports Pages*. New York: Harper and Row.

Atwell, Nancie. 1987. *In the Middle*. Portsmouth, N.H.: Boynton/Cook.

Brooks, Bruce. 1984. *The Moves Make the Man*. New York, Harper and Row.

Buchwald, Emilie and Ruth Roston, (Ed.). 1987. *This Sporting Life*. Minneapolis, Minn.: Milkweed Editions.

Conroy, Pat. 1987. *The Great Santini*. New York: Bantam.

———. 1987. *The Prince of Tides*. New York: Bantam.

Cormier, Robert. 1974. *The Chocolate War*. New York: Dell.

Dillard, Annie. 1987. *An American Childhood*. New York: Harper and Row.

Donelson, Kenneth L., and Alleen Pace Nilsen. 1989. *Literature for today's young adults*. Glenview, Ill.: Scott Foresman.

Feelings, Tom. 1981. *Daydreamers*. New York: Dial.

Fox, Mem. 1988. *Koala Lou*. Orlando, Fla.: Harcourt Brace Jovanovich.

Ghigna, Charles. 1989. *Returning to Earth*. Livingston, Ala.: Livingston University Press.

Glenn, Mel. 1982. *Class Dismissed!*. New York: Clarion.

———. 1986. *Class Dismissed II*. New York: Clarion.

Hinton, S. E. 1967. *The Outsiders*. New York: Dell.

Janeczko, Paul B. 1989. *Brickyard Summer*. New York: Orchard.

Janeczko, Paul B. (Ed.). 1983. *Poetspeak: In their work, about the work*. New York: Bradbury.

Kleinbaum, N. H. 1989. *Dead Poets Society*. New York: Bantam.

Knudson, R. R., and May Swenson, (Ed.). 1988. *American Sports Poems*. New York: Orchard.

Little, Jean. 1986. *Hey World, Here I Am*. New York: Harper and Row.

Mazer, Norma Fox. 1989. *Waltzing on Water: Poetry by Women*. New York: Dell.

O'huigin, Sean. 1985. *Atmosfear*. Windsor, Ont.: Black Moss Press.

Wigginton, Eliot. 1985. *Sometimes a Shining Moment*. Garden City, N.Y.: Anchor.

Yes! Sometimes We All Read the Same Book

> **S**ome books are to be tasted, others to be swallowed, and some few to be chewed and digested.
>
> Francis Bacon, *"Of Studies"*

We read fine literature in many different ways: kids choose their own books, we read different books to each other, and sometimes we read the same book together. I think we need to get at reading from all those angles. I like choosing books to read by myself most of the time. I also like reading to the students and being read to. But occasionally I like to read the same book a number of other people have read, or even the whole class has read, and get into a good discussion about it.

We read the play version of *The Diary of a Young Girl* (Frank 1967) together as a class. First, however, I spend several days on the history of the period by reading over and discussing the events in Germany that led up to the Holocaust. The book *Friedrich* by Hans Pieter Richter (1970) has a simple but powerful chronology of the systematic removal of freedoms and rights from the Jews.

After going through the chronology I have the students choose a card from a bucket. About one-third of the cards have large blue dots on them, the other cards are blank. The rules are simple. Those with blue dots must pin them to their shirts and wear them for twenty-four hours. The Blue Dots must try to abide by the same laws the Jews had to abide by, based on what we discovered in the chronology in *Friedrich*, such as: they may no longer attend plays, movies, concerts and exhibitions; they must hand over securities and jewelry; they must hand in their radios to the police; they may no longer have telephones, etc. Those who are not Blue Dots are just to go about business as usual. That's all.

After twenty-four hours we find out that that's *not* all. I have the students write out everything they did, saw, felt, and thought. We discuss what happened.

I discovered how Stacey and Marie reacted to each other when I read their pieces. They each wrote from their own point of view and had no idea how well their contrasting views fit together (Figure 5.1).

101

DURING LUNCH I WALKED AROUND AND LOOKED FOR BLUE DOTS
DOING THINGS THAT WERE FORBIDDEN. "A-HA! I'M GOING TO REPORT
YOU!" I SAID TO MARIE.

> I looked up. Stacey had a mean look on her face. "What are you talking about?"
> I asked.

"THE BLUE DOT. YOU'RE NOT WEARING IT!"

> The blue dot. I had forgotten about it. That dumb dot! It made me feel so guilty all
> day long.

I HAD TO REPORT HER. IT WAS MY DUTY FOR BEING A NO DOT.

> That morning we were told to staple the dot to our shirts. It was humiliating. It made
> me feel different. So I put it in my pocket. No one would know. Until now.
>   As I taped (I wasn't going to staple it) the Blue Dot to my shirt pocket I thought
> to myself, I can take this easily. It's just a dumb dot. I'm going to follow all the rules
> with no trouble. It will be like camping.

BUT EVEN THOUGH I DID THIS, DEEP IN THE BACK OF MY MIND I FELT
REALLY GUILTY BECAUSE I WAS IN A SENSE "SPYING" ON MY FRIEND'S
PRIVACY.

> Walking down the hall, the faces of other Blue Dots lit up and smiled as I passed,
> even people I had never spoken to before.
>   It's funny though how being restricted changes your point of view about lots of
> people. Even after only ten minutes of not being able to talk to people without Blue
> Dots, I began to feel inferior. I kept to myself all through our last yearbook meeting,
> and didn't enjoy myself much at all.

I GUESS I REPORTED HER, TOO, BECAUSE I WAS FEELING LEFT OUT OF
THE GROUP OF FRIENDS WHO WERE ALL BLUE DOTS. IT WAS A WAY OF
GETTING BACK AT THEM.

> At first I thought being a blue dot would be fun, but I soon realized my fun was
> limited.

IN A WAY I FEEL THE BLUE DOTS JUDGED ME BEFORE THEY EVEN KNEW
IF I WAS FOR OR AGAINST THEM. AFTER I HAD WRITTEN PEOPLE UP, I
JUST FORGOT ABOUT IT. THEY REALLY DIDN'T DO ANYTHING TO ME, SO
I DIDN'T WANT TO THINK ABOUT IT. NOTHING WAS GOING TO HAPPEN TO
THEM ANYWAY.

> Lunch was after English, where we had all gotten our roles. What can happen at
> lunch? I thought. Well, everything can happen at lunch when you're a Blue Dot. Like
> Stacey yelling at me for not wearing it. I sat with a bunch of other Blue Dots. Since
> we were not able to communicate with people who were not Blue Dots, it was
> better if we all sat together.
>   Sitting together was the easiest part of it. I got out my ham sandwich and
> chocolate milk. I realized my friend Melissa was staring at me like I was from
> another planet.
>   "What's the matter?" I asked.
>   "You're eating meat and drinking milk!"
>   Oh no! What was I supposed to do? I continued eating, looking around for any
> sign of anyone not wearing a Blue Dot who would report me.

AT LUNCH I WATCHED THE BLUE DOTS EATING. I WAS JUST WAITING FOR
SOMEONE TO BUY MILK OR FRUIT. IT'S A GOOD THING THEY COULDN'T
EAT GRAPES. THERE WEREN'T ENOUGH TO GO AROUND.

> At recess, everybody came up to us, saying things like:
>   "Oh, you're a blue dot! Stay away from us!
>   "Pick this up for me, Blue Dot!"

"NO BLUE DOTS ON THE SOCCER FIELD!" I YELLED AT RECESS. I
COULDN'T BELIEVE MY OWN EARS.

Figure 5.1 Stacey's and Marie's response to "Blue Dots"

How humiliating. I felt like my own friends were betraying me. Of course, there were the people in other classes who didn't know what was going on. What was I supposed to do with them? It was pretty annoying, having your own friends telling on you just because they didn't have a stupid Blue Dot pinned on their shirt.

How was I supposed to get through the afternoon without being able to talk to anybody but other Blue Dots. I wanted to ignore the Blue Dot, but I felt too guilty. In science I took my regular seat. Stacey sat down. She had no dot. I stood up. I couldn't sit next to her.

"WHO'S GOING TO FIND OUT?" she said.

But I didn't trust her. I moved to sit with another Blue Dot, even though I didn't like her much.

"HEY, MARIE. YOU ARE NOT SUPPOSED TO TAKE A BUS HOME, REMEM-BER?" I SHOUTED.

"I am not going to walk ten miles!" I said. What was the matter with me? I had just told myself to take this more seriously, and there I was, breaking another rule. But the teacher couldn't have expected us all to walk home! Why was I feeling so guilty?

I figured I was safe on the bus. I took the Blue Dot off. I felt ten pounds lighter. I relaxed on the way home, not having to worry about anything. I realized this plaything was really getting to me. I was taking it more seriously than I thought.

I FLIPPED ON THE TV, GLOATING OVER THE FACT I WOULD KNOW WHAT HAPPENED ON GENERAL HOSPITAL AND MARIE WOULDN'T. I POURED MYSELF A LARGE GLASS OF ICE COLD MILK AND GRABBED A HANDFUL OF CHOCOLATE CHIP COOKIES.

As I fixed a snack, I became aware of the reflex to turn on the radio. It seemed so natural. But I resisted. No television, either. I decided to take my dog for a walk. But then I remembered Blue Dots weren't allowed to have animals. I went by myself. I wore my thinnest jacket instead of my woolen winter coat. At last, by myself, no one could discriminate against me.

I cut the walk short and hurried home. I had to do my homework before the sun dropped too low. I started to run as the sun sank lower and lower in the sky. I couldn't use electric lights.

THIS WASN'T FUN, NOT HAVING ANYTHING NOT TO DO. IF I COULDN'T BE A BLUE DOT, I'D FIND SOMETHING TO DO.

At home, was I supposed to avoid my parents too? I thought about it and said to myself, girl, if you are a Blue Dot, your parents must be Blue Dots too. As I opened the door for my mother and brother, I jumped up and down, shouting, "Look, I'm a Blue Dot. I'm a Blue Dot." Why was I suddenly so enthusiastic? I thought.

After dinner (tuna casserole, not meat) I went back to my homework. I was trying to figure out one of the algebra problems when the phone rang. I casually got up and answered the phone.

"MARIE! YOU'RE A BLUE DOT! YOU KNOW YOU'RE NOT SUPPOSED TO ANSWER THE PHONE. I'M REPORTING YOU IN THE MORNING.

Of all people, Stacey, my best friend. Still she tricked me. I felt embarrassed, ashamed, and most of all, guilty.

The next morning I taped my Blue Dot to the front of my light cotton coat and waited coldly for the bus. I felt like everyone was staring at me. But I thought, I can certainly handle a little Blue Dot.

At school the next day the No Dots were really getting into their roles. They began to discriminate, throw insults, and make fun of us. Jeff was a little too earnest in his attempted gesture at saving all Blue Dots. He tried to convince us all to take them off. Jared got so angry he tore his Blue Dot up. "I'm not playing this stupid game!" he said.

Friends wouldn't sit near me in classes and made jeering remarks about the Blue Dots. I wanted so badly to hit anyone who whispered, "NASTY LITTLE BLUE DOT" in my ear . . .

We discover that a lot happens in one day. Some of the Blue Dots do exactly as they're told, passively accepting their fate; others argue, rebel, refuse to participate. Some wear the dots proudly; others hide them and hope no one will discover them. Others, like Jared, tear the dots off and refuse to play "the game."

The students with no dots discover they are not passive participants. For reasons unknown to them it is not "business as usual." A few pay little attention to the Blue Dots. Some of them turn Blue Dots in for violations, others protect friends, trying to hide them from harassment. A few even get angry when they find out there is no punishment for a violation. The kids are surprised at how they react.

We view the movie *The Wave*, a true account of a high-school classroom in California in the 1970s. The teacher attempted through an experiment (the forming of a group they called The Wave) to answer the students' questions: "How could so many people know what was going on and not do something to stop it? How could so many people be pulled along to participate?"

I ask my students to follow one character throughout the film—what is the character like at the beginning, how does he or she change, and what causes that change?

"I can't believe how quickly they got sucked into the group," Julie said.

"I can't believe how quickly *I* got sucked into the Blue Dots!" noted David. "That's even scarier."

And then we read *The Diary of a Young Girl* (1967). The students are ready. They have a better understanding of the people and the times. They are immersed in the text before we even open the book.

Why *The Diary of a Young Girl*? Many reasons. Because eighth graders can relate to Anne, Peter, and Margot—perhaps not to their situation, but certainly to their feelings. Because it was a horrific time in our history, and we should not forget. Because kids need to know and talk about prejudice as it exists today for so many minorities: for Blacks, for Hispanics, for women, for Native Americans, for the poor, the disabled. Because I want the students to be sensitive, empathetic human beings. Because we don't want anything like this to happen again. Because reading a book together lets us discover new meaning through the views and interpretations of others.

Why *any* book like this? Toshi Maruki says it best when she talks about her book *Hiroshima No Pika*: "It is very difficult to tell young people about something very bad that happened, in the hope that their knowing will keep it from happening again."

Some years I don't use *The Diary of a Young Girl* as the common reading. I might use several books, have multiple copies, and give the kids more choice. Some years we read *"Flowers for Algernon"* (Keyes 1967). Some years we read *Romeo and Juliet* or *MacBeth*. I especially like reading plays with the students, as many of them can take parts, they can practice well ahead of the reading, and they can use their voices to act.

Whatever we read together as a class it is literature that I love and I think the students can relate to. When I stop being surprised or affected by a certain piece of literature, I stop teaching it. It must be fresh for me, so it can be fresh for the kids.

Sometimes I choose a book based on a theme I'd like to explore: generations, human rights, the environment, prejudice, and so on. Sometimes I choose the theme based on what's happening in the students' lives or in the world around them. Sometimes the choice is based on the experiences they bring to the classroom. Always, the choice is based on the fact I like the book. If I'm not passionate about the book and what it says, I will not pass on that love of learning from reading.

I still expect the students to produce five pages of writing a week and to read and respond in their journals for a half hour each night. I read numerous excerpts aloud to them from various novels. We read a variety of essays, poems, short stories, and children's books, based not only on the Holocaust, but on the theme of prejudice. We view and respond to several movies.

I give them a recommended list of books. They read at least three books from the list. I never give my students a recommended list until we are well into the year and they trust me as a reader.

The writing they do during this time span may bear a direct relationship to what we are reading. It may not. I invite them to try certain pieces such as, what happened? what were you thinking? what were you feeling? during the Blue Dot simulation. Most students write well over five pages in response. However, the piece may never make it to final draft. The choice is always theirs.

Monica wrote:

HOPE?

"Hope," mutters the woman to herself. "Hope is one thing." I sit there studying this woman, who I believe is crazy. What am I doing sitting in this dumb old nursing home on a Saturday anyway? I should never have taken this job. "Remember," says this crazy woman. "Remember is another."

Remember what? I think, looking this woman over. Remember a birthday? An anniversary? Her first love?

Nice slippers, I think to myself. This woman is strange: crazy uncombed white hair, wearing a brown terrycloth bathrobe with big, fluffy orange slippers. No wonder she talks to herself! She's nuts!

The woman sits, rocking slowly back and forth, whispering to herself. "Hush, hush, keep silent. They're coming. Hush, hush. Filthy Jew . . . no better than a dog. Keep quiet. Mama, mama, where'd you go mama?"

The woman rocks faster now, holding herself tightly, sunk deep into a corner. I'm glued to her. All my attention is on what she is saying.

"No, they can't be dead. They went to take a shower, only a shower. You'll never see the end of the war. You'll all be dead by then . . . deportation train's going to hell . . . let me out . . . mama, don't let them hurt me, mama . . . "

The lady screams louder and louder. I don't understand what's going on. Why is she saying these things? What's this all about? I don't dare to move, or even breathe. Why doesn't someone come to help this woman?

"All Jews must leave for deportation. Deportation? No, mama, no . . . it's a trick . . . women and children to the left, men to the right . . . move it you filthy Jews! It's a body . . . no mama, don't leave me here alone. MAMA!" the woman screams, as tears fall slowly down her face.

Heavy footsteps can be heard coming down the hall. "No, no, they're coming . . . help me . . . " she says, looking at me. What do I do?, I think to myself. "Run, run, run!" I hear in the back of my mind, but I can't. I'm glued to this woman.

Finally a man in a white coat appears. "Rachael," the man says softly. "Come, it's time to go." The woman tries to hit the man, screaming, "No, I want to live, please don't hurt me," she cries, throwing a last effortless punch. She collapses, crying softly, "You hurt them . . . you killed my family . . . " The man reaches out, pulling her into an embrace. It is then that I see the numbers. The blue tattooed numbers imprinted in white flesh. The constant reminders that won't let her go.

Jill wrote "Theresienstadt, 1941" and a letter to Adolf Hitler.

THERESIENSTADT, 1941

Terror strikes my city
Terror strikes my home.

Soldiers fill the streets.

Anger fills my mind.
Anger fills my heart.

Guns fill the streets.

Hunger fills my stomach.
Hunger fills my mind.

Killings fill the streets.

Silence strikes my home.
Silence strikes my city.

To: Mr. Adolf Hitler

Dear Mr. Hitler,

I have been contemplating the Holocaust for quite some time now, and I thought I would write to you, the creator of this tragedy, in hopes of my questions being answered.

"The law of existence requires uninterrupted killing, so that the better may live." A quote from you. What were you thinking when you said that? Did you truly believe in it? If so, who taught you these ideas? Or were you brilliant enough to develop them yourself?

When you put people into gas chambers, broke babies' appendages repeatedly, and just plain shot people, did you realize the torment you were causing? The brutality, and severity of what you were doing? Death is forever.

Did you really think that Jews weren't human? They were human, they are people. Until you took away their being, souls, and pride. First, kosher butchering was forbidden. Then Jews couldn't participate in parliamentary elections. Next, all Jewish businesses were closed. All German Jews were required to wear the Star of David, along with the Polish Jews, who already had to. The Jewish schools were closed down. And finally, Jews were not allowed to buy meat, eggs, or milk. In the midst of all these things, you somehow found the time, to take away the Jewish peoples' namesakes and make them numbers.

No one, including I, can change what you did, or the fact that you took six million Jews' lives. You may still be proud of that, I don't know. But because of what you did, we, the rest of the world, have learned, and progressed, and realized that prejudice is wrong. We will continue to progress, until the quote "All men are created equal," bears true. I guess, maybe, I thank you, for teaching me, and the world, the difference between right and wrong. But I hope that a person such as you, may never rise to power again.

Jill explained in her log how she came to write her pieces:

I tried earlier in the year to write a piece on the Holocaust, but didn't succeed. I felt that I had given up, by not working to a finished piece, and thought this would be the perfect time to work towards a finished piece.

I was taking a walk around the neighborhood, and the first few lines of "Theresienstadt" just came to me. I had been thinking about the story *Rose Blanche*, and what the city that it took place in must have been like during the war. I went home and wrote the first few lines down, and the rest went onto paper by itself. I then revised it: took out some repetitive lines, rearranged the order of some of the lines, and changed a word from mouth to mind . . .

It came to me to write to the creator of the tragedy, Hitler. Mostly, I wrote down my feelings, but I did refer back to some of the things I had written during my first attempt to write about the Holocaust. I used a quote I had written down. I conferenced, then revised on my own. I added in some significant dates regarding the Jewish people and their freedom. I changed a few words to stronger ones.

I read *Rose Blanche* by Innocenti (1985), *Friedrich* by Richter (1970), and *Black Like Me* (1961) by Griffin. *Rose Blanche* gave me the idea to, and made me, write "Theresienstadt, 1941." *Friedrich* helped me write my letter, and I am now working on a piece that *Black Like Me* inspired.

While Jill was reading *The Hiding Place* (ten Boom 1971) she wrote in her log

During the war so many people lived their lives as secrets, and in secret. I hate secrets. They also lived their lives as lies, and in lies. I hate lies, too, even little white lies. I couldn't stand living in lies and secrets. I guess it would be better than going to a concentration camp, or dying. But to me, the three are the same. Their dignity and pride have been stolen, and that's something no one should be able to take away.

This book shows that during the war there were some people who stood up for what they believed in, and were willing to do what they felt was right. The war teaches us death, the ten Booms teach us life. They teach us that life is sharing, giving, working to benefit others—they cared for the world before themselves and their family . . .

What Carrie and her family went through is unimagineable to me. We live in such a sheltered world. We know nothing of poverty, loss of freedom, or war. We think that we know about these things but we have such a limited view. We read about them, and watch the news to try to learn, but we have not experienced any of these things, therefore, we don't know the truth of them.

Jill also responded to *Friedrich*.

So many times in this book Jews are referred to as "dirty rotten Jews," "good-for-nothing Jew-boy," "cheats and sneaks" . . . Whatever happened to "all men are created equal"? . . . If we would only take time to remember this quote, and live it, so many people would hurt so much less. Jews, Blacks, handicapped, deformed people, they would all live their lives with less pain . . . Friedrich's whole family dies, including Friedrich. I still don't understand how this could have happened. How people could think that way. Today, in America, we think we are way above and beyond anything like that happening, or thinking of that sort. There are still people who are prejudiced, because they know no better. We deny it, but it's there. You would think that after the holocaust people would learn what prejudice can do, but so many people have not yet learned their lesson.

My reading response to Elie Wiesel's *Night*, my letter to Wiesel, and my poem "Waiting for Her to Die" (all in chapter 1) were written during this time period in class. Some of Sandy's writing in chapter 1 and in Appendix A (Portfolios) shows the influence from these readings. Jay, Nahanni, and Sarah all read *Night* after hearing a few excerpts. That book and the Holocaust became the focal point for their Art of Literature project described in chapter 8.

On the board I write:

> Despite everything, I believe man is good at heart.
> —*Anne Frank*

> My mission is to destroy and exterminate.
> —*Adolf Hitler*

I ask the students to respond for ten minutes. In my log I wrote:

> . . . What allows humanity to be so inhumane? How could all those people have been killed without someone stopping it? It makes me think that perhaps TV and pushy journalists are not so bad after all. Little can be hidden. People should get angry . . . Everyone is so quiet. I look up and everyone has more than a page of writing. It seems apparent they have a lot to say. I wonder what each really thinks? I wonder what peoples are being persecuted today? what religious group in the Middle East? which Blacks in Africa? which Moslems in the Philippines? . . . I wonder how awful the concentration camps must have been to make grown men weep at the recollection? I wonder about the grandparents of all these students—what did they endure to make life so free, so easy for their grandchildren? And I wonder if Sandy will ever get an answer to her questions—in her log last week she wrote," . . . it really makes me wonder, when Americans come together, we write letters, make phone calls, demonstrate, petition and protest when Coke changes their formula, but we can't end apartheid in South Africa."

Yes, sometimes we read the same books, because if a few are "chewed and digested" it might make a difference in some child's life. It might make a difference in the world.

Works Cited

Frank, Anne. 1967. (Dramatized by Frances Goodrich and Albert Hackett 1956.) "The Diary of Anne Frank." In *Counterpoint*, ed. Frances Goodrich. Glenview, Ill.: Scott, Foresman.

Griffin, John Howard. 1961. *Black Like Me*. Boston: Houghton Mifflin.

Innocenti, Roberto. 1985. *Rose Blanche*. Mankato, Minn.: Creative Education.

Keyes, Daniel. 1967. "Flowers for Algernon." In *Counterpoint*, ed. Frances Goodrich. Glenview, Ill.: Scott, Foresman.

Maruki, Toshi. 1980. *Hiroshima No Pika*. New York: Lothrop, Lee and Shepard.

Richter, Hans Pieter. 1970. *Friedrich*. New York: Viking Penguin.

Shakespeare, William. 1971. *MacBeth*. New York: Penguin.

Shakespeare, William. 1970. *Romeo and Juliet*. New York: Penguin.

Strasser, Todd. 1981. *The Wave*. New York: Dell.

ten Boom, Corrie. 1971. *The Hiding Place*. New York: Bantam.

Wiesel, Elie. 1960. *Night*. New York: Bantam.

Further References (Historical Fiction and Nonfiction)

I find it difficult to recommend books. My list is not comprehensive. It includes the books which have had the strongest impact on my students and on me. There are many others relating to these themes, and hundreds of other themes. I teach the literature which speaks to my students. I still think the best way to fill our classrooms with books is to first take the recommendations from our students and friends. Then, sit in a bookstore and look through the books. It's worth the time.

Holocaust:

Adler, *We Remember the Holocaust* (Henry Holt)

Arnothy, *I Am Fifteen—and I Don't Want to Die* (Scholastic)

Arrick, *Chernowitz* (Signet)

Frank, "The Diary of a Young Girl" (Pocket Books)

Frank, *Tales from the Secret Annex* (Washington Square Press)

Frankl, *Man's Search for Meaning* (Washington Square Press)

Gehrts, *Don't Say a Word* (Macmillan)

Gies, *Anne Frank Remembered* (Touchstone Books)

Greene, *Summer of My German Soldier* (Bantam)

Hautzig, *The Endless Steppe* (Harper Keypoint)

Hayden, *The Sunflower Forest* (Avon)

Hochschild, *Half the Way Home* (Penguin)

Holm, *North to Freedom* (HBJ)

Holman, *The Wild Children* (Russian Revolution) (Puffin)

Houston, *Farewell to Manzanar* (Bantam)

I never saw another butterfly. (Schocken)

Kerr, *Gentlehands* (Bantam)

Kerr, *When Hitler Stole Pink Rabbit* (Dell-Yearling)

Kuchler-Silberman, *My Hundred Children* (Dell)

Lasky, *The Night Journey* (Puffin)

Lasky, *Prank* (Dell)

Leitner, *Fragments of Isabella* (Dell)

Lowry, *Number the Stars* (Dell-Yearling)

Meltzer, *Never to Forget-The Jews of the Holocaust* (Harper & Row)

Orgel, *The Devil in Vienna* (Puffin)

Reiss, *The Upstairs Room* (Harper Key point)

Rogasky, *Smoke and Ashes-The Story of the Holocaust* (Holiday House)

Sender, *The Cage* (Macmillan)

Shirer, *The Rise and Fall of Adolf Hitler* (Landmark-Random House)

Siegal, *Upon the Head of a Goat* (Signet Vista)
Spiegelman, *Maus* (Pantheon)
Steinbeck, *The Moon is Down* (Penguin)
Yolen, *The Devil's Arithmetic* (Puffin)
Zar, *In the Mouth of the Wolf* (JPS Phila.)
Zyskind, *The Stolen Years* (Signet)

Movies

Night and Fog
The Wave
The Attic
Anne Frank, The Diary of a Young Girl
Shoah by Claude Lanzman
    Shoah Guide, WNET/Thirteen
               Publishing Dept.
               356 W. 58th St.
               New York, NY 10019

Japanese (internment, the atomic bomb)

Coerr, *Sadako and the Thousand Paper Cranes* (Dell)
Kogawa, *Obasan* (Godine)
Maruki, *Hiroshima No Pika* (Lothrop)
Morimoto, *My Hiroshima* (Viking)
Takashima, *A Child in Prison Camp* (Tundra)
Tsuchiya, *Faithful Elephants* (Houghton Mifflin)
Uchida, *Journey to Topaz* (Creative Arts)

Blacks (prejudice in America and South Africa)

Adoff, *All the Colors of the Race* (Lothrop)
Angelou, *I Know Why the Caged Bird Sings* (Bantam)
Angelou, *All God's Children Need Traveling Shoes* (Vintage)
Fox, *The Slave Dancer* (Dell)
Golenbock, *Teammates* (Gulliver-HBJ)
Gordon, *Waiting for the Rain* (Bantam)
Haley, *Roots* (Dell)
Hamilton, *M. C. Higgins, the Great* (Dell)
Lanker, *I Dream a World: Portraits of Black Women Who Changed America* (Stewart, Tabori & Chang)
Lee, *To Kill a Mockingbird* (Warner Books)
Lester, *Long Journey Home* (Scholastic)
Lester, *To Be a Slave* (Scholastic)
Mathabane, *Kaffir Boy* (Plume)
Meltzer, *The Black Americans: A History in Their Own Words* (Crowell)
Meyer, *Voices of South Africa* (HBJ)
Naido, *Journey to Jo'burg* (A Trophy Book)

Rochman, *Somehow Tenderness Survives* (Harper Keypoint)

Taylor, *The Cay* (Avon)

Taylor, *Roll of Thunder, Hear My Cry* (Bantam)

Taylor, *Let the Circle Be Unbroken* (Bantam)

Turner, *Nettie's Trip South* (Macmillan)

Two Dogs and Freedom-Black Children of South Africa Speak Out (Rosset and Co.)

Walker, *The Color Purple* (Washington Square Press)

Winter, *Follow the Drinking Gourd* (Knopf)

Wright, *Black Boy* (Harper & Row)

Vietnam

Ashabranner, *Always to Remember: The Story of the Vietnam Veterans Memorial* (Putnam)

Bunting, *The Wall* (Clarion Books)

Lopes, *The Wall: Images and Offerings from the Vietnam Veterans Memorial* (Collins)

Mason, *In Country: The Story of an American Family* (Harper & Row)

Myers, *Fallen Angels* (Scholastic)

Palmer, *Shrapnel in the Heart* (Random House)

Paterson, *Park's Quest* (E. P. Dutton)

Scruggs and Swerdlow, *To Heal a Nation* (Harper & Row)

Schaeffer, *Buffalo Afternoon* (Ivy Books)

Williams, ed., *Unwinding the Vietnam War: From War to Peace* (The Real Comet Press)

Native Americans

Bierhorst, ed., *In the Trail of the Wind* (Farrar Straus Giroux)

Caduto and Bruchac, *Keepers of the Earth* (Fulcrum)

Carter, *The Education of Little Tree* (New Mexico Press)

Ehle, *Trail of Tears: The Rise and Fall of the Cherokee Nation* (Anchor-Doubleday)

Freedman, *Indian Chiefs* (Holiday House)
———, *Buffalo Hunt* (Holiday House)
———, *Children of the West* (Clarion)

Hobbs, *Bearstone* (Atheneum)

Hudson, *Sweetgrass* (Scholastic)

Kroeber, *Ishi-Last of his Tribe* (Bantam)

McLuhan, *Touch the Earth* (Touchstone)

Miles, *Annie and the Old One* (Atlantic)

Neihardt, *Black Elk Speaks* (University of Nebraska Press)

Rowland, *Sacajawea, Guide to Lewis and Clark* (Dell)

Silko, *Ceremony* (Penguin)

Stewart, *Letters of a Woman Homesteader* (Houghton Mifflin)

Wood, *Many Waters* (Doubleday)

Yue, *The Pueblo* (Houghton Mifflin)

**Yes! Sometimes
We All Read the
Same Book**

Baker, *Where the Buffaloes Begin* (Puffin)

Baylor, *Moon Song* (Scribner)
————, *If You Are a Hunter of Fossils* (Aladdin)
————, *When Clay Sings* (Aladdin)
————, *The Way to Start a Day* (Aladdin)
————, *The Other Way to Listen* (Scribner)

Blumberg, *The Incredible Journey of Lewis and Clark* (Lothrop)

dePaolo, *The Legend of the Indian Paintbrush* (Putnam)
————, *The Legend of the Bluebonnet* (Putnam)

Driving Hawk Sneve, *Dancing Teepees* (Holiday)

Eagle Walking Turtle, *Keepers of the Fire* (Bear and Company)

Fritz, *The Good Giants and the Bad Pukwudgies* (Sandcastle)

Goble, *The Girl Who Loved Wild Horses* (Bradbury)
————, *Star Boy* (Bradbury)
————, *The Great Race of the Birds and Animals* (Bradbury)
————, *Beyond the Ridge* (Bradbury)
————, *Death of the Iron Horse* (Bradbury)
————, *Buffalo Woman* (Bradbury)

Grossman and Long, *Ten Little Rabbits* (Chronicle)

Holling, *Paddle to the Sea* (Houghton-Mifflin)
————, *Tree in the Trail* (Houghton-Mifflin)

Jakes, *Susanna of the Alamo* (HBJ)

Levitt and Guralnick, *The Stolen Appaloosa* (Bookmakers Guild)

Locker, *The Land of Gray Wolf* (Dial)

Lopez, *Crow and Weasel* (North Point Press)

Toye, *The Loon's Necklace* (Oxford University Press)

McDermott, *Arrow to the Sun* (Puffin)

Prusski, *Bring Back the Deer* (Gulliver-HBJ)

Whitehead and McGee, *The Micmac* (Nimbus)

Miscellaneous

John Clavell, *The Children's Story* (Dell)
Ann Durell and Marilyn Sachs, eds., *The Big Book for Peace* (Dutton)
Dr. Seuss, *The Butter Battle Book* (Random House)

Reader's-Writer's Project

Anne produced a pamphlet on child abuse. Her opening paragraph reads: "Abuse. It's an awful thing done to many people of all ages. Physical, sexual, psychological abuse and neglect are the different types of abuse. You might think that only people you don't know will abuse you, or be abused. But, it could happen to you, and by someone that you love and trust. It's always good to be prepared."

The pamphlet describes each kind of abuse and gives the reader questions to ask him or herself. It concludes with numbers to call in Maine or New Hampshire for help. Anne hoped to earn money during the summer to copy the pamphlet professionally and distribute it to elementary schools in the area.

Chris studied the Vietnam War. Holly researched the mentally handicapped. Sara and Jen collaborated on two children's books after studying children's book illustrators Trina Schart Hyman and Jan Brett.

For the last six weeks of the school year I ask the students to look back on everything they've written or read. What surprises them or what do they want to know more about? They are to choose one author, one genre, one theme, or one topic to learn about in depth. They are to research their topic in at least three different ways (three different genres of writing, personal interviews with people, study of films, plays, etc.) and present their findings in three different genres (letter, poem, essay, video, storytelling, children's picture book, mime, drawing, rap, song lyrics, etc.). They are to show a range and depth of knowledge on the chosen topic. They are to prove expertise. I encourage the use of primary sources for information—people, literature, stage productions, and the like.

In addition to the three pieces of "writing," they must also write a *process paper*, which describes what they did and how they went about doing it (what their original plan was, where they got their ideas, what they chose to read and write—the what, how, and why of what they did). They may use their logs to take notes, respond, react, or draft ideas as they research their topics.

Each student also has to give a five-minute oral presentation on his or her topic. They might choose to read a piece of writing, give a presentation, bring in an author or guest speaker, or simply let the class interview them. The format for the oral is up to the student.

I ask the students to begin their project with several questions:

- What or who do you really want to know more about?
- What are the ideas that pop into your head?
- How will you go about researching your topic?
- Where will you start?
- What are the questions you want to answer?

Each week I ask: "How's the research going? What have you done? What have you discovered? What are your plans?"

Two Sample Projects

Tricia studied the Indians of Western America. She presented what she found through a benediction, a letter, and a cover design (see Figure 6.1).

Figure 6.1
Cover design from
Tricia's "Indians of
Western America"
research

Benediction

The weariness in my body pushes me to sleep, but my heart is thinking many thoughts.

Indians are one soul.

We learn through the old who have known the soul of the earth the longest. The earth has the oldest soul of all.

My inside self sings loudly when the ground is decorated with water bringing it life. It's often hard to hold it inside so I don't show how happy I am.

But by the fireside you can see my soul dance with happiness in the shadows. I'm the happiest though when the sun sets the cliffs on fire and warms the stones from the cold of night. The sheep laugh as they walk up the path. The land is clean and wonderful. My soul answers the earth's soul when it speaks like this, hoping for praise.

We talk without a human word, but we say, "I am Indian." My veins pump with the same red clay of the hills. My baby sleeps the quiet sleep of the mountains and becomes strong. The land is my friend and without her, I die.

DEAR MS. BAYLOR,

. . . In many ways I envy the Indians. They had close families, were close to nature, knew exactly who they were, and didn't have stress . . .

At first your style surprised me and I didn't think I'd be able to get past how different it and the illustrations are, to hear what you had to say. But I read *Moon Song*. The pictures were more what I'm used to so I listened to the words more. I loved it. I was still thinking about it the next day. After that I decided to just listen and learn.

Your book *The Desert Is Theirs* is the best I've read in a long time. I could relate with so many things in it. I think that what makes me love a book is when I can put it together with something in my life.

The part where people ask if the Indians get lonely reminds me of when people ask how I can like where I live. I live in a mobile home park about 10 minutes out of town. It's away from any stores, but the kids are closer to each other than any who live in town. People don't understand that. There are small things that make us love where we live, like we all love the sound of the rain on our metal roofs. Possible I babbled on a bit but I think you'll understand what I meant.

Maybe I've interpreted the book differently from how you intended, but I'll tell you anyway.

"They'll say they like the land they live on so they treat it well—the way you'd treat an old friend." We should all learn to do that. The world would be in such better shape. I envy how you can teach people a lesson like that in your writing. Or "No matter what happens he won't give it up. He won't trade it for easier wishes." But you don't make the reader feel guilty, only like they need to excel to meet that goal.

The other thing I really warmed up to in the book was "They have to see mountains and have to see deserts everyday or they don't feel right." That's the way I basically feel about living in New England. In fifth grade I wrote a piece about how I feel about N.E. I'll enclose a copy for you. Do you think everyone feels that way about where they live?

There are a few questions I'd like to ask. I hope you don't mind. (Especially since you've only just met me in this letter.)

What was it that made you write in the style you do? Did you originally do the usual idea of poetry, or was it that after writing for awhile words just came out that way? Do you write for a certain amount of time per day, or is it just when something inspires you? Why did you start writing about the Indians?

Thank you for your time. I hope you continue writing. And thank you for writing things that make us think. "May the wind be always at your back, may the sun shine warm upon your face."

Sincerely,
Tricia

Process paper

It was through talking with my mom that I decided upon the American Indians. She gave me the idea of Arizona, then my mind wandered. The Indians' culture and beliefs would be interesting and fun to learn about. I wanted to know what they believed life was.

I then started reading books that Mrs. Rief brought in. With each book I read I made notes either on facts to remember, or I'd write down quotes so I could look back later and have a feeling of a style that fits the discussion of Indians.

The first book was *Sing Down the Moon*. It gave me information on how Indians felt toward their sheep and elders. While reading the novel I was also reading at least one short book per day. The first was *Dancing Teepees*, a collection of poems by Indians. I copied down poems like this one by Virginia Driving Hawk Sneve.

I watched an eagle soar

Grandmother,
I watched an eagle soar
high in the sky
until a cloud covered him up.
Grandmother,
I still saw the eagle
behind my eyes.

From this book I learned the value of the spoken word, and the amount of respect and marvel for nature.

The next novel was *Stream to the River, River to the Sea*. I learned about the Indians' knowledge of who they are, and how marriage was a political matter without the input of the females.

In the next month I spent every class reading and taking notes. During that time there were some books I got little or no information from: *When Clay Sings, Hawk, I'm Your Brother, If You Are a Hunter of Fossils, Desert Voices*, all by Byrd Baylor.

But there were many good books. The best of these were *Moon Song* and *The Desert Is Theirs*, both by Byrd Baylor. *Moon Song* taught the most about the folklore and legends. The other focused on their sentiments toward the land they live on.

Another helpful one was *Annie and the Old One*. Lines like, "My children when the new rug is taken from the loom, I will go to the mother earth." really give the reader a feel for the people. The book also did a good job showing the peoples' strength. "She would always be a part of the earth, just as her grandmother had always been, just as her grandmother would always be, always and forever.

I read many books that left me with a feeling but nothing easily written on paper . . . All notes and drafts are in my reading log.

Books read:

Byrd Baylor: *A Hunter of Fossils*
               Desert Voices
               Hawk, I'm Your Brother
               If You Are a Hunter of Fossils
               I'm in Charge of Celebrations
               The Desert Is Theirs
               The Other Way to Listen
               Moon Song
               When Clay Sings
Collective?:   *Dancing Teepees*
Scott O'Dell: *Sing Down the Moon*
               Stream to the River, River to the Sea
Diane Siebert: *Mojave*

Donaldo, who received tutorial help daily from the Resource Room in support of academic classes studied the art of storytelling. He wrote a paper on the history of storytelling, took the lead role as storyteller in the school musical *Heads or Tales*, interviewed the writer and director of the musical, attended the performance of professional storyteller Odds Bodkin, and wrote Bodkin a letter and questionnaire. Writing was extremely difficult for Donaldo. He was a natural on the stage, however, and had no trouble learning even a major part.

For his oral presentation, in addition to playing the lead in the musical, Donaldo brought the writer and director, Dave Ervin, to class to answer questions about the development of the play and Donaldo's role as storyteller.

The art of storytelling

Storytellers are people who develop their own kind of art. Storytelling is used for passing on news, history, culture, religion, and for the enjoyment of entertainment. As time passed on, storytelling mixed truth with myth.

The storytellers were regarded with respect. Storytelling existed in all civilizations. There is recorded evidence that storytelling existed as early as 2000 B.C. Greek myths were told by storytellers.

In Europe, minstrels were a kind of storytellers. Minstrels used instruments to tell their stories. When the printing press was invented in 1450, that's when stories began to be printed. The printing press assured that the stories would be passed without change.

Recently I was in a school musical called *Heads or Tales*. I played the part of the storyteller, whose purpose it was to tell stories and keep them alive by ensuring that they were retold.

I interviewed the writer and director of that musical, David Ervin. The way he developed the character of the storyteller was from very hard reading and research. Over the summer he spent weeks and weeks trying to find books with his character. He wanted the storyteller to be a medieval storyteller. He didn't stick to only one character, he had to change the character of the storyteller many times.

That amazed me. The way Mr. Ervin had to create the storyteller. It would seem to me that creating a storyteller would be easy, but it is very complicated.

Then I got fascinated in storytelling. Mrs. Puffer, a teacher from the Resource Room, told me about Odds Bodkin and how he would be in Portsmouth performing *Tales from the Wisdom Tree*. The stories he performed were absolutely amazing. He would actually play the role of each one of the characters in his story.

I then wrote a letter to Odds Bodkin with questions which are on the following pages.

Dear Mr. Bodkin,

Hello, my name is Donaldo _____. I'm an eighth grade student at Oyster River Middle School in Durham, New Hampshire.

I recently played the role of a storyteller in a musical written and produced by David Ervin, my school's music teacher.

Last week my English teacher assigned a writing project and left the choice of the topic up to me. Because of my fascination with the role of storyteller, I chose to do my project on storytelling.

I was fortunate enough to attend your performance at the Bow Street Theater on May 2nd. Now I would like to interview you to find out how you became interested in storytelling.

Enclosed is a questionnaire. I would appreciate it if you could take time to fill it out and return it to me by Friday May 9th. I have enclosed a self-addressed stamped envelope for your convenience.

Sincerely,
Donaldo

1. Why did you become a storyteller?

2. What kind of training did you go through?

3. How long have you been a storyteller?

4. Can you support yourself as a storyteller?

5. Do you have a family?

6. Where do you get your stories?

7. How long does it take to learn a story?

8. Do you have to play an instrument to be a successful storyteller?

9. How far do you travel to put on a performance?

10. Why did you settle in New England and why Bradford?

11. Do you get nervous in front of an audience?

12. Do you ever get lost in a story?

13. Is Odds Bodkin your real name?

Conclusion

The research I put into storytelling, I learned another world. I never knew that storytelling could be so hard. . . . It took Odds Bodkin seven professional years just to get to where he is today. The art of storytelling is one of the most complex arts existing today. Without storytelling I don't know where we would be today.

Tricia's benediction was accepted for publication in *Merlyn's Pen*. Donaldo tried out for and won the lead role in the school musical. Nahanni, Sarah, and Jay were asked to exhibit their *Night* mural (see chapter 9) at the University of New Hampshire. Sara and Tricia are writing articles for a professional publication describing their use of children's literature in an eighth-grade classroom.

All of these projects brought together what these students knew and wanted to know more about as writers and readers. All that they know reaches far beyond the classroom walls.

Works Cited

Baylor, Byrd. 1982. *Moon Song.* New York: Scribner's.

———. 1980. *If You Are a Hunter of Fossils.* New York: Scribner's.

———. 1981. *Desert Voices.* New York: Scribner's.

———. 1976. *Hawk. I'm Your Brother.* New York: Macmillan.

———. 1975. *The Desert Is Theirs.* New York: Macmillan.

———. 1972. *When Clay Sings.* New York: Macmillan, Aladdin.

Miles, Miska. 1971. *Annie and the Old One.* Boston: Little, Brown.

O'Dell, Scott. 1986. *Stream to the River, River to the Sea.* New York: Ballantine.

———. 1970. *Sing Down the Moon.* New York: Dell.

Siebert, Diane. 1988. *Mojave.* New York: Thomas Y. Crowell.

Sneve, Virginia Driving Hawk. 1989. *Dancing Teepees.* New York: Holiday House.

Evaluation

Chapter 7

Where I Am, Where I Want To Be

On a Friday in early October we received the package of tests from the district hired testing service. The writing test was to be administered on Monday. Twenty-one school districts would be giving the same test to their eighth graders. The pieces were to be graded holistically by a gathering of teachers representing each district.

I read the "rules." Students had one 45-minute period to write on the subject given. They could use dictionaries as they redrafted, but they could not talk to anyone. I opened the booklet. There was a picture: two muskrats, one dressed as a man, one as a woman. The muskrats were posed on a hill, a castle in the background, a picnic blanket spread on the ground before them. The female muskrat was posed seductively on the blanket, the male muskrat standing over her. Both held what appeared to be champagne glasses, toasting each other. Eighth graders were asked to respond to the question: "Create a story based on this picture. Ask yourself, 'What do you think will happen next?' "

I was furious. My colleagues were appalled. We cut the picture and question out of each test booklet. On Monday morning we handed out the writing sample test and posed our own question: "Write about anything you care deeply about. Try to convince the reader how much you care." We complied with the other conditions.

I don't know what happened at the scoring session. I was too angry to participate. We were not fired. Our students were not blocked from entering the high school. I did write a letter to the testing company. I argued vehemently about the inadequacies and inappropriateness of such a test, with suggestions about better ways for testing actual instructional methods and beliefs of what real writers do. No one responded to my letter.

That was several years ago. I vowed I would never allow a test like that in my classroom. But with that vow came the realization that I had to find a better way to show what kids can do as writers and readers—that I had a responsibility to come up with the evidence.

I started with a question: What happens when I consider the students the primary evaluators of their own writing? I wanted to see if students could evaluate any writing as well as even professional writing teachers and I wanted to know how well the students could evaluate their own writing.

I put together a packet of twenty-two pieces of writing, written in a variety of genres, all from my own students. I shared the writing with all my students and hundreds of teachers, kindergarten through college. I asked them to rank the writing from most effective to least effective and talk about the reasons for making their decisions. I made several discoveries:

- It is far easier to identify poor writing than good writing.
- Writing is subjective.
- Students immersed in writing are as effective at evaluating as teachers are.

Out of hundreds of responses to the same pieces of writing, the same three pieces were identified as the three least effective. It was far more difficult to reach consensus on the most effective pieces. Every piece but the bottom three ranked as the top for one or more readers at some time.

In discussing the criteria for effective writing, readers revealed that they bring all kinds of biases, experiences, and likes or dislikes for a topic or genre to a piece; these affect their reading.

The students who were immersed in writing were as effective at identifying the most effective pieces and the criteria that made those pieces good, as the most experienced writing teachers.

Based on the results of the informal study, I wondered what kids would be able to do with their own writing. I asked them periodically to arrange their writing from most effective to least effective and to discuss their reasons for the ranking order. At first, kids had a hard time looking at their own writing. Janet said, "I can't. I've written it so many times I don't know what a reader thinks when they first read it. And this means so much to me I'm not sure it will mean as much to another reader."

But the more they were asked to look at their own writing, one piece against another, the better they got at it. What started out as a "wondering how well kids can evaluate their own writing" is now an integral part of my classroom. I *know* kids can evaluate their own writing; I know that the evaluative process helps them make their writing better; I know that evaluation of writing *in progress* is as important, if not *more* important, than the final product; and I know how to help them learn to do that.

Where I Am

Conferences: Evaluation in Progress

Conferencing with students is an integral part of my day and part of the evaluation process. I've adapted what I've learned from Paula Fleming in the New Hampshire Writing Program and Peter Elbow in *Writing Without Teachers* (1973) to fit the process that works best for me.

Students have their own conference sheets (Appendix N) which stay in their working folder. The students fill out the portion that asks: How can I help you? As I kneel down beside the student who has requested a conference (they sign up daily on the board), he or she hands me the sheet and tells me how I can help. I ask the student to read his or her piece or a portion of it. I jot down what I liked or heard or what stuck with me, write down any questions, and add a suggestion or two after the reading. I try to keep all my responses focused on the student's response to "How can I help you?" That question alone asks the student to evaluate the strengths and weaknesses of the piece before a conference even begins.

So often during a conference I hear myself saying to students, "What do you think?" as they try to get me to make decisions about their pieces. I want the students to make their own decisions. Good conferences seem to be confirmations of what the students already know and have to hear themselves say. I will certainly make suggestions, but only when I know they're ready for them.

I know good writing takes time and patience. I expect good writing from my students. I won't let them get away with mediocrity. I push all of them like I pushed Craig. Figure 7.1 shows his first draft of a piece.

All those corrections? That's what I *used* to do, when I first began teaching, because that's what was done to me. I did not actually do that to Craig's piece. I stopped writing comments like that when I realized they were no help to me. What helps me? Someone pointing out what

Figure 7.1 Craig's first draft

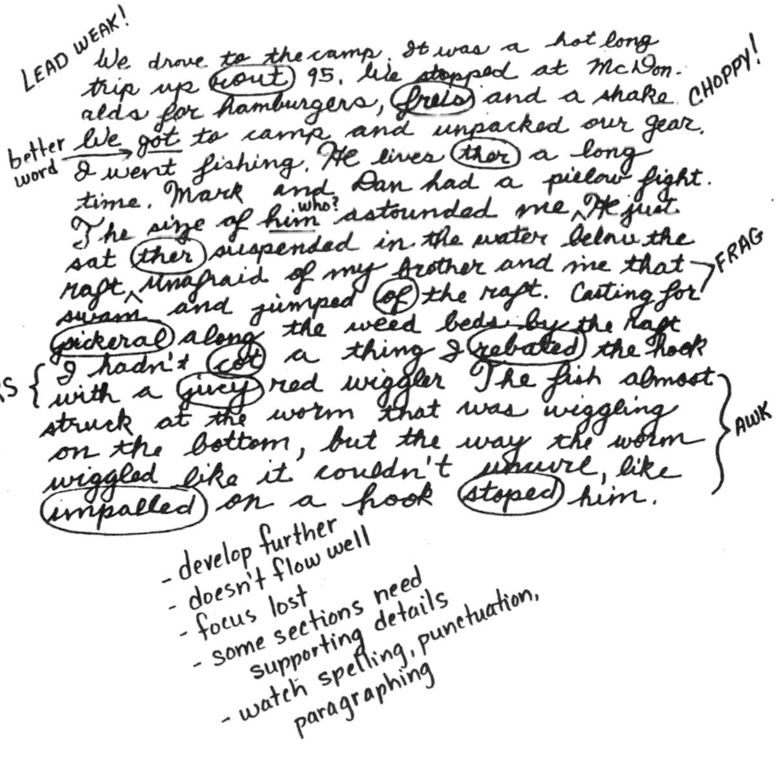

stuck with them, asking me questions about things they really want to know more about, and giving me a suggestion or two about ways to enrich the piece.

In actuality, I pointed out words and phrases to Craig that really stuck with me: " 'astounded,' 'suspended in the water below the raft,' 'impaled on the hook,' 'juicy red wiggler' . . . You sure know a lot about fishing . . . tell me more about that . . . What did that fish look like? How big was he? What happened? Did you catch him? How'd you do that?"

I talked with Craig at least three times about the content of the writing, pushing him each time to tell me more about the fish. He added and organized lots of information about fishing, without deleting any other parts for a while.

Sometimes kids will recognize the focus of a piece on their own and delete unnecessary information. Sometimes they need to be nudged. "Craig, you talked about camp, McDonald's, a pillow fight, and fishing. If you had to pick the most important part, which would it be?" Craig spent weeks commiting words to paper about fishing. He knew what part was most important and was able to take out all the irrelevant information.

Lastly, I helped him edit the piece. He had already read it to me and his peers so many times that he had corrected a lot of the periods, commas, and simple spelling mistakes ("there," "off," "stopped"). I did have to teach him paragraphing, missed periods and commas, and several difficult spelling words. He underlined "pickerel," "caught," "rebaited," and "juicy" after I asked which words he wasn't sure of. I gave him those words and also "impaled," which he had missed.

It took Craig four weeks to finish the piece. It was worth the time, the effort, the patience.

The fish had lived there a very long time, probably five or six years. He sat there suspended in the water below the raft. We could see him through the cracks in the boards. Even when my brother and I jumped in and swam off the raft, the old bass remained there, unafraid.

His size astounded me. The bass was longer than any fish I had seen before, almost as large as the fifty gallon drums that kept the raft afloat . . .

. . . I kept thinking about that old bass as we drove to camp.

Mark, Dan and I arrived at the camp at dusk, unpacked and put our gear in the musty loft that sheltered bats and mice. After stretching and unpacking, we went fishing. I had forgotten about the monster fish.

I began casting for pickerel along the weed beds and was without luck. I rebaited the Eagle Claw hook with a juicy red wiggler. Then I cast off under the raft and let the worm sink slowly to the muddy bottom. This time a fish struck, almost ripping the rod from my hands. It dove to the bottom, then boiled to the surface and jumped in an explosion of fury, trying to rid itself of the hook.

As I battled the fish closer to shore, I put as much pressure on the light tackle as I dared. The battle raged on with the angry fish diving and jumping. After many minutes of intense excitement, the fish lay there gasping on the surface. I then realized this was the monster fish from under the raft.

After the old bass was beached, we put him in a galvanized wash tub and went into the camp. That night I lay there thinking about the old bass. I thought about how long the fish had lived under the raft and how we had watched him grow to his huge size.

Then I heard him splashing about in the tub. Tiptoeing down the creaky stairs, I went out into the cool night.

The next morning I was awakened by my brother's shouts, "Your fish is gone!" I just smiled to myself.

When conference questions turn the decisions back to the writers they are forced to make evaluative judgments. These decisions made in progress help them better the writing. If kids can be taught how to detect and diagnose strengths and weaknesses and how to come up with strategies for dealing with those problems through conference questions and suggestions, then we have taught them not only how to become better writers, but how to be independent writers along the way.

In addition to the individual conferences, I model response using my writing with the whole class. I respect their opinions and questions. I show them how their response helps me revise the piece.

Grades: Evaluation of the Final Product

I don't like assigning a letter grade to students, but every six weeks I have to do it. Those letter grades come from their writing and reading.

Writing

Before I grade any writing, I take the students through a process similar to the informal research project I described at the beginning of this chapter. It occurred to me that students need to look at other students' writing also, in order to learn to look at their own more objectively. I collect samples of writing from previous years and other districts. I select twenty pieces based on varying degrees of strengths and weaknesses. I ask the students to read each piece, assign a holistic number (1 = ineffective, 2 = ineffective but salvageable, 3 – effective, 4 = most effective), and jot down three reasons for a particular score.

We then discuss the papers. In small groups I ask the students to share their criteria for the most effective pieces and reach consensus on the criteria for effective writing. Lindsay's list of criteria for effective pieces included: "kept my attention, ending strong, vivid description, left me feeling something, leads pulled me in . . . "

When asked to summarize her own ideas about effective writing, Lindsay wrote:

> If I were to characterize pieces with certain features as being "effective" or "ineffective," I would say that a strong lead is important. The first lines of "Mailbox Surprise" say: "I opened the mailbox, expecting to only get once in a lifetime offers and instant millionaire contests." This makes me want to find out what's in the mailbox; it pulls me into the story. I like endings with a twist; endings that are not what I was thinking of. Peaceful endings also are effective to me: "I clicked the flag on the ropes with my nervous, shivering fingers, and slowly raised the flag to half mast." But the stranger endings are more fun: "I bet I'm the only one Bert has injured in his time. He's pretty rough."

We came up with the following synthesis of opinions on what makes writing effective:

1. Good lead.
   – Attention grabbing, catchy, unexpected.
   – "Pulls you in!" "Hooks you!" "Sucks you in!"
   – Makes you want to read it to find out or know more.
   – Gives you an idea or direction for the writing.

2. Topic appeal for reader.
   – Can relate to or identify with it.
   – Reminds reader of an experience in his or her own life.

3. Appeal for writer.
   – You can tell the writer likes what he or she wrote because it's so real.
   – Writer is honest, knows what he or she is saying.

4. Lots of detail, description.
   – You get a good picture in your mind of what is happening, the scene or what the character is like.
   – Images, word pictures, metaphors are so realistic, you feel like you're part of the story.
   – Awesome but simple detail.
   – Descriptions grab your attention, then the words hold it ("I could see myself there watching and hearing, but powerless to do anything but continue to read").

5. Good middle.
   – Keeps the reader in it/keeps the reader reading.

6. Good ending.
   – Ends just as well as it begins.
   – Leaves you knowing, or thinking, or feeling.
   – Unexpected.
   – Sense of closure, but you want to read it again.

7. Nice use of words.
   – Language is simple, clear, not flowery or stiff.
   – Great vocabulary—the strongest nouns and verbs.
   – It sounds like the author is just talking to you comfortably.
   – It doesn't repeat words too often, like "she said . . . "
   – Balance of dialogue, thought, and description.

8. Style.
   – Uses a new, different, or surprising perspective or viewpoint to write about something common.

9. Purpose.
   – Something happens rather than wandering aimlessly.
   – Makes you think, or tells you something, or entertains you, or makes you feel something.

10. Feelings/emotions.
    – Reader is affected with the same emotions as the writer or the characters: scared, hopeful, upset, confused, humorous, relieved.

11. Flows well/consistent.
    – Makes sense.
    – Moves nicely, steady pace, one thought follows another.
    – The mood and voice sound like the writer or character all the way through.

12. Overall appeal.
   – The writing stays with you even when you're done.
   – You can relate to it, understand it, feel it.

13. Mechanics.
   – Appealing title.
   – Legible.
   – Correct spelling, punctuation, paragraphing, and usage.

This synthesis for effective writing is condensed even further on a handout the students staple into the inside covers of their working folders (Appendix O). Each year the list varies slightly in wording. They have ownership in what we consider the characteristics of effective writing. Most importantly, they do exactly what I hope they will. They begin looking at their own writing more carefully. I hear comments like: "Let me work on the lead a little more before I read this to you. It really doesn't pull you in." "This story doesn't even make any sense. I need to start all over." "I've got a good beginning and middle, but I really need to work on the ending." "There are no surprises here. It's boring. I don't even like reading it."

To reinforce conferencing techniques, I ask the students to take eight of the twenty pieces and respond to them (Appendix P), the same way I want them responding to each other's writing, and the same way I will respond to them. I cannot separate revision and evaluation, any more than I can separate reading and writing. They are inextricably interwoven.

In discussing their responses to similar pieces we discover there are similarities and differences in the way we respond. So be it. Writing is subjective. That is why I should not be the sole evaluator of their writing. I bring my own biases to a piece. I am only one responder. The best evaluator/responder is the writer. The better they become at that evaluation process, the better the writing will become.

We also discuss what kind of response helps best and why: how we all need to hear first what works well or stays with the reader, so we know we did something well and know what to keep; how questions let us know where more information is needed; and how suggestions give us ideas to think about.

When the students hand in their own final pieces, I ask them to write out a case history or process paper about the piece of writing and to grade their own writing first. Then I grade it and comment (Appendix Q). On the process paper I ask: "Tell me everything you can about how this piece of writing came to be. What do you want me to know about this piece that I would never know from just reading the writing?"

The student's grade carries as much weight as my grade. We each give it three grades: a process grade, a content grade, and a mechanics grade.

Process grade   The process grade is based on attitude, effort, and what Tom Romano (1987) calls "good faith participation" in the class and on the particular piece. I ask the students to write five rough draft pages of writing per week and to select the two pieces going best during each six-week grading period to take to final draft. The two pieces they select are the ones taken to conference with at least three peers and me. (The writer takes the Conference Sheet—Appendix N—to those conferences, to be filled out by their peers and by me.) The rough draft pages are not graded. I don't even read them. I just count them—done or not done. The rough drafts count toward good faith participation. The two final drafts (which might be several poems and a ten-page short story, for example) are the only pieces graded.

Content grade   The content grade is based on the quality of the piece. We take into consideration the characteristics of effective writing. We comment on the qualities of the piece.

Mechanics grade   The mechanics grade takes into consideration legibility, punctuation, spelling, usage, and so on.
Tom Romano notes:

> When I evaluate papers, I bring to bear my history as a writer, my tastes in reading, my prejudices and moods, my ever-developing understanding of teenagers, and my perceptions of how a particular student will be affected by what I say. Yes, *who* the student is helps determine what grade I give, what response I make. It cannot be otherwise. Each of my students is an individual. (1987, 113–114)

> Writing is the writer. "It embodies her voice, her passion, her thinking, her intellect, her labor, and, on some occasions, her very soul. . . . Our responses and grades should nurture. (1987, 125)

I agree. Certainly our grades and comments should be honest, but they must keep the writer writing. The way I respond to Shawn, who comes out of the Resource Room for the first time in many years and has never written more than a sentence, is *not* the same way I respond to Kyle, who writes complex novels on the scale of Tom Clancy.

By the time we've reached the final product, the students have received so many comments from me in conferences along the way that written response does not have to be complex.

> To Jenn and Sara: I like the way you attempted something very different—a children's picture book. The two of you collaborated so well on this as you researched, planned, and wrote. I like the way the illustrations "play with the words," and the examples you use for the child getting the mother's attention. Nice story about sibling rivalry, and how it eventually turns around. You both have wonderful families. It's clear how much you love them. (Don't forget how to spell congratulations and paid.)

> To Scott: Beautifully done piece. You capture a couple of minutes in a hockey game so well. Strong verbs carry us through the action with you: skated, hit, flashed, sandwiched, crumbled, rammed, squirmed, dove, etc. At the * please note that you change from the past tense to the present. Present tense makes action feel stronger (example: "The whistle blows . . . ") so you may want to use that. Tense must remain consistent throughout.

Reading

I base their reading grades on "good faith participation" (Romano, 1987) as readers: they have a book with them and they read it on silent reading day, they share reading excerpts voluntarily in whole-class discussions, and they read for a half hour each night and maintain a reading list.

I also grade their reader's-writer's logs based on quantity of response and quality of response. Often I ask them to assign the quantity/quality grade, and I write just a written response.

I don't like the idea of assigning a grade to what is essentially "thinking," but that's the point I'm at right now. I count the number of entries or times read and the number of pages of response for a particular time frame. I expect ten entries (or times read) and six to ten pages of writing in the log (for a B for quantity) during a two-week period. Some kids write a lot more, others less. I try to individualize my expectations of each student based on what they tell me they are capable of doing. A grade/comment in a log might look like this:

$$10/6^+$$

10 entries

6 pages of writing

B for quality

To Steve: $8/4^{\checkmark}$ Nice writing. I understand what you're learning about dirt bikes from the articles you're reading. Are you still reading *Dune*? It's the only book on your list. Have you read anything, besides dirt bike magazines, this trimester?

To Jay: $16/17^+$ It sounds like Alice Bach is talking down to teenagers. What could she have done to make this a more believable book? Have you read *A Tree Grows in Brooklyn* or *One Child*? I think you'd find them more challenging than this.

To Holly: $9/12^{++}$ This book makes me so sad. I wonder if Charlie wouldn't have been better off being left alone. Dr. Nemur infuriates me—he seemed to "experiment" without feeling—all for his benefit—to increase his discoveries in the name of science for his name—not for the betterment of fellow humans. Your discoveries make me wish we had all read this book and taken a trip to Great Bay Training Center. How lucky *we* are!

To Scott: $3/3^+$ I like the way you are trying to get your thoughts into entries. Especially liked how you mentioned *humiliation* and *gullibility*. In what other ways do those words fit this book?

To David: $3/0^{-0-}$ You are a prolific reader. I see you reading Farley Mowat books, Patrick McManus, Jack London. It's obvious you really like adventure and the outdoors. It's also obvious you are *not* crazy about writing in this log (0 response in 6 weeks). Can we find a way of you teaching me about these books, about your ideas for writing, about your observations of the world—since this log is obviously not working for you? How about—you tape response? Just tell me on tape about what you've read and what you think about it. Let's try it . . . or do you have other ideas?

Ironically, my written comments in the logs and on the student's writing seem to carry far more importance than the grades. When students rank pieces from most effective to least effective at the end of the year, I notice that the grade on the piece has little to do with its placement. Appeal to the writer based on topic or commitment to the piece, and appeal to a reader based on audience feedback carries far more weight in those evaluative judgments than a letter grade. If I only write a quantity/ quality grade in the log, I will frequently hear, "Where's your note to me?"

The overall grade then is based on writing and reading, and their goals set and achieved. They respond to a series of self-evaluation questions (see chapter 8) every trimester, which includes response to writing, reading, and goals set and achieved. I look at the student's portfolio of finished writing, the reader's-writer's log and reading list, and evaluation responses to determine a grade.

I cannot separate the child from the grade. There is too much that enters into my process in determining the grade that is not objective. How much has this child grown? What risks has he or she taken? What risks hasn't he or she taken because of school or home circumstances? What kind of new challenges has this child taken on? No matter what I do, evaluation is subjective.

I am not the editor of a national magazine rejecting or accepting a piece of writing. I am a teacher trying to nurture growth. I am a teacher trying to help the child become better at using language. I am a teacher trying to help the child like reading and writing. I am a teacher trying to help the child become an independent learner. Grades get in the way. Unfortunately, I have to give them. And I have to give them every six weeks. This is where I am for the moment.

Keeping Parents Informed

Most parents were not taught reading and writing this way. Until our students become parents, we will have to keep them informed about the differences, in methods, theory, and philosophy. Keeping them aware of what happens in the classroom starts with a letter home at the beginning of the year and continues with invitations into the room and to readings. It also means informing them about their child's progress throughout the year. I am still struggling with a way to do that beyond a cursory letter grade, a few one-word comments, or a checklist.

In chapter 8 I describe problems I encountered when writing a narrative on every student. What seems to be working better is a letter to parents (see Appendix R).

Self-Evaluation Throughout the Year

We are on a trimester basis. For the first two trimesters, I ask the students to look at their writing and reading in depth and answer the questions listed in chapter 8. For the last trimester, I ask them to respond to a Self-Evaluation for the entire year (Appendix S). These self-evaluations become part of their portfolios.

Throughout this book I have shared students' evaluative statements about themselves as readers, writers, and learners. The portfolios in

Appendix A are especially rich with their comments. Coupled with their writing and reading it is easier to get a better picture of who these students are, what they are able to do, and how connected all their learning is.

Where I Want To Be

I want to do away with grades on the individual pieces of writing and in the logs. I want to sit down with each student and take a look together at the working folder of works in progress, the portfolio of finished pieces, the reader's-writer's log, the reading list, and the student's self-evaluation. I want to look at goals set, and goals achieved. I want to base the grade on attitude, effort, growth, and "good faith participation." I want the students to evaluate themselves. Sitting down with 125 students to do this takes a lot of time. More time than I have, but time I must try to find.

As teachers I think we have to fight for what we believe is good teaching, and the conditions under which good teaching can happen.

When I vowed not to let "two muskrats" back into my room to test what my kids knew as writers, I took a stand as a professional. I will continue to take a stand: on testing, on class size, on good teaching practices.

Until we realize that the student is the best evaluator of his or her own learning, we will never know what our students really know or are able to do.

Until we reduce the number of students to a manageable size (four classes of no more than twenty students each), we will seldom be able to individualize our curriculums to meet the needs of the diverse children we have in our classrooms.

Until we (classroom teachers) are trusted to know what to teach and how to teach it, we (the taxpayers) will spend needless millions of dollars trying to design the "perfect" curriculum of mediocrity and the "perfect" objective test to measure it.

We must listen to our students. Craig struggled with reading and writing, unless he had something important to say—then he worked hard to make sure his message was heard.

Mr. John C. Quinn
1000 Wilson Blvd.
Arlington, VA 22209

RE: *USA Today* Tuesday December 9, 1986
    "Kids Need to Write if They're to Succeed", p.12A

Dear Mr. Quinn,

I am an eighth grade student at Oyster River Middle School in Durham, New Hampshire.

When I saw the title of the article in your paper I was really pleased, because many people in the United States do not think that it is necessary for people in our country to know how to write.

As I read your article I had some disagreements with your thoughts and statistics. If I had written the piece of writing you published next to a cartoon where you made students in the United States look like they had the I.Q. of this paper, I really would not feel too much like writing another piece of writing in my life, especially if after it came the words, "Dreadful! It's a travesty when a child falls so short of coherent expression."

Even after you stated that, you said, "even worse, this example is typical." I know that isn't true. My eighth grade English teacher once brought some of our classes' best pieces to a teachers' workshop and read them. When she was done many of the teachers there asked how old the students in her class were, because they couldn't believe that thirteen and fourteen year olds could write so well. Many of them said we wrote better than they did.

Your statistics state: four percent of fourth graders, fifteen percent of eighth graders, and twenty-three percent of eleventh wrote letters that were considered "adequate" or better. In my class that would mean only three students could come up with an adequate or better piece of writing. Everyone in my class can come up with a better than average piece of writing.

I also really think students should be tested differently. Students should be writing the best pieces of writing they can all year long and at the end of the school year they should hand in their best piece of writing to be graded.

There is just one more thing; I sensed a real negative tone of voice as I read this article. I think you should concentrate a little more on peoples' feelings rather than criticizing them so fast. Thank you for reading my letter.

Sincerely yours,
Craig

Craig's letter was never published. He received no response. It's time we responded.

Works Cited

Elbow, Peter. 1973. *Writing Without Teachers*. New York: Oxford University Press.

Hayden, Torey. 1981. *One Child*. New York: Avon Books.

Herbert, Frank. 1987. *Dune*. New York: Ace.

Romano, Tom. 1987. *Clearing the Way*. Portsmouth, N.H.: Heinemann.

Smith, Betty. 1968. *A Tree Grows in Brooklyn*. New York: Harper & Row.

Finding the Value in Evaluation

Chapter 8

Portfolios

~~~~~~

"To be nobody-but-yourself—in a world which is doing its best, night and day, to make you everybody else—means to fight the hardest battle which any human being can fight; and never stop fighting."

e. e. cummings (*Murray, 1990, 37*)

"We must constantly remind ourselves that the ultimate purpose of evaluation is to enable students to evaluate themselves."

Arthur L. Costa, "*Re-assessing Assessment*"

Sarah was adamant. "They don't know me as a person and a writer. They don't know how I've improved." I watched as Sarah read the one mark on her writing sample—a 7. On a test mandated by our school district, Sarah had received a 7 out of 8, certainly a good score. It didn't matter. "What does this tell me?" she continued. "I'm one less than an eight and one more than a six. So what?"

Sarah was right. The writing sample didn't show who Sarah was as a person or a writer, and the response she received didn't help her. No one who had read her piece knew where she'd been, so how could anyone tell how she had grown? How sad, I thought, especially when all the evidence was right here in the classroom.

Portfolios: A Wealth of Information

The evidence was in Sarah's portfolio. In my classroom, the portfolios have become the students' stories of who they are as readers and writers—rich with the evidence of what they are able to do and how they are able to do it. Each portfolio is a collection of each student's best work.

I impose the *external* criteria for the portfolios—each student's two best pieces chosen during a six-week period with all the rough drafts that

133

went into each piece, trimester self-evaluations of process and product, each student's reading list, and, at year's end, a reading-writing project.

The students determine the *internal* criteria—which pieces, for their own reasons. I invite them to work on reading and writing from other disciplines and to include them in their portfolios, if they think their efforts are some of their best. Joel's portfolio has a piece entitled "The King and His Achievements," a paper written for social studies but worked on in a remedial reading class and English. Sara and Jennifer each included a children's book about a little boy investigating tidal pools along the coast of Maine. It was written for science, based on their field trips while studying marine biology. In English class Jen wrote while Sara, studying the work of Trina Schart Hyman and Jan Brett, drew the illustrations.

A Changing View: Teacher as Learner

My classroom has evolved slowly. When I first started teaching I used to make all the decisions about what the students read and wrote and what they learned from that reading and writing. I tested them on all the information.

But times have changed. I've turned much of the responsibility for learning over to them. They choose what they write, what they read, and what they need to work on to get better at both. I invite them to try different genres of writing and I share a variety of literature that I love with them. They used to keep writing folders and were judged on all their writing. Now *they* select the best pieces to revise and rework. The portfolio of best pieces is separate from the working folder of works in progress.

I keep a portfolio also. If I don't value what I ask the students to do, they seldom value it either. We have to trust ourselves as learners first, before we can understand the trust we put in students. My portfolio this year has an educational article entitled "Seeking Diversity," a poem, a personal narrative about my mother's sewing, and a letter written to the governor nominating our parent group for a state award for support of our Arts in Education program. I keep a writer's/reader's notebook, which includes my list of books read. I begin my writing and reading with my students. I share drafts in progress and reflections on or reactions to books read. I have to trust my own possibilities as a learner if I'm going to trust and value my students as learners.

Evaluation: The Last Holdout

For years, however, I was the final decision maker about how well each student did. But then I began to wonder what would happen if I considered my students the best evaluators of their own reading and writing, both in progress and as a final product? I not only have them choose their own topics, but I have them choose the pieces that are going best. So that they will have a selection, I ask them to write at least five rough draft pages a week. Further, I not only have them choose their own books, but I have them select the ones they want to respond to.

Over time, the quality of the students' work has changed. I see more diversity and depth to their writing, their reading, and their responses to literature. I have discovered that the students know themselves as learners better than anyone else. They set goals for themselves and judge how well they reach those goals. They thoughtfully and honestly evaluate their own learning with far more detail and introspection than I thought possible. Ultimately, they show me who they are as readers, writers, thinkers, and human beings.

As teachers/learners we have to believe in the possibilities of our students, by trusting them to show us what they know and valuing what they are able to do with that knowledge. The process of turning the learning back to the students, from choice of topic or book all the way through to the evaluation of their own processes and products, produces students like Nahanni. I chose Nahanni, not because she had the best portfolio, not because she was the most articulate, but because she was a little better able to reflect on what she did. Through Nahanni's portfolio I want to show the possibilities in diversity, depth, growth, and self-evaluation. This kind of evidence shows the value in evaluation. This kind of assessment *matters*.

One Student's Story

Nahanni's portfolio is typical of the diversity and depth of writing and reading I am now seeing in my students. I think several things have contributed to that change: the students have been immersed in reading and writing in more and more elementary classrooms, they are more sophisticated and articulate as language users, and I am actively seeking and expecting good writing and reading.

For the year she has ten final drafts: three poems, a personal narrative, a character sketch, a letter, a pen-and-ink drawing, a play, an essay, and a picture of her final project for the year—an acrylic collage representing her interpretation of the book *Night* by Elie Wiesel. Everything that contributed to the final draft—lists, rough drafts, sketches of ideas—is attached to the final draft of each piece. Nahanni's reading list and her evaluations of herself as a writer and reader are also in the portfolio, just as they are for all my students.

As a final project for the last six weeks of the year, I want the students to synthesize what they know about themselves as readers and writers (see chapter 6). I ask them to review all their reading and writing with two questions in mind: What surprises you? What do you want to know more about? Next, I ask them to investigate that one theme, one topic, one genre, one author—whatever it is—by reading and writing in three different genres. Nahanni's project, an in-depth study of the book *Night*, is also in her portfolio.

Nahanni ranked "Looking Across Rows of Music Stands" as her best piece for the year.

Looking across rows of music stands
  I see an ordinary face—
    a quarter note on a sheet of paper.
I know the face has laughed and cheered,

and laughed some more,
  and cried.
Violins raise, bows sway, and the ordinary
  quarter note frowns and comes alive.
The lone quarter note is gently carried off
  the page and lightly danced onto the air.
Hidden behind Beethoven's Fourth, one little
  second plays along.
A royal theme, issued from the staff,
  fades, then crescendos to a wail.
My loving ear fills up with sounds
  of staccato arpeggios, an Allegro,
    and a triste.
And so, to my dear little—quote—
  unimportant second,
  I give my ballad of soft support.
Bring your laughter and your cheers,
  your sorrows, and your fears,
We'll laugh together, heal the pains,
  use up some tears.
    We'll share in both.

Nahanni reflected on her poem in her trimester self-evaluation and why she chose it:

> My best piece is my best . . . because it meant the most to me. I used music to portray a friendship. I think there are layers of understanding, and to enjoy the poem you don't need to understand all the layers.
>
> The piece that is the least effective means nothing to me. It was an assignment. (An assignment from me: an essay entitled "MacBeth and Lady MacBeth, the Perfect Marriage." Why? Because every few years my past history as a learner haunts me and all the essays assigned to me hover like ghosts wailing "There must have been a good reason . . . " But after reading two of the hundred plus essays, I'm bored to tears and the ghosts are put to rest for a few more years.) That's why the pieces that are my best are so important. I chose topics that were important to me.
>
> Now I know that in order to write something well, you have to care about it. The first important thing is that you like a piece of writing, then you worry if anyone else likes it . . . I've learned to add detail to get something across. I've learned to care about my writing and write it in order to resolve things, because even when I read a piece over, I always learn something.
>
> I think writing is like visiting places; you see different things each time. You read a piece of writing and you think you read it and saw it, but then you go back and read it again and see new things . . .
>
> **Writing isn't just a school subject. It's part of how we think. I write because I need to figure out what I'm thinking.**

Nahanni's final version of the poem, however, is a far cry from her first draft, which looked like this:

> Letter to my best friend—You are my very best friend and I think you know I don't mean that in what you'd call a "superficial" way . . . It feels like we are always one person. I want you to know how much I care about you. Never forget me, I will NEVER forget you. The times with you will never leave me.

How did she get from this early draft to such a different final version? Nahanni's second draft looked like this:

> Two worlds collided and they could never ever tear us
> apart.
> Two hearts that can beat as one, there aren't a single
> thing we can't overcome. We're indestructable.
> The light of a billion stars pales
> As we comfort each other . . .
> . . . the universe and a billion stars—and you and me.
> Crowded there with others or alone.
> At times it seems—we will always be together
> So bring your laughter and your cheers
> your sorrows, and your fears.
> We'll heal the pains, use up some tears,
> we'll share in both.
>   poem or letter (She asks herself.)

Nahanni's first breakthrough on this piece came when she answered my question, "Where did you first think of writing this to your friend?" She said she looked up in music class across rows of music stands and saw her friend—just an ordinary face. "What does an ordinary face look like?" I continued.

"Like . . . like . . . a quarter note on a sheet of paper . . . just an ordinary note." And Nahanni was off. She made lists of musical terms. Played with them in phrases. Literally cut the phrases apart and made two piles—the ones that related best to what she was trying to say about friendship and the ones that didn't feel right to her. She read drafts in progress to her peers and to me for responses to what was working, to hear the questions we had, and to get suggestions.

Although questions are at the heart of conferences, I always point out first what I liked or heard or what stuck with me. The questions then help the writer decide what to add, what to delete, what direction the piece might take in terms of focus and format, and even how to organize the ideas. Timing is key. When I know my students well, I begin to know when I can ask questions that push their thinking—push them to make evaluative judgments as they write.

In a conference I asked Nahanni to elaborate on the poem's development ("Tell me more about how you did this."):

> In my first draft I didn't tie in much about music to my theme of a
> person—or face. I made a quarter note be the face. I use a piece of music
> to interpret what the face fears and what I feel about that face. I did away
> with a rhythm and pattern I had in each stanza of some of my earlier drafts.
> I feel now that what I have to say is more important than trying to shape
> and twist the words into specific phrases and lines. This is one of my best
> pieces because I've worked so hard, making all the words say exactly what
> I mean.

At the beginning of the year, Nahanni set three goals for herself: to try writing poems, to write longer pieces, and to send a piece off for consideration for publication. To meet her second goal she wrote two prose pieces. One was entitled "Melted Wax":

At the time, all I was worried about was what to say to Emily. At the time I was too nervous to know. Now I know . . . this is for Emily.

Our knobby knees and skinny, six-year-old legs stuck out from underneath the little baby table that Emily and I sat at. Dinner was over and her father, Steve, brought us the cake and lit the six colored candles.

Emily had made her wish and was just about to blow out the candles, when Steve came in and told us her mother would have to be taken to the hospital. Linda had been suffering from cancer for some time. She didn't feel especially well tonight, and thought it best to be in the hospital.

I remember watching the scene, trying to look unafraid. Unafraid for Emily, her mother, and myself. How would I comfort my friend? As I sat in the little chair, I watched with fascination as the six candles sank into the cake, spluttered and went out, leaving melted wax and small burnt holes in the frosting. I wanted to tell Emily to hurry and blow them out before it was too late, but I didn't.

"My mother made this for me when she was in the hospital." We were in Emily's room. She was showing me a needlepoint design.

Linda died that night on her daughter's birthday. Those six candles had been too short. They hadn't even given Emily the chance to make her wish come true. Whatever the wish had been, it just spluttered and burned out until nothing was left except for small burnt holes in her cake.

Nahanni's second piece was entitled "Unable to Forward":

He leaned against a tree in his "garden" of driftwood branches. My father stepped out of our shiny, white Buick and walked hesitantly up the highway with outstretched hand. The old man shook his hand heartily and, patting him on the back like an old acquaintance, said, "I'm glad you stopped in, friend, for one reason."

He instructed my dad on how to "plant" a dried up tree stump in the orchard. He had the stump roped to a live tree and had been struggling to lift it into the ground. Lucky for him that my dad came by, because that stump was HEAVY.

My brother wasn't so keen on joining my father in meeting the old hermit, but he followed as I quietly climbed out of the car. We arrived just in time to see my father and the old man heave a dead tree stump into an upside-down, mushroom-like position.

We picked our way down his rocky pathway into a low-ceilinged room that was more like a cave. I stepped carefully in and then quickly withdrew. My brother and I cringed and waited at the door in the sunlight. My father, no hesitation in his voice now, turned to us and said, with a gleam of amusement in his eye, "Come on in, kids, this is great."

The gnomelike man showed us his house. It didn't take long. It consisted of two rooms, bedroom and kitchen. My eyes slid quickly over the old photographs, the sink overflowing with green wine bottles, and the cat food strewn across the floor.

My father crouched, hands on knees, squinting to see a picture on the cover of *National Geographic*. "Hey, that's you!"

"Yeah, that's me!"

"This is great!" My dad was looking at a black and white photograph of the hermit sitting in the corner of his house. In the picture he held a plaster figure of the Virgin Mary, which I noticed above a chair in the same corner.

I escaped, dragged my brother out the door into fresh air. My father stayed in the dusty cave, pouring over pictures from the man's life. I watched, from outside in the cool breeze, as my father crept at a snail's pace along the wall, exploring details of paintings, photographs, revealing the spark of memory in the old man's eye with each question. I thought, amused, I can picture my father living here.

Reluctantly, my father pulled himself away. Up the path, through the orchard, back to the highway we climbed. The old man told us about his land. It wasn't really his, but, he said, he "claimed" it. He lived here, no one else. A seventy-year-old man by himself on a mountain in the middle of Nowhere, California. He gestured, his arms encompassing his whole mountain. "As far as I'm concerned, it's all mine. No one bothers me. Who cares?" His voice floated out to us, "Who cares?"

My father took pictures of the old man in his orchard, which we enlarged and sent to his P.O. Box. Weeks later, they came back, stamped in red ink, "UNABLE TO FORWARD-DECEASED."

Who cared? I thought sadly. Who cared?

At first she steered away from writing long pieces because she was "afraid of losing peoples' focus, or maybe I'm afraid I'll lose my own focus." She discussed how "Melted Wax" came together:

I was brainstorming all the positives and negatives in my life . . . trying to think of a way to write down what I thought of Linda's death. One of the things that makes the piece good is the layers of depth that people can find. Most people can come up with the point of the candles . . . but if they don't, they can still enjoy it. I changed the lead all around. My first lead was dull and didn't say much. Now it starts with a quote and pulls people in better.

I think I'm learning more to do—as you say—showing, not telling. Before now I don't think I was really aware of the difference. My writing really changes a lot from first to last draft.

Nahanni also accomplished her third goal: to send a piece for publication. The poem she selected went through as many changes as her first poem.

Through the night window I imagine what could be hidden

A delicious horizon,
    like smoked white-fish on deli-wrap,
    beckons to me.
A Great-Blue Heron breezes on grayish icy water.
Straggling, gnarled fingers struggle for a place
    to clutch the faulty ground halfway across the bay.
Houses scatter across the distant shore,
    bread crumbs for the birds.
Tiny diamonds of light pinpoint
    quickly darkening gray-blue shadows.
I turn . . . delicious.

Nahanni recounted what happened when she submitted the poem for publication:

I like this poem, but it doesn't have that much meaning to me. I sent it to *Merlyn's Pen* and that's what they told me. I knew they would. I think I'll make it how they'd like it since if it's changed it's not a problem for me because I don't think it's perfect the way it is. It's not for me that I want to keep it that way. Like the poem I wrote for my friend, "Looking across rows of music stands." I wouldn't change that no matter what anybody thinks. It's for me—and my friend.

One thing I'm not sure I will change in "Night Window" is the image of smoked white-fish on deli-wrap. The editor said he didn't think it fit. My dad and I think he must not be Jewish. If he was he'd know how well it does fit.

This year Nahanni started reading thirty books and finished twenty-nine. Her favorites were *The Little Prince* (de Saint-Exupery 1971), *One Child* (Hayden 1980), *The Chocolate War* (Cormier 1974), *Beyond the Chocolate War* (Cormier 1985), *Night* (Wiesel 1960), and *The Princess Bride* (Goldman 1973). She formed some definite opinions on the subject of reading:

> In order to be a good reader I think you must read a lot and think about what you are reading—how it relates to you, what the writer wants you to think, versus what you really get out of it. I think the responses I got to my reactions to books were written or asked in such a way that I feel my ideas are important and that makes me think more.

In responding to *The Runner* (1985) by Cynthia Voigt, Nahanni wrote:

> Sometimes reading this book really frustrates me. It's not that it isn't interesting, but sometimes I just don't *get* it. Bullet seems to have dealt with things in a strange way, burying himself. Burying himself from everyone else . . . Bullet and Katrin have learned to sort of remove themselves and not get too involved in things they care about, so they never get hurt . . . I think I'd rather live through some of the bad times than never see any good times. Sort of like in *The Little Prince*, the taming of the fox. I'd rather love, lose, get hurt and go away with memories than never love and never know the difference between happy and sad. What words! Happy? Sad? When I was little I thought happy meant a smile and going to the circus. Sad was a frown and sitting home on a rainy day.

Jay, a peer, responded to Nahanni, in her log confirming and extending what they both knew:

> I never thought of Bullet burying himself, I guess I thought of him isolating himself, in the boxes he always talks about. Isolated, being able to see out, but no one can see in. Do you think people, like Bullet and Katrin *really* know what happy and sad are? In their boxes? I guess you and me would rather have loved and lost than never to have loved at all. There is a point though where the lost overcomes the loved and you (collectively, not just you, Nahanni) end up feeling like cuckies. At least for me—Have you read *Sons from Afar?* By Cynthia Voigt. That, and I guess *Homecoming*, too, show how similar Sammy is to Bullet. Them and their boxes. Scot is like that too. I guess he has good reason. His parents both left him in the lurch. But he doesn't let himself get too involved, because he always got hurt. Whatever flics your Bic, I guess . . .

Discussing her reading goals, Nahanni said, "I wanted to read faster . . . I don't think I accomplished this, except that now I read more than I did before. I don't think that [reading faster] was a really important goal. I wanted to understand more, and I think I've done that."

Nahanni not only understood more, but she connected what she understood with her own life as she questioned, evaluated, criticized, and analyzed.

Of *The Little Prince* (1971) she said:

I love the part in this book about the fox who believes in taming something unknown. When the little prince sees all the roses that look exactly like his flower, he is angry. When the fox is finished with him, he returns to the roses and is happy. He knows that no other rose in the whole universe is like his . . . because he's cared for it . . . What is the point in life if you are afraid to get involved? People want to be strong so that they will not be sad when they have to leave someone. How do they think they can be strong? They have to be sad at one point in order to really learn . . .

From the *Chocolate War* (1974) she learned "what peer pressure and a crowd of kids can do to people." She wrote:

Archie, and people like him . . . are always in the darkness. They do things to other kids because it makes them feel superior and happy with themselves . . . For a second, at the end of the book I thought Archie won. But he didn't. He just thought he did . . . The vigils live in darkness. I thought vigil was supposed to mean watchfull? VIGIL—dictionary definition—a purposeful or watchful staying awake during the sleeping hours. Is this name appropriate? . . . I guess it is sort of appropriate. The vigils do think that they are watchful, sort of keeping an eye on things. Oh, here's a good analogy: The cord—I forget what it's called—that goes to the brain from the eye, is broken or just not there. So the image that is focused on the retina is upside down. There is no cord going to the brain so the image remains upside down. I think the vigils have an upside down picture of the world and people's place in it . . . in the last paragraph of the book Archie and Obie sit and walk around in darkness. They might think they are vigilant, but who can see clearly in the dark?

Through reading *One Child*, (1980) Nahanni not only discussed Hayden's influence on her choice of professions, but how the book opened her mind to diversity.

Reading Torey Hayden's books, I always think I want to do what she does. I would love to do something to help people. I used to think that maybe I'd be a psychiatrist but I don't think that I'd really want to help the people who could come to me. To make a real difference I'd want to help the people who really couldn't go to look for help themselves . . . I wonder if Boo really does live in another world. Actually, I guess we all do. We've all got separate ideas of what the world should be. It's like we're all sitting in a building looking into a circular field. Everybody has their own window and everybody looks out onto the same field but from a slightly different angle . . . you can never really see the field from exactly their angle.

Nahanni believes that if a book is really good, you can learn a lot about what the author thinks about life. In a good book, the meaning is made clear so that "I think I've discovered it on my own, not so it was told to me." Her responses to books showed she was wondering deeply about life and discovering how we all fit into the bigger scheme.

She knows too there are some books just to read for the pure pleasure. "I loved *The Princess Bride* (1973) because it was so funny and a break from a book where you are expected to think!"

Nahanni doesn't think she consciously connects reading and writing, but listen to what she consciously did:

> I think I've been looking a lot more for metaphors and hidden meanings in my reading. This is a result of my change in writing—or vice-versa. I think that now I look for things that aren't so obvious. I've discovered that those things mean so much to me. The words that are unwritten teach much stronger than words on paper. . . . [In books] I can understand the author's point of view and think about my own views.

She not only reads as a writer, trying to find her own views, but she observes as a writer. While driving to New York, Nahanni was reading a book, yet decided to respond to what she saw, instead of what she was reading.

> At the side of this thruway is an Hispanic man looking into the hood of a pale green, rusty, dented old station wagon. In the lane to the right of us is a shiny black new Oldsmobile with shaded windows. It purrs along . . . Here we are in the middle, pretty neutral—a Toyota. Why do the wealthy people like to tint their windows so we can't see in? Are they embarrassed that we might find out they look exactly the same as we do . . . grafiti . . . why do people write on walls? Central Park . . . people on park benches. On rocks. In bushes. Their pantry is a garbage can.

By the end of the year Nahanni even recognized she had become the observer of everything around her. She wrote in her log:

> I'm just lying here in the grass with sun on my face. Outside the UNH library on the lawn . . . Monica is working with Dave on his MacBeth essay. Now, finally, just feeling like an observer, I can just listen and watch and realize how wonderful it is. Dave is perfectly able. He just needs Monica to help him hold his concentration and put his pen on paper. Dave sits there yelling and grumbling. But he's got a smile on his face . . . That's how it was with Jeff and me. He always had the ability. Sometimes he caught onto things faster than I did. What he didn't have was the patience to sit down and work on something. He needed the trust in himself and to believe that he could do it . . . One great thing about Jeff is his art. He always encouraged me in art, just like I always did for him . . . Every night I sit up and draw. I never showed anyone or told anyone but when Jeff comes back I'll show him so he can encourage me again.

Nahanni told me constantly—in her writing, in her reading responses, and in her evaluations of herself as a learner—that she learns from her own discoveries, not from someone telling her what she is supposed to know. Isn't that what real learning is all about? Isn't that what real teaching should be about?

For her reading/writing project, Nahanni chose to focus on the Holocaust. She read *Night* (Wiesel 1960), *The Wave* (Strasser 1981), *The Hiding Place* (ten Boom 1971), *Rose Blanche* (Innocenti 1985), and *Faithful Elephants* (Tsuchiya 1988), just to name a few. Her original plan was to write a children's book in the form of a diary. It would be from the perspective of a little girl who is swept up into the Nazi movement believing she is wrong to have Jewish friends. But along the way Nahanni shifted her focus. She collaborated with Jay and Sarah on an art project that portrayed the impact of all they were feeling from immersing themselves in literature of the Holocaust (see chapter 9). They read essays from a similar art/literature project in Brooklyn, N.Y. They looked at slides of art work using the same medium. The art teacher, Beth Doran, and I worked together with the students, discussing writing and visual arts with the same words. Nahanni described what she, Jay, and Sarah did:

> We brainstormed what the book [*Night*] was making us feel . . . it was difficult because we don't have much practice brainstorming pictures. After we succeeded in getting our feelings portrayed in shapes and pictures, we put together some of our ideas: a triangle, upside down to show rising growth that finally got out of control, a growing sense of disorder and helplessness. We worked to get the image artistically interesting and show contrast, not just in the art form, but in the book. The black and white. The full pages to the ashes . . . We decided to burn pages of *Night* and fasten them to a triangular board. In the bottom of the triangle the pages were orderly and gradually got more burnt and unorganized. We traced our hands and made them reaching up from organization to chaos. We revised our original idea of the clenched fist. People weren't fighting back. We skipped the people and the barbed wire that we had originally planned because we decided it would be distracting and split the picture, stop the movement.
>
> In the book I saw Wiesel going from not believing anything would happen, even after he was warned by an old man, to almost becoming that old man . . . I think the painting means something different to each one of us . . . but all the ideas worked together and are related. The important thing is that people can look at our painting and feel something . . . for their own meaning . . . Art, literature and music all mean something different to everyone . . .

In addition to doing the mural/collage she wrote a letter to Elie Wiesel and a process paper on what she did as she researched the Holocaust. Her project was a synthesis of what she wanted to know more about and who she was as a reader and writer by the end of the year.

Self-Evaluation, Self-Discovery

I don't have to be the sole evaluator of Nahanni's reading and writing. She, like the rest of my students, is far better at it than I am. With 125 students the only possible way of keeping such extensive notes on each student is to have them keep them themselves. And the better Nahanni, as with all the students, knows her own process as a writer and reader, the better she becomes at both.

But I *can* evaluate Nahanni's growth, as well as any of my students, as a writer and reader, if I have to. I have all the evidence in front of me in their drafts of writing from rough to final, in their response to what they're reading, in their self-evaluations of themselves as writers and readers, and in my responses to them on their writing and in their reading logs. As teachers, we must listen first to the perceptions our students have of themselves and address what they think they can and cannot do.

From what Nahanni showed me, I wrote the following narrative about her growth as a learner:

> Nahanni has become an independent learner who reads and writes for her purposes because she wants to become better at what she does. Her reading

and writing show me she is engaged in the excitement of learning. She is a keen observer of life around her, often finding topics because she is always looking. She has an acute sense of detail, able to see, hear, smell, touch, and feel things that many of us miss, all with a sensitivity to the human factor. She is a caring, sensitive young woman who leaves an impact.

As a writer Nahanni knows how to find a topic, how to play with words until they say exactly what she means, how to seek help for revision, how to use a variety of resources, how to ask questions, and how to answer them. She is not afraid to take risks with new genres. She has tried poetry this year, while still attempting a variety of prose pieces. There is always a meaning to her writing—a reason to read it. She has a message for her reader, because the message is always for her first.

Nahanni has an acute sense of audience. After receiving a rejection from one magazine I asked her if she wanted to revise the piece and resubmit it. "No," she said. "This wasn't for the magazine anyway." Another piece she will revise because it wasn't for her alone.

Her images are fresh and vivid. She searches for precise words. Her dedication is evident in her willingness to revise until pieces say exactly what she means, in the appropriate format. She grapples with big issues: friendship, control, love, prejudice, hate, fear, uniqueness . . .

She is a thoughtful reader who reads a variety of books (historical fiction, nonfiction, adolescent fiction, classics, poetry, realistic fiction, etc.) for a variety of reasons: some to make her think, others just for fun. She knows how to take meaning to, and meaning from a book. She relates her own life experiences to the experiences she draws from books. She comprehends, analyzes, reflects, contrasts, compares, synthesizes, wonders, questions, criticizes, and enjoys literature.

I believe her writing has changed her perceptions as a reader. At first she wondered why authors did certain things; now she reflects on how she would have done it differently. Nahanni writes and reads to find out what she is thinking.

I wrote extensive narratives like Nahanni's for more than fifty of my students. (I gave up after two classes and three sleepless nights.) It took about twenty to thirty minutes per student. Of the more than fifty I wrote, I received response from only one parent, even though I asked specifically for feedback as to what this kind of response showed them about their son or daughter. Several students told me their parents were "too busy" to read the evaluations, let alone respond in writing to them. Several other parents asked, "What's the bottom line?"

Will I do such extensive evaluations again? No. But I know I *can* do them if I have to. I have more information about what my students can do as writers and readers than any number of standardized tests or writing samples could ever show.

I also know I can show someone very quickly and succinctly how any one of these students has grown. Simply by taking Nahanni's self-selected "most effective" piece I can see growth by looking at two areas I value in writing: verbs and leads.

Verbs in first rough draft		Verbs in final draft	
are	go	looking	hide
think	wish	see	issue
mean	were	know	fade

call	feels	laugh	crescendo
had	want	cheer	fill
be		cry	give
		raise	bring
		sway	laugh
		frown	heal
		carry	share
		dance	

Lead in her first draft: You're my very best friend and I think you know I don't mean that in an ordinary way.

Lead in final draft: Looking across rows of music stands I see an ordinary face—a quarter note on a sheet of paper.

Certainly the verbs chosen for the final piece are stronger and carry more vivid images than the first draft. The two leads are also very different—the first bland and nondescript, the final clearly metaphorical, rhythmic, and far more appealing even in its design.

As teachers, if we think about what we value most that our students can do, we can look at those areas as the students move from draft to draft or as they respond to literature in their journals. Simply looking at verbs or leads takes less than five minutes, yet I know that I can discover other strengths in a similar manner. I can look at metaphorical language, dialogue, topic choices, genres of books chosen, types and changes in reading response. I can use what I discover as minilessons in addition to evaluative information.

If students keep their rough drafts attached to their finals, all the evidence for showing growth is right in front of me. If students keep their responses to literature in some kind of journals, who they are as readers is documented for the entire year.

Searching for the perfect record-keeping system for 125 students has always been a cumbersome and impossible task for me. When I started looking at portfolios, working folders, logs, reading lists, and self-evaluation sheets, I realized all the information was there and the students were keeping the records themselves. It is far more comprehensive evidence than any checklist I have ever attempted to develop.

Reading, writing, speaking and listening are the tools students work with to create meaning for their own purposes. I value students who are able to communicate, think, create, and reflect with those tools. Portfolios become the evidence for what we value in our classrooms. The act of putting together a portfolio is a reflective act in itself, as students choose what to put in there and why. That reflection on where they've been, where they are now, and how they got there, is what real learning is all about.

Building Toward Portfolios: What Can We Do?

Will the portfolio concept work in every classroom? Yes, but certain conditions must be present. First, students must be immersed in reading, writing, speaking, and listening. Second, they need to be given time in large blocks. Third, they need to be allowed choice as to what they are writing and reading—for their reasons, their purposes. And fourth, they must receive positive response to their ideas.

Once these conditions exist, we can introduce the concept of portfolios as places where students collect evidence of who they are.

- as readers and writers (their best pieces of writing, their reading lists, and most effective responses to literature)
- as learners (all their rough drafts and ideas for getting from one draft to the next attached to their best pieces)
- as *reflective* learners (self-evaluations and reflections on what they've done and how they've done it)

And we have to keep and share our own portfolios. The more I discover what I can do, the higher my expectations are of what kids can do. They seldom disappoint me.

Each trimester I ask the students to arrange their writing from most effective to least effective and to evaluate it considering the following questions:

- What makes this your best piece?
- How did you go about writing it?
- What problems did you encounter?
- How did you solve them?
- What makes your most effective piece different from your least effective piece?
- What goals did you set for yourself?
- How well did you accomplish them?
- What are you able to do as a writer that you couldn't do before?
- What has helped you the most with your writing during this trimester?
- What are your writing goals for the next twelve weeks?

As readers I ask them to consider similar questions about the books they have chosen to read.

- What's the best book you've read this trimester?
- What makes this one of the best you've ever read?
- How did you go about choosing this book to read?
- What's the most significant thing you learned from this book and/or discovered about yourself as you read it?
- What were your reading goals at the beginning of the trimester?
- How well did you accomplish them?
- In the past twelve weeks, how have you changed or grown as a reader?
- What has helped you the most with your reading?
- What are your reading goals for the next twelve weeks?
- In what ways are your reading and writing connected? How does one affect the other?

The Kind of Evaluation That Matters

Nahanni is not the exception in my class. She is becoming the norm. She is motivated and persistent, and she cares about learning. She reads and writes for real reasons. Through her portfolio and her reading log I know her as a reader, writer, thinker and human being, not as a 5 or a 7. Through their portfolios of writing and reading, I know all my students. They, like Nahanni, are articulate learners because they continually *practice* discussing what they know and how they know it: by sharing with me, their peers, the community, and other grade levels; by teaching teachers at writing workshops; and by publishing locally and nationally.

Learning to make meaning in writing and reading is not objective, as our evaluation systems would seem to indicate. We must become more flexible in our assessment of students' work. When kids are given choices in what they read and what they write, and time to think about what they are doing, their writing and reading get better. When we trust them to set goals and to evaluate their learning in progress, we will begin to realize that they know much more than we allow them to tell us through our set curriculums, our standardized tests, our writing samples.

If our goals are to keep students writing and reading, to help them get better at both, and to help them become independent learners, then we *must* nurture self-evaluation of writing and reading in progress *and* as a final product. This is the kind of evaluation that matters because it is for Nahanni, for Joel, for Sarah. Who else is evaluation for?

Works Cited

Cormier, Robert. 1974. *The Chocolate War.* New York: Dell.

———. 1985. *Beyond the Chocolate War.* New York: Dell.

Costa, Arthur L. 1989. "Re-assessing Assessment." *Educational Leadership* 46,2 (April).

Dillon, David. (ed.). "Evaluation of Language and Learning." 1990. *Language Arts* (entire issue). 67, 3 (March).

Goldman, William. 1973. *The Princess Bride.* New York: Ballantine.

Goodman, Kenneth S., Yetta M. Goodman, and Wendy J. Hood, eds. 1989. *The Whole Language Evaluation Book.* Portsmouth, N.H.: Heinemann.

Hayden, Torey. 1980. *One Child.* New York: Avon.

Innocenti, Roberto. 1985. *Rose Blanche.* Mankato, Minn.: Creative Education.

Murray, Donald. 1990. *Shoptalk.* Portsmouth, N.H.: Heinemann.

Saint-Exupery, Antoine de. 1971. *The Little Prince.* Orlando, Fla.: Harcourt Brace Jovanovich.

Shepard, Lorrie A. 1989. "Why We Need Better Assessments." *Educational Leadership* 46:2 (April): (pages 4–9).

Strasser, Todd. 1981. *The Wave.* New York: Dell

Taylor, Denny. 1990. "Teaching Without Testing." *English Education* 22, 1 (Feb.)

ten Boom, Corrie. 1971. *The Hiding Place.* New York: Bantam.

Tierney, Robert J., Mark A. Carter, and E. Desai. 1991. *Portfolio Assessment in the Reading-Writing Classroom.* Norwood, Mass.: Christopher-Gordon.

Tsuchiya, Yukio. 1988. *Faithful Elephants.* Boston: Houghton Mifflin.

Voigt, Cynthia. 1981. *Homecoming.* New York: Fawcett-Juniper.

———. 1985. *The Runner.* New York: Fawcett-Juniper.

———. 1987. *Sons from Afar.* New York: Fawcett-Juniper.

Wiesel, Elie. 1960. *Night.* New York: Bantam.

Wolf, Dennie Palmer. 1987/1988. "Opening Up Assessment." *Educational Leadership* (December): 45:4 (p. 24–29).

———. 1989. "Portfolio Assessment: Sampling Student Work." *Educational Leadership* (April): 46:2 (pages 35–39).

Chapter 9

The Art of Literature/ The Language of the Arts

January

Anne and Amy paint silently.

Nahanni sketches in depth lines.

Sarah and Jay glue ashes to masonite.

Jennifer, Todd, and Ben negotiate colors, delete symbols, change shapes.

June

"Oh wow!" Ben says, standing in front of the newly hung murals. "That's awesome! I had no idea we did that well."

"This looks like a real museum," adds Charlie. "Like—this is real art."

"Why can't we take these up to the high school with us?" Amy asks. "I want to look at them every day."

We all stand back, open-mouthed, silent, losing words to our surprise. The kids are right—these murals are "awesome."

I look for ways in my classroom to allow students to respond to their world in ways other than *just* "the word." Writing is extremely important. But it is not the only way for literate human beings to show what they know. I respect the fact that the visual arts and the performing arts are as important and as valuable as the literary arts.

Too often, when we refer to the arts, or label something "creative," we demean its importance. Why are the arts the last disciplines funded, or the first disciplines cut in a budget crunch? Why do people continue to ask me, "Well, yes, your kids do some incredible *creative* writing, but when are they going to learn *real* writing?" Creating *is* the highest form of intellectual development.

Beth Doran, our art instructor, and I often talked about the similarities of her process as a visual artist and my process as a writer. "We'll have to do something together," we said. We stopped talking and called our something "The Art of Literature."

I shared an article from the *New York Times* (Glueck, 1988) with Beth about a teacher in the South Bronx. Tim Rollins works with Hispanic kids who might otherwise drop out of school. As an alternative curriculum, he has these students create painting-collages based on literary classics. Their "signature" is to use the pages of the books as the background for their creations. We were not only intrigued by what these students did, but also by what our kids might do with a similar idea. We had never done anything like this before and didn't know what would happen. We would have to learn with the students.

The only scheduled time we could coordinate was our schoolwide activity block, during which each teacher designs an elective that meets once a week for forty minutes. We went into the activity with only an idea.

Fifteen eighth graders signed up. Eleven stuck with it. As in our classrooms, we had a heterogeneous mix of kids—some who are successful at most everything, others who find little success in anything academic.

The first day we met we gave the kids copies of the articles we read about the New York project: "From Dead End to Avant-Garde" (Nilson 1988) and " 'Survival Kids' Transform Classics to Murals" (Glueck 1988). We talked about what these high schoolers did.

Our students each listed five books that had made the strongest impression on them. They shared their lists in an attempt to find a common book. Monica and Jay felt so strongly about *Night* (Wiesel 1986) that they convinced Sarah to read it before reaching consensus on which book to represent. Sarah agreed on the book's impact. "It's a real story, which really scared me: thinking 'Oh my God, this man is still alive, writing and teaching, after all he has been through. He must be so strong . . . to be able to get his story across like he does.' "

After each member of the group read his or her chosen book (Beth and I read the books also), we free-wrote for ten minutes about what came to mind when we thought of the book. One group had chosen *Night*, another *Go Ask Alice* (Anonymous 1978), and the third group *The Outsiders* (Hinton 1967).

One list for *Go Ask Alice* read: "She got into drugs. She got hooked. She turned into a different person. . . . She died. The honest and simple truth. That's the way it happens. It goes so fast. You don't even know what hits you. One day everything is fine. Then you're dead."

A list for *Night* read: "dead babies, people on shovels, flames, lips—mean, cruel, fleshy, terror, losing track of all time, shoes—possessions—keep them, stay together as long as possible, teaspoon, a cup of barbed wire, hope, dreaming—am I dead?"

The students looked back at anything that surprised them in their lists of images or ideas from their free-writing. Sarah said, "What I noticed was that we had all, in our own way, gotten a feeling of hopelessness . . . " Beth showed slides of murals by Rivera, Benton, and Picasso. We spent time looking carefully at Picasso's *Family of Saltimbanques* and *Guernica*. We talked about how the Picasso murals originated, specifically with regard to the history of the time, asked what feelings they evoked, why he might have chosen certain symbols, how the shapes and lines contributed to a certain effect, what effect certain colors had, and how they related their own lives, their own experiences, to what Picasso might have been trying to get across.

Before they drafted designs, Beth and I asked: "What images come to mind when you look at what you wrote? What feelings? What shapes? What colors?"

Everyone brainstormed ideas and then shared their thoughts and their sketches (Figures 9.1, 9.2, 9.3)". "At first it was difficult," Nahanni said, "because we don't have much practice brainstorming pictures."

The students continued drafting and designing ideas, while Beth and I gathered materials. Rollins and his students in the Bronx use only the highest-quality materials. He believes that "this is another way to foster the sense that their art is important, lasting, worthy of respect" (Nilson 1988). Beth made sure the students had top quality acrylics and the best brushes. We wanted the students to know their art was important right from the start.

Students shared their separate ideas and negotiated a collaborative piece. They sketched out their ideas on paper (see Figures 9.4, 9.5, 9.6) before even tackling the masonite. One of our first questions to them was: "What shape will best serve the feelings you want to generate in the viewer?" (See color plate for the final murals.)

Those students working on *Go Ask Alice* chose a circle, "because with a circle there's usually no way out."

Night students chose an inverted pyramid because it emphasized the growing power of the Nazis and the horror of the Jews' situation. "It just grew worse and worse, more horrifying, more chaotic . . . " they said.

The group working on *The Outsiders* felt a diamond shape best reflected good and bad: "Both sides are the same . . . It's not clear sometimes who's good, who's bad."

Their compositions evolved as they worked on them. About her *Night* project, Sarah said:

> I had an idea that if, as the pages got more and more disorganized and chaotic, the edges, and finally entire pages, could be burnt. The whole book came down to fire and flames, and Holocaust actually means fire. . . . We started out having a fist near the bottom of the triangle. Once this was on, we decided it was too small and just didn't fit with the feeling we wanted. We wanted that small sense of hope, that reaching for something . . . We decided the swastika, which was going to be a marionette holder of shadow figures hanging from barbed wire should stand alone, the most powerful single object in the art piece.

Like the process of writing, through which the writer discovers new thoughts, new ideas, new directions, the process of painting revealed new images, new feelings. The breakthrough for all the students came as they moved from the realistic to the abstract. Beth often led them to more abstract thinking with her questions. "What will this figure do for the painting? What do you want a viewer to understand from this design? Of all the images you have here, what's the one you think holds the most power? What made you choose this line? How does repetition contribute to the effect you want? How could toning down this color change the feeling?"

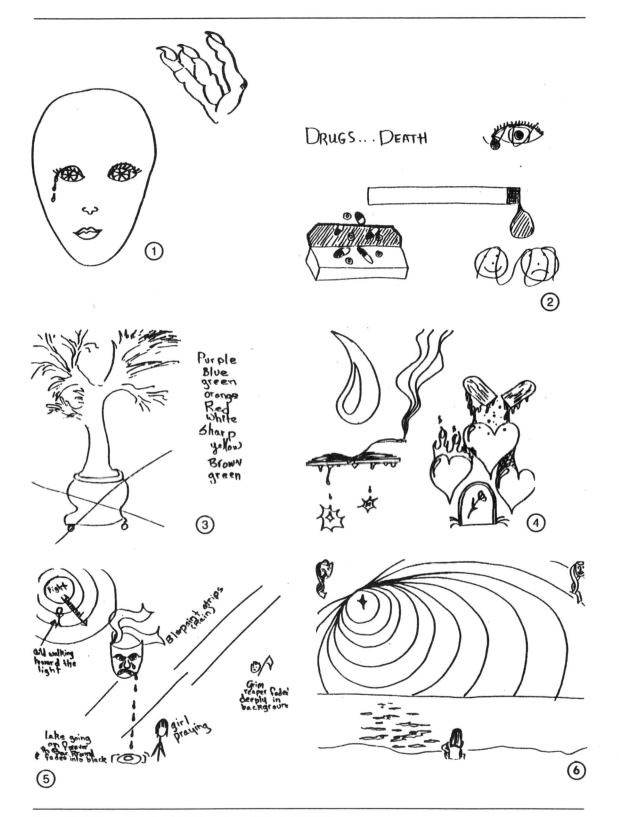

Figure 9.2 Brainstorming sketches from "Go Ask Alice"

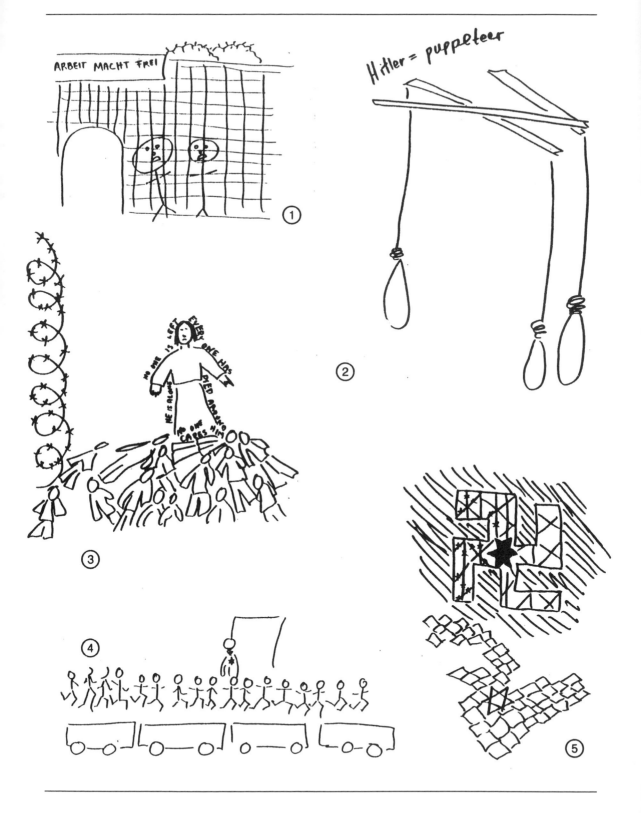

Figure 9.3 Brainstorming sketches from "Night"

Figure 9.4
Final sketch before
"Go Ask Alice" mural

Beth's questions were so similar to what I ask writers (What do you want a reader to get from your writing? What does this section do for the piece as a whole? Of all the things you've talked about, which one is most important? Which one best describes what you're trying to say or do? What sentence conveys the strongest image? What made you think of this image?) that I no longer worried about asking the "right" questions. She and I moved from group to group, pointing out what we liked (what images, colors, designs had the most impact for us) and asking questions, constantly turning the thinking back to the artists.

Ben, Amy, Charlie, and Anne painted silently, often moving intuitively from section to section. "We chose the design by each brainstorming ideas and then voting on them," said Ben. "We liked the tunnel

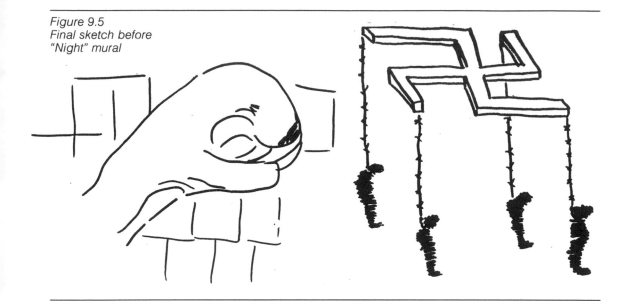

Figure 9.5
Final sketch before
"Night" mural

effect because we wanted it to portray infiniteness, like some universal tone of infinity. Drugs is like being sucked into a tunnel. But we didn't want it to look like just drugs. It could be anything bad in the society that sucks you in."

"As the piece evolved it got less complicated," added Charlie. "We took out a lot of stuff because we didn't want anything to detract from the picture by making it too crowded. . . . We took the idea from the story but turned it into a piece of artwork where everyone can understand what we mean. They can get their own feeling from it." Readers often take their own meaning to and their own experience from a text. Charlie recognized that same potential in his art.

"I think the tunnel effect pulls the viewer right in, too," said Amy, working on *Go Ask Alice* (Anonymous, 1978). "The pinpoint is the light at the end of the tunnel. It's so hard to get there once you're caught inside."

Jennifer, Todd, Jeff, and Angie, in representing the book *The Outsiders* (Hinton, 1967), negotiated colors, deleted symbols, and shifted ideas. "We tried to show good/bad, love/hate, using a sunrise/sunset," Todd said. "And we wanted to show 'nothing gold can stay.' The whole idea was to show we alternate all the time between good and bad."

The Process

Like a piece of writing, the students had to get everything down on paper first to see what worked, what didn't. As the murals evolved they often discovered they had too much detail that took away from the power of simplicity. They revised their text: deleted, reorganized, changed colors, shifted images. They were constantly asking themselves: What do we like best about this? How does this image make us feel? How can we best present this to get our feelings across to a viewer?

Figure 9.6 Final sketch before "The Outsiders" mural

These are the same questions writers ask of themselves, writing first to find out what they know and what else they want to know. Lizzie, an eighth grader talking about her writing with me one day in English class, said she had too many words in her piece and that took away from the sadness. She chose stronger, simpler words. Nahanni, Ben, Anne, and Sarah discovered in their art that the fewer the images, the stronger the impact.

Technique and color played important roles, as the group from *Night* (Wiesel, 1986) explained: "We added black 'ash' or 'smoke' at the top, the final destructions. We used black paint on a sponge to get an 'ash' effect, and this was really hard to get right, because if you were too careful it didn't look natural, and if you weren't careful enough it looked like black blotches, spaced over the top. We picked the power color red, for the swastika, and gave it dimensions (depth) in dark black. We wanted the one symbol, the swastika, to look strong and hard and overpowering."

The students working on *Go Ask Alice* (Anonymous, 1978) chose the color blue because it gave a sense of security, a sense almost of lulling the viewer, like the drug user, into a sense of false peacefulness.

Bright reds, oranges, and yellows were chosen for *The Outsiders* (Hinton, 1967) because of their harshness—no muted colors, to portray nothing in between good and bad. Across the pages at the bottom of their diamond shape the students used a gold wash that disappeared in flames— "that's because 'nothing gold can stay,' and that's what this book is about."

157

Time

"I can't even begin a painting," Beth told me, "unless I have at least three hours. And then I don't want to know what time it is." Because Beth is an artist, not just an art instructor, she knows the importance of time. The students were obsessed with completing what they had started. When we realized they couldn't possibly complete this in one trimester, we continued into the second.

Artists need time to gain a sense of control so they can develop their ideas and their skills. With time we saw the kids gain competence in making critical and aesthetic judgments, in developing their imaginations, and in mastering the medium of acrylics and collage. They learned new techniques and skills: laminating, applying transparent washes, using sponges or brushes to achieve the desired effect, and using retarder to slow down the drying process to achieve a better blend. They began to understand the impact of varying shape, proportion, color, intensity and order.

Writers master skills in context—when they need the skill and want to know it. Like all creators, they discover style, variety, usage, and mechanics once they know what they want to say and because they want to make what they create their best. Like artists, writers begin to understand the impact of choices.

Most importantly, these students had time to get a sense of becoming part of their creations. "This was the first time I saw kids lose themselves in a painting," Beth said. Kids take ownership in writing when it matters to them. This art mattered.

Artists' Responses

Before the works were hung in the middle school lobby for exhibition, another teacher in the building suggested to several students that they write a paragraph or two saying why they did what they did and what it meant. They decided against the suggestion. Sarah explained:

> The whole purpose of the piece is to let people interpret it as they will. Will they see it as we do, or find a new, hidden meaning, that perhaps we haven't even discovered? . . . For me, I wanted to get a piece that is able to be interpreted. If someone who had never heard of World War II and the Holocaust, saw this artwork, I would want them to get the feeling of chaos (and hope) blocked by a single force. If a survivor saw it, then I hope that we have done a good enough job so that a survivor can see something in our art that moves them, that makes them stand still.

Both artists and writers know they can accept or reject reader response, depending on their intent. The suggestions may work for them, they may not. When writers take responsibility for their pieces, they decide what the reader needs to know. Sarah knows they *showed* what they meant and felt and therefore didn't need to tell or explain what it was about. They are also aware that viewers or readers bring their own experience to a text and take their own meaning from it. They hoped that would

Art of Literature murals with a close-up of "Go Ask Alice"

Art of Literature murals with a close-up of "Night"

happen. Sarah hopes that the art "moves them . . . makes them stand still," the same way a piece of wonderful writing makes you laugh or cry or stop to think.

Sarah knows O. Henry as a famous American short story writer. She probably doesn't know him as an artist. O. Henry and Sarah share common purposes as artists. "Some of us fellows who try to paint have big notions about Art," said O. Henry. "I wanted to paint a picture some day that people would stand before and forget that it was made of paint. I wanted it to creep into them like a bar of music and mushroom there like a soft bullet" (Hjerter 1986, 67).

Exhibition

These students worked passionately to complete their murals and make them the best they could. They had choice in what they did. They felt commitment and responsibility to the project because they had a sense of pride about what they were doing.

We wanted the students to know we valued their efforts and had great respect for their products. We chose a wall in the school that gave the murals the greatest exposure to the student body. We spent hours laying out the most aesthetically pleasing design for exhibition. Alan Stuart, the shop instructor, made sure the murals were mounted on solid wooden frames. Our custodian, Don Mone, took time bolting the murals securely to the wall, patiently maneuvering them "higher," "lower," "a little to the left," to make sure they were just right. Next to each mural, we hung a professionally framed copy of the book jacket and laser-printed names of the artists.

Evaluation

We didn't grade or test the students. We did ask them to take a look at what they had done and reflect on the process and the product. We asked them to respond to the following questions:

What made you choose the book? How did you go about designing your project? How did you go about reaching decisions as your design evolved?

Monica was determined to do a project about the Holocaust. When we discovered *Night* on several brainstormed lists, she talked the rest of us into reading it. We agreed that if we could represent the book artistically it would be a powerful piece.

—*Sarah*

We brainstormed what the book was making us feel . . . It was difficult because we don't have much practice brainstorming pictures. After we succeeded in getting our feelings portrayed in shapes and pictures we put together our ideas: a triangle, upside down, to show rising growth that finally got out of control, a growing sense of disorder and helplessness. We worked to get the image artistically interesting and show contrast, not just in the art form but in the book. The black and white. The full pages to ashes . . . We decided to burn pages of *Night*—in the bottom of the triangle the pages were orderly and gradually got more burnt and unorganized. We traced our hands and made them reaching up from organization to chaos. We revised our original idea of the clenched fist. People weren't fighting back. We skipped the people and barbed wire that we originally planned because we decided it would be distracting and split the picture, stop the movement.

—*Nahanni*

What did you hope to get across to an audience from the book through this piece of artwork? How well do you think you succeeded?

I think the painting means something different to each one of us. We all had different ideas, but they all work together and are related. I want people to see that in the beginning no one thought all the rumors could be true. Everyone was denying that terrible things could happen. They wouldn't let themselves prepare for disaster because they didn't want to believe how awful people can be. The hands reaching out are because the people just sat and waited, figuring the situation would fix itself. If you wait for something to fix itself, it's almost as if you are accepting whatever might happen . . . The important thing is that people can look at our painting and feel something, it doesn't have to be a message that we thought of, but their own meaning.

—*Nahanni*

We hoped to get across the feelings—the fear, the terror, the hopelessness, the ignorance. We wanted to show that no matter what you thought, it was worse . . . This book showed me what the government, the media and American culture didn't tell us.

—*Jay*

When you stand back and look at what you did, what surprises you? What do you like that really worked?

All pieces of art mean something different to everyone. It's the same with literature and music for that matter . . . The important thing is, we somehow managed a terrific piece of artwork that means *so* much to us. We learned how to put an emotion into a painting, how to make it look good artistically, and how to make it mean something to our audience. It was very exciting to watch it grow out of our six hands.

—*Nahanni*

What have you learned or discovered about this book, about pieces of art, about yourself, with respect to the content of what you read and the design you created?

I learned a lot about the process behind creating a work of art. It takes a *long* time. A *lot* of attention must be paid to detail. Things that you learn only by doing. I also learned that I *am* an artist. Just as I *am* a writer.

—*Jay*

I learned a lot about the sensitive sides of Jay and Nahanni. We had very serious discussions about how to do a certain part—the design, paint, color—would this help? Hurt? How does this make you feel? Working together on such a serious project really let me get to know both of them in a different way.

—*Sarah*

What might you have done differently to accomplish your goals?

> I don't feel that it could have gone much better. I'm not saying that it went perfectly. If nothing had gone wrong we wouldn't have learned anything. If we didn't want to learn anything we could have just plowed through . . . But that would not have been very rewarding. I guess the end justifies the means . . . I'm very impressed. —*Jay*

You spent a lot of time on this project. In what ways was it worth the time and effort?

> Every time I look at this piece, I kind of have to hold my breath. I think because I'm Jewish . . . it's sort of a let down to know it's over . . . I want to keep working together on this project. —*Sarah*

What kind of thinking did these students participate in as they created these murals? These students came to know, came to thinking *intensely*, shuttling back and forth through every cognitive process: comprehending, applying, analyzing, synthesizing, and appreciating. Their thinking evolved from the literal to the inferential to the critical. Their talk was laced with the words "we *decided*," as they brainstormed, collaborated, negotiated, restructured, and evaluated. They took the language, thoughts, and perceptions of an author and interpreted them visually through, and for, themselves. They found and made meaning for others, as they found and made meaning for themselves.

These students created startling new ways, ways beyond words, to look at three books—to look at the human experience in these books through their own personal experiences and personal knowledge. And they made meaning so effectively that viewers bring their own meaning to, and take their own meaning from, the experience.

Viewers' Reactions

On the first day after hanging the murals, Steve LeClair, our assistant principal, said he wished he'd had a camera to record the other students' faces and reactions. Kids started past the wall, stopped, turned, backed up, often speechless, staring open-mouthed, reading the artists' names ("Eighth graders did these?!"), eyes moving back and forth from piece to piece. These twelve-, thirteen-, and fourteen-year-olds are treating these pieces and their creators with the greatest respect.

Trusting and respecting our students may be the best models we provide for them in creating culturally healthy environments in our schools.

In addition to the pleasure and respect for art these pieces are providing for their makers and their viewers, they provided a bridge into other disciplines. Teachers and students began discussing the Holocaust, drugs, books, collages, and murals. And other kids are reading. By the end of that first day of exhibition students had checked out every copy of *Go Ask Alice* (Anonymous, 1978), *The Outsiders* (Hinton, 1967), and *Night* (Wiesel, 1986). These pieces were proving William Bennett, former secretary of education, right: "The arts convey what it means to be human . . .

and give coherence, depth and resonance to other academic subjects" (1987/88, 4). Indeed, just as Charlie said when he saw his mural on exhibition for the first time, "This *is* real art."

Summary

Jay, Nahanni, and Sarah constantly evaluated and revised their design as they recognized the visual impact of what they did. A fist became an open hand, an unheeded plea, because the Jews were not defensive. The swastika stood alone, without the originally planned human shadows dangling marionette-style from barbed wire. The students recognized that simplicity made the symbol even more horrific. They knew when a symbol worked, because it surprised them.

Charlie, Anne, and Ben deleted all the concrete symbols that took away from the power of the one image they emphasized—the tunnel, their abstract symbol for being sucked in with little hope of escape.

Jay said she learned that a lot of attention had to be paid to detail—"things that you learn only by doing."

As reading and writing teachers we share the best literature and writing of the professionals and of our students, guiding them through the process of recognizing the similarities and differences, the qualities that make the pieces good, the books effective. We offer kids options—topics of their choice, books they like, and varied ways of presenting what they know. Donald Murray, in discussing the process of teaching, says teachers must give their students all of these opportunities: "The student must be given four freedoms—the freedom to find his own subject, to find his own evidence, to find his own audience, and to find his own form" (1982, 142).

When students make art—visual or literary—and are led to look at it with respect to their own *and* others' pieces, we foster expression and impression. We often saw Sarah or Anne or Charlie standing on a chair to gain some distance on their piece, to answer some of those same questions. The students looked also at what Picasso and the students in the Bronx had done. We talked about the qualities in both. They looked introspectively into their own.

This self-evaluation process is the same in reading and writing. When students write and read often, and are asked to look critically at what they have done compared to previous works of their own and the work of others, they develop an eye toward making their own work better.

While discussing Picasso's work, Beth pointed out the relation of symbols to the culture of the time. In context, they made more sense and led to a greater understanding of Picasso's interpretation of society. Drugs and identity are ingredients in the adolescent culture today. The students began with graphic images of drugs, needles, coffins, and the like, but moved to the abstract with the tunnel—the feeling of being sucked in, "by anything." It made sense to them culturally. It allowed Anne, Ben and Charlie to talk openly about the problems adolescents face today, as they interpreted their thoughts and feelings.

All of these students were trying to answer their biggest questions: Who am I? Where do I fit in this world?

Sarah, Nahanni, and Jay talked with a rabbi who viewed their work. He was most impressed with what they did, yet puzzled. Shaking Jay's hand, the rabbi said, "But you're not Jewish."

"But I still care," Jay said.

Whether it's my regular classroom or a special project, I want my students participating in the process of creating real art, visual or literary, for their purposes. "Pictures as well as words," says researcher-teacher Ruth Hubbard, "are important to human beings in their communication; we need to expand our narrow definition of literacy to include visual dimensions, and in so doing answer the call of researchers for the recognition of multi-literacies . . . " (1989, 150).

Tom Newkirk reiterates that belief: "An exclusively word-centered view of literacy may not prepare students for the graphic and design opportunities that will be available to them . . . we may need to alter our view of 'writing' to include these opportunities. We may need to think of composing . . . as symbol-weaving" (1989, 65–66).

The recognition of multiliteracies not only prepares students for design opportunities, but gives them chances to develop more acute sensory abilities for observing and processing the world around them.

When I recognize and value these multiliteracies, so do the students: "The rhythm that [the teacher] establishes in the structure and environment of the classroom, the values that are both explicit and implicit, the theories on which he or she bases day-to-day decisions [of a child's abilities and of how people learn], and his or her own literacy behaviors . . . have enormous influence on the emerging literacy of the child" (Hubbard 1989, 150).

I try to be the advisor who offers models and guidance, so the students can determine for themselves what literacies they are best at. When the learning is for them, a sense of responsibility and pride drives the process and betters the product.

I think too often I forget the name of the course I teach—language *arts*, not language *art*. Too often I forget that there's more to the arts than the written word.

I want to encourage and celebrate my students' multiliteracies. I need to foster the intelligences for which they are, in the words of Howard Gardner at Harvard, "at promise" (Walters and Gardner 1985).

I'm looking for a balance, a way to integrate the visual and performing arts with the reading and writing, without sacrificing the integrity of either one. I want my English classroom to be a laboratory for the language arts. I have to listen to my students as they tell me what language arts they are good at, so I can help them become better.

Yet I don't want to balance too carefully. I want surprise. I want to stand open-mouthed, speechless, like Ben, and Charlie, and Amy, at all the new ways kids can create, when they're allowed their own voices. I want to hear the voices of students like Nahanni in postcards during the summer. She wrote: "I just got home from a music camp in Orono, Maine. I was in a string quartet there and we played a piece by Mendelssohn. I wish you could have seen (heard) it. Playing it reminded me exactly of our Art of Literature project. The music would go *so* perfectly with that painting. Mendelssohn is a particularly dramatic composer and this piece went back and forth from sad, crying, wailing, *anger* . . . "

I want to hear the combined voices of Sarah, Jay, and Nahanni as they wrote to Elie Wiesel—an invitation to the "showing" of their work.

What struck us [about *Night*] (Wiesel, 1986) was the sense of gradual and unidentifiable growth from the beginning, into something so horrible that once it was recognized, it was uncontrollable. In the beginning, the old prophet who warned the people of such a terrifying future was a sharp contrast to the peoples' false sense of security.

During the year we learned a lot about ourselves as people through the writing process, and while working on this project we found the same problems and ways of solving them as in writing.

Over these months, the painting became less and less of a school project and more a part of our feelings and a challenge to get those feelings out on paper the way we wanted to . . .

What can be done on a more sustained basis to incorporate the arts into a language arts classroom?

- Fill the classroom with posters/prints of favorite artists. Display the students' work beside the professionals. Paint their poetry and their pictures on the walls of the school. Send their best artwork beyond the classroom, to national publications like *Merlyn's Pen*.
- Allow students to convey what they know about language through more than just written response; respect their desire to create through the visual and the performing arts.
- Fill the classroom with real books, models for students to learn from:
  - nonfiction books that focus on visual arts
  - children's literature, those books that contain the richest language and the finest illustrations
- Look for opportunities for interdisciplinary projects, especially with the art, music, or drama teachers.
- Contact your state arts council or the National Council on the Arts for opportunities for bringing artists into the classroom.

The painter Georgia O'Keeffe, once explained that she found she could say things with color and shapes that she had no words for.

I need to remember to give my students the opportunities to say things in ways they have "no words for." When students, like Jay, find work "less and less of a school project, and more of a challenge to get real feelings out," then we will all stand open-mouthed, silent, losing our words to surprise, over the extraordinary things kids can do.

Works Cited

Anonymous. 1978. *Go Ask Alice*. New York: Avon Books.

Bennett, William J. 1987/88. "Why the Arts Are Essential." *Educational Leadership* 45, 4 (Dec./Jan.): 4–5.

Boyer, Ernest L. 1987/88. "It's Not 'Either Or,' It's Both." *Educational Leadership* 45, 4 (Dec./Jan.): 3.

Brandt, Ron. 1987/88. "On Discipline-based Art Education: A Conversation with Elliot Eisner." *Educational Leadership* 45, 4 (Dec./Jan.): 6–9.

Glueck, Grace. 1988. " 'Survival Kids' Transform Classics to Murals." *New York Times,* Nov. 12.

Hinton, S. E. 1967. *The Outsiders.* New York: Viking.

Hjerter, Kathleen J. 1986. *Doubly-Gifted: The Author as Visual Artist.* New York: Abrams.

Horgan, Paul. 1988. *A Writer's Eye.* New York: Abrams.

Hubbard, Ruth. 1989. *Authors of Pictures, Draughtsmen of Words.* Portsmouth, N.H.: Heinemann.

Murray, Donald. 1982. *Learning by Teaching.* Montclair, New Jersey: Boynton/Cook.

National Endowment for the Arts. 1988. *Toward Civilization: A Report on Arts Education.* Washington, D.C.: U.S. Government Printing Office.

Newkirk, Thomas. 1989. *More Than Stories: The Range of Children's Writing.* Portsmouth, N.H.: Heinemann.

Nilson, Lisbet. 1988. "From Dead End to Avant-Garde." *ART NEWS* (December): 133–137.

Sorrell, Walter. 1974. *The Duality of Vision: Genius and Versatility in the Arts.* New York: Bobbs-Merrill.

Walters, Joseph M., and Howard Gardner. 1985. "The Development and Education of Intelligences." In *Essays on the Intellect,* ed. Frances R. Link. Alexandria, Va.: (A.S.C.D.) Association for Supervision and Curriculum Development.

Wiesel, Elie. 1986. *Night.* New York: Bantam.

Further References

Nonfiction Books

Arnosky, Jim. 1982. *Drawing from Nature.* New York: Lothrop, Lee and Shepard.

Exley, Richard, and Helen Exley, eds. 1985a. *My World Nature.* Lincolnwood, Ill.: Passport Books. (Thoughts and illustrations from the children of all nations.)

———. 1985b. *My World Peace.* Lincolnwood, Ill.: Passport Books. (Thoughts and illustrations from the children of all nations.)

Isaacson, Philip M. 1988. *Round Buildings, Square Buildings, and Buildings That Wiggle Like a Fish.* New York: Knopf.

Koch, Kenneth, and Kate Farrell. 1985. *Talking to the Sun.* New York: Henry Holt. (Poetry combined with artwork from the Metropolitan Museum of Fine Arts.)

Lopes, Sal. 1987. *The Wall: Images and Offerings from the Vietnam Veterans Memorial.* New York: Collins. (Photography, sculpture and writing from the Vietnam Veterans Memorial.)

Reading Is Fundamental. 1986. *Once Upon a Time.* New York: Putnam's. (Illustrated collection of true and fictional anecdotes and stories by well-known children's authors and illustrators about reading.)

Sills, Leslie. 1989. *Inspirations: Stories About Women Artists.* Niles, Ill.: Albert Whitman. (Discusses the lives and art of Georgia O'Keeffe, Frida Kahlo, Alice Neel, and Faith Ringgold, with color reproductions of their work.)

Wright, Peter, and John Armor. 1989. *The Mural Project.* Santa Barbara, Calif.: Reverie Press. (photography by Ansel Adams)

Children's Picture Books

Bjork, Christina, and Lena Anderson. 1987. *Linnea in Monet's Garden*. New York: Farrar Straus Giroux. (Linnea, a little girl, visits the painter Claude Monet's garden and sees many of his famous paintings in Paris.)

Collins, Judy. 1989. *My Father*. Boston, Mass.: Little, Brown. (A shared dream carries family members out of their drab life into a finer world of music and travel, a dream later fulfilled by the youngest daughter when she becomes a parent herself.)

De Paola, Tomie. 1989. *The Art Lesson*. New York: Putnam's. (Having learned to be creative in drawing pictures at home, young Tommy is dismayed when he goes to school and finds the art lesson there much more regimented.)

Fleischman, Paul. 1988. *Rondo in C*. New York: Harper and Row. (As a young piano student plays Beethoven's *Rondo in C* at her recital, each member of the audience is stirred by memories.)

Kesselman, Wendy. 1980. *Emma*. New York: Doubleday. (Motivated by a birthday gift, a seventy-two-year-old woman begins to paint.)

Lionni, Leo. 1967. *Frederick*. New York: Knopf. (While his relatives gather food and seek shelter, a small mouse gathers images and dreams to sustain his family through the long winter.)

Rylant, Cynthia. 1988. *All I See*. New York: Orchard Books. (A child paints with an artist friend who sees and paints only whales.)

Wordless Picture Books

Collington, Peter. 1987. *The Angel and the Soldier Boy*. New York: Knopf.

De Paola, Tomie. 1981. *The Hunter and the Animals*. New York: Holiday House.

Ormerod, Jan. 1981. *Sunshine*. New York: Viking Penguin.

Select list of illustrators (authors in parentheses)

Aliki:
A Medieval Feast. 1983. New York: Harper Trophy.
Mummies Made in Egypt. 1979. New York: Harper Trophy.
The Two of Them. 1979. New York: Greenwillow Books.

Graeme Base:
Jabberwocky (Lewis Carroll). 1989. New York: Abrams.
My Grandma Lived in Gooligulch. 1988. Davis, Calif.: Australian Book Source.
The Eleventh Hour. 1988. New York: Viking Kestrel.

Jan Brett:
The Mitten. 1989. New York: Putnam's.
Beauty and the Beast. 1989. New York: Clarion Books.
The First Dog. 1988. San Diego, Calif.: Harcourt Brace Jovanovich.
Goldilocks and the Three Bears. 1987. New York: Dodd, Mead.
Annie and the Wild Animals. 1985. Boston, Mass.: Houghton Mifflin.
Fritz and the Beautiful Horses. 1981. Boston, Mass.: Houghton Mifflin.

Marcia Brown:
Shadow (from the French of Blaise Cendrars). 1982. New York: Charles Scribner's.

Barbara Cooney:
Hatti and the Wild Waves. 1990. New York: Viking Penguin.
Louhi, Witch of North Farm (Toni de Gerez). 1986. New York: Viking Kestrel.
Miss Rumphius. 1985. New York: Viking Penguin.
Ox-cart Man (Donald Hall). 1979. New York: Viking Kestrel.

Tomie de Paola:

The Mountains of Quilt (Nancy Willard). 1987. San Diego, Calif.: Harcourt Brace Jovanovich.
The Quilt Story (Tony Johnston). 1985. New York: Putnam's.
The Legend of the Blue Bonnet. 1983. New York: Putnam's.
The Good Giants and the Bad Pukwudgies (Jean Fritz). 1982. New York: Putnam's.
Now One Foot, Now the Other. 1981. New York: Putnam's.
The Kids' Cat Book. 1979. New York: Holiday House.
Nana Upstairs and Nana Downstairs. 1973. New York: Putnam's.

Tom Feelings:

Now Sheba Sings the Song (Maya Angelou). 1987. New York: Dutton.
Daydreamers (Eloise Greenfield). 1981. New York: Dial Books.
Something on my Mind (Nikki Grimes). 1978. New York: Dial Books.
jambo means hello (Muriel Feelings). 1974. New York: Dial Books

Stephen Gammell:

Dancing Teepees (Poems selected by Virginia Driving Hawk Sneve). 1989. New York: Holiday House.
Airmail to the Moon (Tom Birdseye). 1988. New York: Holiday House.
Song and Dance Man (Karen Ackerman). 1988. New York: Knopf.
Old Henry (Joan Blos). 1987. New York: William Morrow.
Waiting to Waltz—A Childhood (Poems by Cynthia Rylant). 1984. New York: Bradbury Press.

Paul Goble:

Beyond the Ridge. 1989. New York: Bradbury Press.
Death of the Iron Horse. 1987. New York: Bradbury Press.
Buffalo Woman. 1984. New York: Bradbury Press.

Trina Schart Hyman:

Swan Lake (Told by Margot Fonteyn). 1989. San Diego, Calif.: Gulliver Books.
Canterbury Tales (Selected, translated and adapted by Barbara Cohen). 1988. New York: Lothrop, Lee and Shepard.
St. George and the Dragon (Retold by Margaret Hodges). 1984. Boston, Mass.: Little, Brown.
Little Red Riding Hood. 1983. New York: Holiday House.
How Does It Feel to Be Old? (Norma Farber). 1979. New York: Dutton.
The Sleeping Beauty. 1977. Boston, Mass.: Little, Brown.
Snow White (Translated from the German by Paul Heins). 1974. Boston, Mass.: Little, Brown.

Susan Jeffers:

Hiawatha (Henry Wadsworth Longfellow). 1983. New York: Dial Books.
Stopping by Woods on a Snowy Evening (Robert Frost). 1978. New York: Dutton.

Steven Kellogg:

Is Your Mama a Llama? (Deborah Guarino). 1989. New York: Scholastic.
Johnny Appleseed. 1988. New York: William Morrow.
Best Friends. 1986. New York: Dutton.
Pecos Bill. 1986. New York: William Morrow.
The Island of the Skog. 1973. New York: Dial Books.

David Macaulay:

Castle. 1977. Boston, Mass.: Houghton Mifflin.
Pyramid. 1975. Boston, Mass.: Houghton Mifflin.
Cathedral. 1973. Boston, Mass.: Houghton Mifflin.

James Marshall:

Goldilocks and the Three Bears. 1988. New York: Dial Books.
Red Riding Hood. 1987. New York: Dial Books.

Wendell Minor:
    Heartland (Diane Siebert). 1989. New York: Thomas Y. Crowell.
    Mojave (Diane Siebert). 1988. New York: Thomas Y. Crowell.

Peter Parnall:
    Quiet. 1989. New York: Morrow Junior Books.
    Feet. 1988. New York: Macmillan.
    Apple Tree. 1987. New York: Macmillan.
    Desert Voices (Byrd Baylor). 1981. New York: Scribner's .
    The Other Way to Listen (Byrd Baylor). 1978. New York: Scribner's.

Ted Rand:
    Once When I Was Scared (Helene Clare Pittman). 1988. New York: Dutton.
    Knots on a Counting Rope (Bill Martin, Jr. and John Archambault). 1987.
    New York: Henry Holt.
    Barn Dance! (Bill Martin, Jr. and John Archambault). 1986. New York:
    Henry Holt.
    The Ghost-Eye Tree (Bill Martin Jr. and John Archambault). 1985. New
    York: Holt, Rinehart and Winston.

Dr. Seuss (A. S. Geisel)
    You're Only Old Once!. 1986. New York: Random House.
    The Butter Battle Book. 1984. New York: Random House.

Chris Van Allsburg:
    Swan Lake (Mark Helprin). 1989. Boston, Mass.: Ariel Books.
    Two Bad Ants. 1988. Boston, Mass.: Houghton Mifflin.
    The Polar Express. 1985. Boston, Mass.: Houghton Mifflin.
    The Wreck of the Zephyr. 1983. Boston, Mass.: Houghton Mifflin.
    Jumanji. 1981. Boston, Mass.: Houghton Mifflin.

Julie Vivas:
    The Very Best of Friends (Margaret Wild). 1989. Orlando, Florida: Harcourt
    Brace Jovanovich.
    The Nativity. 1986. San Diego, Calif.: Gulliver Books.
    Wilfred Gordon MacDonald Partridge (Mem Fox). 1984. New York: Viking
    Kestrel.
    Possum Magic (Mem Fox). 1983. Nashville, Tennessee: Abingdon Press.

David Weisner:
    Free Fall. 1988. New York: Lothrop, Lee and Shepard.
    Kite Flier (Dennis Haseley). 1986. New York: Four Winds Press.

Don Wood:
    Heckedy Peg (Audrey Wood). 1987. San Diego: Harcourt Brace Jovanovich.
    King Bidgood's in the Bathtub (Audrey Wood). 1985. San Diego, Calif.: Harcourt Brace Jovanovich.
    The Napping House (Audrey Wood). 1984. San Diego, Calif.: Harcourt Brace
    Jovanovich.
    Moon Flute (Audrey Wood). 1980. San Diego, Calif.: Harcourt Brace
    Jovanovich.

Professional Publications/Organizations

Hickman, Janet, and Bernice Cullinan, eds. 1989. *Children's Literature in the
    Classroom: Weaving Charlotte's Web.* Boston, Mass.: Christopher-Gordon.

Huck, Charlotte S., ed. 1979. *Children's Literature in the Elementary School.*
    New York: Holt, Rinehart and Winston.

Moir, Hughes, Melissa Cain, and Leslie Prosak-Beres, eds. 1990. *Collected Perspectives: Choosing and Using Books for the Classroom.* Boston, Mass.:
    Christopher-Gordon.

O'Sullivan, Colleen. 1988. *The Challenge of Picture Books.* Melbourne, Australia: Thomas Nelson Australia.

The New Advocate
Christopher-Gordon Publishers
P. O. Box 809
Needham Heights, MA 02194-0006
(A magazine "for those involved with young people and their literature." Frequently publishes articles about, or by, authors and illustrators.)

Merlyn's Pen
The National Magazine of Student Writing
P.O. Box 1058
East Greenwich, RI 02818
(Publishes students' artwork as well as writing.)

National Endowment for the Arts
Arts in Education Program
1100 Pennsylvania Ave., N.W.
Washington, D.C. 20506

Director
Alliance for Arts in Education
The John F. Kennedy Center for the Performing Arts
Washington, D.C. 20566
(Both of the above federal agencies offer numerous programs in the arts for students and teachers at the local, state, and national levels.)

Chapter 10

Apprenticeship

At Four or Fourteen

I slipped into the window seat, buckled my seat belt, and waited for the remaining passengers to board. A little boy and his father sat down next to me. The boy crawled up into the middle seat and waited for his dad to buckle him in. I smiled, then turned back to the window.

"Are we off the ground?" the little boy asked me.

"Not yet," I answered. "What's your name?"

"Jimmy," he said. "I'm four." And he held up four fingers. "Are we in the sky yet?"

"Not yet," I said.

When we reached cruising altitude he watched as his dad unbuckled his own seat belt. Jimmy unbuckled his, climbed down, and maneuvered his way in front of me so he could see out the window. I asked his dad if it was okay. He nodded, leaned back against the cushion, and closed his eyes.

Jimmy asked questions, seldom waiting for an answer. "Do the wings flap? What holds us up? Where's the pilot? What are clouds made of? Are we there yet? Are we above the trees? Above the biggest mountains? Why's the sky blue?"

When he tired of my answers he returned to his seat and told me his story. His mom and dad were " 'vorced." He'd been visiting his mom in Boston. He could count to one hundred. He and his mom had built a snowman in her yard.

I held my journal in my lap. When Jimmy asked me about it I explained it was a notebook in which I wrote down things I heard or saw—things that mattered to me, things I wanted to remember. "Would you like to write too?" I asked. When he said yes I ripped several pages from my journal and handed them to him with a pen. He wrote quietly, while I wrote. When I looked over he was finished (Figure 10.1).

"Can you tell me about your writing?" I asked.

"This is the snowman I built with my mom," Jimmy said. "This is the head," he continued, pointing to the round circle with two dots and a mouth. "These are the bodies. This is the nose," he added, pointing to the carrot. When he tried to explain what they used for eyes, I couldn't understand the word. In frustration he began to draw. He pointed to the tree on the paper and said, "From oak trees. You know, the seeds of oak trees."

Figure 10.1
Jimmy's snowman
story

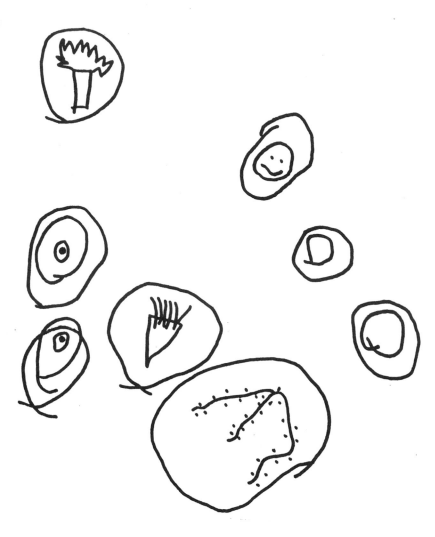

"Oh, *acorns*," I replied.

"Right," he said.

"Could you write the word *snowman*?" I asked.

"Sure." Jimmy drew an M, covered it with dots, framed it with a circle, like his other story parts, and handed the piece to me.

"Nice," I said. "What are these dots?"

"Snow," he said, looking at me as if I should know that. "*Snowman*," he explained, tracing the flake-covered M with his finger.

"Of course," I said.

Jimmy continued drawing while I wrote in my journal. Between drawings, he'd look up and say, "I know how to count to a hundred. Want me to count to a hundred?"

"Sure," I'd reply, not paying much attention. He must have counted to a hundred a hundred times. Each time I'd say, "Wow! That's great. You

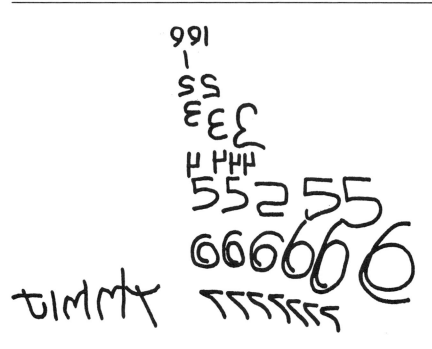

Figure 10.2
Jimmy's poem

sure know how to count to a hundred." He'd lean back into his drawing and start counting all over again.

I bent over and pulled a book of poetry from my briefcase: Paul Janeczko's *Strings: A Gathering of Family Poems.*

"What's that?" Jimmy asked.

"It's a poetry book."

"What's 'pote-tree?' "

I explained that poetry was a special kind of writing. That poets wrote what they knew or wanted to know more about and shaped the words the way they wanted them to be read. I showed Jimmy several poems, pointing out the titles, and telling him that's what poets called their poems. He turned back to a blank piece of paper and was quiet for a long time. The plane began to descend on its way into Minneapolis, so Jimmy stopped writing. When I glanced to my right again, I was surprised at what I saw (Figure 10.2). "What did you write?" I asked.

"Pote-tree," he said proudly. Jimmy read: "One one. Two twos. Three threes. Four fours. Five Fives. Six Sixes. Seven sevens."

"Wow! You got all the way to seven sevens," I said. "That's a wonderful poem about numbers. Have you thought of a title?"

"What do you mean?" Jimmy asked.

"What could you call your poem? What's it about?"

"It's about one hundred," he said, looking at me like, you know that.

"Where would you write that, if you wanted to?"

"Right here," he said, and he added it above his first line.

"Can you sign your name to that, like the poets did with their poems?" I asked. Jimmy wrote his name.

As we taxied to the gate I told him that my sons were in college now and seldom sent me any writing. Would he mind if I kept his drawings for my refrigerator? He agreed. All the way off the plane he kept telling his dad, "She's putting my 'pote-tree' on her refrigerator." I didn't hear his dad's response.

What did Jimmy know by the end of that ride? A lot. He knew how to tell a story. He knew how to write what was important to him: building a snowman with his mom. He wrote what he knew about: numbers. He knew that writing consisted of marks on paper: pictures, numbers, and letters. He knew that the marks made sense to him, the writer, and he knew that sometimes he has to add information so the marks make sense to a reader. He knew how to write a word—his name—and how to combine drawing and letters when he didn't know a whole word (an M covered with snowflakes). He knew what a title was for. He understood the concept of genre: in this case poetry and its unique arrangement on a page. He even knew how to use a drawing to teach someone something they didn't understand—a tree to get at the word "acorn."

Jimmy is a writer, reader, speaker, and listener. Jimmy is four years old.

I met Jimmy five years ago. Every September since then thoughts of Jimmy have helped me to trust my eighth-grade students. Instead of filling my classroom with grammar texts and expurgated literature anthologies, designed by publishers who don't trust kids or teachers to know, I see that I am doing the same thing in the classroom that I was doing on that airplane: immersing myself in writing and reading and sharing it with my students. I trust students to read and write in their own voices, to teach me what they know, and to come up with writing that is worthy of the refrigerator. And I am more aware than ever that what I choose to do in the classroom is a guide for what my students choose to do.

In *More than Stories*, Tom Newkirk argues that "children can adopt a variety of written forms in addition to the story and that they do so by attending to the demonstrations of written language surrounding them" (1989, 24). Jimmy knew a lot about literacy, but I can't be sure what he knew before the plane ride. I am fairly sure he did not know about poetry. But he could attend to the poetic form because of the "demonstration" that he was involved in. Children apprentice themselves to us when we read and write for authentic purposes. That's what I want them to do. That's what Jimmy did—that's what Tricia does.

A month after completing eighth grade, Tricia sent me a packet of her writing. The large manila envelope contained two letters (one a month old), a short piece of fiction, and a play. Tricia's first letter shows she identifies with me as a writer, not just her teacher. She knows how to respond to a writer, noting what especially stayed with her. As a writer she knows she needs to hear that. Tricia's questions to me about her piece indicate she has already detected and diagnosed the problems, and needs only confirmation of her own ideas. She is still searching for strategies. But she also needs response to her writing, to better see what's working or not working (Figure 10.3). She knows I will respond honestly.

June 26

Dear Mrs. Rief,

How is your summer going? I saw your piece in the N.H. Alumnus magasine. I really like the way you started it talking about taking notes on anything and everything. The pieces you included that you had written were great. I think the thing that most people will come away remembering is this line . . . "For a moment, I am the writer who teaches. I can feel the difference."

   The thing that really prompted me to write this letter to you is the fact that I'm stuck. There's a piece I wrote in 6th grade that with some revisions I'd like to send to Merlyn's Pen (Figure 10.3). The problem is I need a reason for the old man to be at the railroad station. Or maybe some inner conflict that he's trying to work out by writing. I'm thinking that possibly I could include the fact that there is obscurity and a strange type of silence that comes from being in a crowd. Since you know the most about writing of the people I know and since you have a head full of ideas I thought I'd ask for help from you. It may be because I had to submit so many pieces to you or for what ever reason but I'm never embarrased to have you read my writing. Actualy I think it's because you never just say "that's a good piece Tricia."

   . . . thank you for your time. If you have any ideas please send them to me. And thanks for your patience reading my scribbles and alterations.

Appreciativly, (is that a word?)
Tricia

In her second letter Tricia has made an important decision. She will be a writer. She is trusting the response she has received from teachers to her writing, that it is diverse and powerful. She is gaining the confidence and courage it takes to be a writer. She is writing. She is the observer of everything around her: comments by teachers, quotes on the board, and thoughts from her own reading. And she is keenly aware of audience. The fiction piece is going to *Merlyn's Pen*. The play might be for just her grandmother. All the writing is for her.

July 21

Dear Mrs. Rief,

Thank you for inviting me to the workshop/class today at UNH. It reinterested me in my own writing. And I've made an *important* decision. Nothing in my life will restrain me from being an active writer. Sounds simple enough, but it also means that I *will* be a *published* writer.

   I have to be. What else could I be when I can sit and talk for hours with teachers (even in the summer) about writing and find it *so* interesting.

   My family (especialy now that my brother has graduated) turns more and more often to me asking what I want to be doing after High School. I always reply that I haven't thought about it. I know it's a lie but I say it. I somehow thought that a writer wasn't a . . . a . . . established enough profession/dream. But *now* I'll answer with my real dream.

   Please thank for me one of the teachers [Brian Adams from Alberta, Canada] in the class today. He was one who had had my folder, he was trying to decide if he wanted to teach 7,8 or 8,9. Physical description? Blond, had a mustache. A coment he made about how I shouldn't quit next year and about how diverse my folder was helped me to decide.

   If I'm going to thank him I should also thank you for all the encouraging things you've said over the year. I should also thank Mr. Schiot. He made a

Ghostwriter *(Does the title even fit?)*

The noise and confusion of the train station was blocked out
How was it blocked out?
from the old man as he walked through the throngs of people
rushing about. A train whistle sounded and people rushed
about jostling each other with luggage and elbows. People *Use "people" too often. Be more specific*
yelled at ticket sellers to hurry up. Paperboys ran around
yelling "Extra Extra read all about it." *Wrong tense?*

The noon sun shone golden through the rose window and a
would the sun shine gold or red?
clock chimed twelve. *too repetetive*

Among all the blusness men and vacationers an old man
walked. He wore a simple outfit, [a pair of] dark pants, a
delete
saltwater blue cablenit sweater and [some] worn black work
boots. Under his arm he had a rather battered Journal. His
hair and beard were thining. His wolf blue eyes were kind
Possibly instead I'll say "his chin was more abundant, with hair than the rest of his head."
and overflowed with laughter. *laughter may not fit with the story*
Who's going to know what wolf eyes look like?

He went over to a bench and sat down crossing his legs
and opened the Journal to a free page. There is something
expectant in a blank lined piece of paper. He took a
mechanical pencil from behind his ear.

The pencil was inscribed Ghostwriter. [And as] he sat
delete
there in [that] busy train station [he held] the pencil erasur
the   *holding*
to his lip, his fingers poised on it ready to drop to the
paper and write.

His brow wrinkled, his [body] shook the tinniest bit as
hand
his eyes appeared to be wrenching his brain for a single
idea. Then his eyes looked glazed seeing but not seeing.
[There were small wrinkles around his eyes] and he seemed
This is unnecessary.
miles away. His lips moved making unheard words. Then his
face looked pained. The thought had gone.
I need something in between these two extremes.
His hand flew up, his face brightened, his eyes were
alert. In an instant all this happened and his pencil was
writing rough scribbles more like bluejay tracks in snow
than writing.

I need to use fewer "his" and fewer "thens."

The old man wrote for [minutes] on end. Then closed the
pages
Journal with a flourish stuck the pencil behind his ear and
walked away triumphantly.

Figure 10.3 Tricia's handwritten revisions as sent to me

comment to my mom yesterday about how my writing had a force to it already that he as an adult hadn't acomplished. All these gave me a confidence and corage to make the decision.

Remember how we once talked about how the Upper Peninsula Of Michigan is full of Scandanavians. Here's a true story of the stubborness of them from when my grandmother went up for a 50 year class reunion this summer.

Family Polotics

A Play . . .

. . . Yes, Mrs. Rief, it actually happened. My grandmother has never liked her ex-inlaws so it became the highlight of her trip . . . That didn't come out bad for a first draft . . . I'll copy it over and do some work on it. If nothing else my grandmother will get a kick out of it.

Anyway, thank you again. I meant to say also that I like the quotes you had on the board. Here's one from a book called *The Flame Trees of Thika*. "But after awhile I think the pursuit of freedom only turns one into a slave." There are so many pursuits of which that could be true.

Appreciatively,

Tricia

P.S. I've just rewritten it and it's already getting better.

In my reply I responded to Tricia's writing, pointing out what I liked and what I wondered about. I asked her questions and confirmed her inquiries about needing a reason for the old man to be at the station. I reinforced what she already knew.

I sent her a copy of *Writing Down the Bones* (Goldberg 1986), a book I use in courses with teachers. And I wrote: "Remember Donald Murray, the professor who was in the magazine with me? He says, 'Good writing makes a reader think or feel. The best writing makes the reader think AND feel.' That's what your writing does, Tricia. He gave me a sign that reads: NULLA DIES SINE LINEA. It's Latin for 'Never a day without a line.' Writing takes confidence and courage—getting published takes patience. It all takes a lot of words on paper. You have it all."

Tricia replied that week:

August 7

Dear Mrs. Rief,

I got your package today. Thank You! I've read one chapter of the book and I'm already excited about it. I'm sure I'll be underlining and scribbling in margins.

Since woodshop in 6th grade I've had a block of sanded wood with a fancy edge on it. Tonite I wood burned the saying in Latin that you gave me into it. It's now on my desk.

I haven't heard from Byrd Baylor. It turns out that all her books aren't in stock in local stores so I'm going to have to order *The Desert Is Theirs*. I *really* (notice the correct spelling) want it now that I've visited Arizona. It turns out I was glad I did that writing-reading project on the West right before going out. I noticed more things that way. After a lot of work I'm going to send *Ghostwriter* to Merlyn's Pen. Thanks for your advice and the book. Till the next time.

Tricia

Tricia is a reader, writer, speaker, and listener. Tricia is fourteen years old.

Jimmy and Tricia taught me not only what they know, but two very important lessons about my own teaching. They taught me to trust my students as readers and writers. They also taught me that I am not a silent partner in my classroom. I guide and direct my students by what I do, and what I choose to immerse them in for reading and writing. In three hours Jimmy had apprenticed himself to me as a learner. He wanted to do what I was doing. Tricia, too, wants to do what real learners are doing. She apprentices herself to those teachers who are reading and writing and learning with her.

I wasn't reading a basal on that flight. I wasn't filling in purple ditto sheets with context-stripped words—*cat, tree, girl*. I had no dot-to-dot books available to occupy Jimmy's time. I shared what I was doing as a learner and for my purposes, not intentionally modeling my learning but sharing my authentic reading and writing with him, trusting that he would get what he could from it. I had no idea he would understand so much.

I do the same thing in my classroom. I had no idea Tricia would understand so much.

References

Goldberg, Natalie. 1986. *Writing Down the Bones.* Boston, Mass.: Shambhala.

Janeczko, Paul B. 1984. *Strings: A Gathering of Family Poems.* New York: Bradbury Press.

Newkirk, Thomas. 1989. *More Than Stories: The Range of Children's Writing.* Portsmouth, N.H.: Heinemann.

Rief, Linda. 1989. "Fragments of Language: A Conversation Between Texts." *New Hampshire Alumnus* 65, 4 (Summer):4–6.

Several years ago at a conference in Albany, New York, I shared my students' portfolios with the participants. One teacher looked up, shook his head, and said, "These kids are all gifted. My eighth graders can't read and write like this."

In Colorado, a teacher asked, "Are these *kids*, or adults?"

In a conversation with Philippa Stratton, Editor-in-Chief at Heinemann, I discussed my discomfort at such comments. "They just don't trust kids," I said.

"You know, Linda," Philippa said, "I love gardening and frequently visit some of the most beautiful gardens on my trips to England. These gardens are so lovely, that I often find myself wishing for a weed or two so I don't feel so awful when I get back to my garden."

There *are* weeds in my garden. My students are not gifted. They are not adults. *Most* of them take learning seriously for all the reasons mentioned in this book. Some do not.

It's difficult writing about the kids who choose to do no work. How do I share an empty portfolio? A missing log? A self-evaluation that is never handed in? How do I write about Nick, who shrugged off failing my class with a handshake and an off-hand, "Well, Mrs. Rief, we've had our differences. But, well, hey, man, it's been fun!"? Do I admit *wanting* to say, "No, Nick, it has *not* been fun. You ruined every class with your rudeness, your arrogance, your incessant demands to be the center of attention at all times, your total lack of respect for anyone and anything."

If students like Nick choose not to participate they will fail. It happens. Students like Nick push me to the edge every day. They are the weeds with which I, too, must contend.

No way of teaching is perfect for every child. I try every alternative I know to help them succeed. But sometimes nothing works. They have different agendas at the moment. No matter how hard I try, I can't reach them. Their failures always feel like my failures.

However, there are too many good stories to let the Nicks drive me from teaching. Shawn, now a part-time freshman at a local university, called the other day. "Remember me?" he asked. Of course I remember him. In eighth grade he struggled with every aspect of language. The most he ever wrote at a time was a paragraph. He seldom read. But he always tried. He struggled with even more difficult issues as he went through high school: a father silenced by his experiences in Vietnam, the death of his mother in an automobile accident, alcohol, and parenthood by the age of 16.

"Would you have time to meet with me?" Shawn asked. "I'd like to talk to you about an idea I have for a book. It's been pretty hard having a daughter when I'm so young. I've been keeping a journal about things she does and things I do to keep from getting mad at her. Games we play and stuff like that. Do you think anyone would be interested in a book for teenage fathers? I think I know some things now I didn't know before ... I've been reading a book called *The One-Minute Manager* and that's kinda the format I've thought about for my book. What do you think?"

I haven't heard from Shawn in five years. After all he has been through, I'm surprised he has turned to writing. I shouldn't be surprised. "I think it's a wonderful idea, Shawn. How about Tuesday?"

The Shawns keep me teaching.

While the question, "Are these kids gifted?" is one I hear frequently in workshops and courses with teachers and administrators, there are several other questions that crop up often. I tried to anticipate and answer questions implicitly between the lines of each chapter. But some need more direct responses.

When do you teach formal grammar?

When I work with teachers and administrators, I frequently ask them to take ten minutes to list those things they value most that they would like their students to be able to do as writers, readers, speakers, and listeners by the time they leave their classrooms. No one has ever listed "identify a gerund," "find indirect objects," or "diagram a sentence."

Formal grammar instruction, identifying parts of speech, is not at the top of my values list either. However, mechanics of the language (spelling, paragraphing, punctuation, and usage) certainly are important to me. They are taught daily within the context of each student's reading and writing, but in their proper place—last, after the student makes the content the best it can be. I use *Writers Inc* to point out standards the students need to know.

Students have sections in their reader's-writer's log for "Notes," "Vocabulary," and "Spelling." The information comes from their reading and writing, or from whole class lessons.

I also give the students a two-day crash course in parts of speech and labels just before they take state-mandated achievement tests. Until we get rid of these senseless tests, I cannot withhold information from the students. I will not let these tests drive my instruction, however. Two days is all I allow.

When I first started teaching, I taught the way I had been taught—read a story, answer the questions, and fill in blanks based on exercises from a grammar text. But it didn't feel right. I couldn't make any connections with real reading and writing. I couldn't make sense of what I was doing for me, let alone for the students in the classroom. I began to wonder why I was doing what I was doing. I had a lot of questions: What am I able to do better after I fill in these blanks? How do I know that's what the author meant? What do real writers do? What do real readers do? How did I learn what I know? What is learning for? As a teacher I feel a responsibility for seeking answers to these questions.

Even as the writing and reading changed in my classroom, I still had questions from parents. "This writing is wonderful and my kids are reading better than ever, but why aren't you doing more formal grammar instruction?"

As a professional I can't say, "Because it doesn't *feel* right." I began reading everything I found about grammar instruction. I started a file folder entitled "Why not formal grammar?" It began to fill up. I found articles that dated all the way back to 1906, the major portion of which confirmed my suspicion that formal grammar instruction does not help students become better writers.

I pulled excerpts from articles by authors I know and respect, putting them together on a sheet to share at workshops and with parents when they ask me, "Why don't you teach formal grammar?"

When one parent asked my principal several years ago why he didn't mandate formal grammar instruction, he said, "Our language arts teachers have a file folder more than an inch thick that proves to me they are professionally aware of the latest research in the field of language arts instruction. Based on our goals of helping students become better writers and readers, my teachers have proven to me that formal grammar instruction doesn't accomplish those goals. If you can show me educational studies that prove differently, I'd be happy to take a look at them." No parent has ever shared his or her own file folder.

If we want to be treated and trusted as professionals, we have to act as professionals. We have to belong to professional organizations like the National Council of Teachers of English. We have to be aware of what research says about learning. We have to attempt to relate those findings to our own situations. We cannot isolate ourselves in closed classrooms. We have to continually question what we do, how we do it, and why we do it. We have to keep our own file folders.

A few years ago, after sharing portfolios and reading response journals with parents, one woman still wasn't convinced. I took her through her daughter's writing, pointing out the level of sophistication and mastery of usage and mechanics in a meaningful context. "But when are you going to teach her the parts of speech?" she demanded. "When are you going to teach her how to diagram sentences?"

"Why?" I asked.

"Damn it," she said, slamming her fist on the table. "I suffered through it in school, and she will, too."

Not in my classroom, I thought. Not in my classroom.

How do you teach poetry?

From every poet I have ever worked with: Gigi Bradford and Jean Nord-hous in Washington, D.C., Roland Flint at Georgetown University, Mekeel McBride at the University of New Hampshire . . . from every poet I have ever heard read: William Stafford, Donald Hall, Marge Piercy, Donald Murray . . . from every poet I have ever read: Robert Frost, Richard Wilbur, Linda Pastan, Georgia Heard, Tess Gallagher, Langston Hughes, Maya Angelou . . . I have heard, "Don't teach poetry. Share it."

In my classroom we read poetry, listen to it, sometimes memorize it, and try to write it when our thoughts and feelings fit the form. I fill the room with poetry anthologies. I type up song lyrics and texts from children's picture books to show the students the poetry of the language. We make copies of poems students request so individually they can own them.

When Jared, a hulking football player, heard Arnold Adoff's sports poems, he wanted his own copy, of the whole book. Several students wanted copies of Shel Silverstein's poems after Julie and Robyn memo-rized and presented them to the class. Lindsay's log is filled with copies of poems she cut from the *New Yorker, Merlyn's Pen,* and *Seventeen.* She has asked for copies of my poetry. She drafts her own poetry in her log, thirty-seven drafts of a poem called "Black Dreams." Josh has an entire notebook filled with his poetry in the style of Jim Morrison. Sarah moves from class to class repeating "Jabberwocky," because "I *love* that poem!"

We talk about sound, appreciation for the strongest words, vivid imagery, and placement of words on a page. We play with language until it says what we want it to say in the format that is most pleasing.

Poetry is idiosyncratic. What one person loves, another hates. I know enough about poetry now not to teach it. We share it. I practice reading it (my own and the professionals) until it feels good in my mouth, the same way I tell kids to shape their poems on paper. "Read it out loud to yourself. Does it feel good in your mouth?"

Jean Little, in her poem "After English Class" (1986, 28), tells us exactly what overanalysis does to her when the teacher tells them all the "hidden meanings" in Frost's "Stopping by Woods on a Snowy Evening."

"It's grown so complicated now that,
Next time I drive by,
I don't think I'll bother to stop."

My students' best-liked books of poetry

Adoff, Arnold. 1974. *My Black Me.* New York: Dutton.

———. 1986. *Sports Pages.* New York: Harper and Row.

Angelou, Maya. 1987. *Now Sheba Sings the Song.* New York: Dutton.

Buchwald, Emilie and Ruth Roston. (Ed.). 1987. *This Sporting Life.* Min-neapolis, Minn.: Milkweed.

Carroll, Lewis. 1987. *Jabberwocky.* New York: Harry N. Abrams. (Illustra-tor: Graeme Base)

Cummings, E. E. 1976. *in Just-spring*. Boston, Mass.: Little, Brown.

Dunning, Stephen, Lueders, Edward, and Smith, Hugh. (Ed.). 1967. *Reflections on a Gift of Watermelon Pickle*. New York: Lothrop, Lee and Shepard.

Eliot, T. S. 1982. *Old Possum's Book of Practical Cats*. New York: Harcourt Brace Jovanovich.

———. 1988. *Growltiger's Last Stand*. New York: Farrar Straus Giroux.

Frost, Robert. 1975. *You Come Too*. New York: Henry Holt.

Glenn, Mel. 1982. *Class Dismissed!* New York: Clarion.

———. 1986. *Class Dismissed II*. New York: Clarion.

Greenfield, Eloise. 1978. *Honey, I Love*. New York: Harper Trophy.

———. 1981. *Daydreamers*. New York: Dial.

Grimes, Nikki. 1978. *Something on My Mind*. New York: Dial.

Hall, Donald. (Ed.). 1985. *The Oxford Book of Children's Verse in America*. New York: Oxford University Press.

Janeczko, Paul. (Ed.). 1981. *Don't forget to fly*. New York: Bradbury.

———. (Ed.). 1983. *Poetspeak: In their work, about their work*. New York: Bradbury.

———. (Ed.). 1984. *Strings: A Gathering of Family Poems*. New York: Bradbury.

———. (Ed.). 1987. *Going Over to Your Place*. New York: Bradbury.

———. (Ed.). 1988. *The Music of What Happens*. New York: Orchard.

———. 1989. *Brickyard Summer*. New York: Orchard.

———. (Ed.). 1990. *The Place My Words Are Looking For*. New York: Bradbury.

———. (Ed.). 1991. *Preposterous: Poems of Youth*. New York: Orchard.

Koch, Kenneth and Farrell, Kate. (Ed.). 1985. *Talking to the Sun*. New York: Metropolitan Museum of Art and Henry Holt.

Knudson, R. R. and Swenson, May. (Ed.). 1988. *American Sports Poems*. New York: Orchard.

Kumin, Maxine. 1989. *Nurture*. New York: Penguin.

Little, Jean. 1986. *Hey World, Here I Am!*. New York: Harper and Row.

Longfellow, Henry Wadsworth. 1983. *Hiawatha*. New York: Dial. (Illustrator: Susan Jeffers)

Mazer, Norma Fox. 1989. *Waltzing on Water: Poetry by Women*. New York: Dell.

Noyes, Alfred. 1990. *The Highwayman*. New York: Harcourt Brace Jovanovich. (Illustrator: Neil Waldman)

Rylant, Cynthia. 1990. *Soda Jerk*. New York: Orchard.

Service, Robert W. 1986. *The Cremation of Sam McGee*. Ontario, Canada: Kids Can Press. (Illustrator: Ted Harrison)

Siebert, Diane. 1988. *Mojave*. New York: Thomas Y. Crowell.

———. 1989. *Heartland*. New York: Thomas Y. Crowell.

————. 1991. *Sierra*. New York: Harper Collins.

Silverstein, Shel. 1974. *Where the Sidewalk Ends*. New York: Harper Collins.

————. 1981. *A Light in the Attic*. New York: Harper and Row.

Soto, Gary. 1990. *A Fire in My Hands*. New York: Scholastic.

————. 1990. *Who Will Know Us?*. San Francisco, Calif.: Chronicle Books.

Sullivan, Charles. (Ed.). 1989. *Imaginary Gardens*. New York: Harry N. Abrams.

Books on writing poetry

Goldberg, Natalie. 1986. *Writing Down the Bones*. Boston, Mass.: Shambhala.

————. 1990. *Wild Mind*. New York: Bantam.

Heard, Georgia. 1989. *For the Good of the Earth and Sun*. Portsmouth, N.H.: Heinemann.

Tsujimoto, Joseph I. 1988. *Teaching Poetry Writing to Adolescents*. Urbana, Ill.: National Council of Teachers of English.

Which professional books do you find most helpful?

In Works Cited at the end of each chapter, the books listed are the ones I use most frequently. Others include:

Nancie Atwell. (1987). *In the Middle.* *

Nancie Atwell, ed. (1989). *Workshop 1: Writing and Literature*.

————. (1990). *Workshop 2: Beyond the Basal*.

————. (1991). *Workshop 3: The Politics of Process*.

Lucy Calkins. (1991). *Living Between the Lines*.

Toby Fulwiler. (1987). *The Journal Book.* *

Donald Graves. (1990). *Discover Your Own Literacy*.

————. (1991). *Build a Literate Classroom*.

Jane Hansen. (1987). *When Writers Read*.

Donald Murray. (1989). *Expecting the Unexpected.* *

————. (1990). *Shoptalk.* *

Thomas Newkirk. (1989). *More Than Stories*.

Brenda Miller Power and Ruth Hubbard, eds. (1991). *Literacy in Process*.

Tom Romano. (1987). *Clearing the Way*.

Frank Smith. (1986). *Insult to Intelligence*.

All of the above books are available from Heinemann Educational Books. (An asterisk indicates a Boynton/Cook title.) In addition, I find the following book extremely helpful:

Donelson, Kenneth L., and Alleen Pace Nilsen. 1989. *Literature for Today's Young Adults*. Glenview, Ill.: Scott Foresman.

185

How do you do all
this in one year
with only 45
minutes a day?

I also belong to the National Council of Teachers of English, from which I subscribe to *Language Arts, English Journal,* and the *ALAN Review,* which publishes extensive reviews and articles on adolescent literature. Their address is National Council of Teachers of English 1111 Kenyon Rd., Urbana, Illinois 61801.

How do you do all this in one year with only 45 minutes a day?

I often don't get to everything in this book in one year. It is my plan. Sometimes kids come to the classroom with more experiences, less experiences, or different needs. I go with them. Sometimes the personality of the class dictates a more structured approach or less structured approach. Sometimes world events shift the students' concerns.

When I have a question, I always ask the kids, and we figure out a solution together. I have faith in my students that they know best what works for them and what they are capable of doing. I have faith in the teachers who read this book that they know best what might work for them and how much they are capable of doing, or want to try.

Appendices

Student Portfolios

In my classroom portfolios have become the students' stories of who they are as writers, readers, thinkers, and human beings. The portfolios are rich with the evidence of what they are able to do and how they are able to do it.

I have shared my students' actual portfolios with many teachers. I can't do that here. I can't show where each student started, how much he or she changed, and everything that contributed to that change. I showed that through Nahanni's portfolio in chapter 8. I can show a few "close-up snapshots" from a range of students' "albums."

During July 1988 I was a teacher-fellow in writing at the Kennedy Center in Washington, D.C. My task was to find a piece of artwork that inspired my writing. When I chose the Vietnam Veterans Memorial, a member of the Kennedy Center staff suggested my writing might include a photo-journalistic display of visitors to The Wall. I tried. But the pictures, no matter how close I moved in on the faces, couldn't capture each person's story. I could only capture the moment. After sitting at The Wall for days, after viewing documentaries on the Vietnam War, after interviewing veterans, and after reading the books and poetry that came out of that war, I could begin to fill in the stories behind the faces.

These portfolios, like photos, are only glimpses into the students' lives as learners. They can't show the whole story. They *can* stand alone, but they *don't* stand alone. Everything that influences each of these students is part of the portfolio: their peers, parents, teachers, the classroom, community, and world. I can't show all the interactions. I can't show all the drafts, all the changes, all the thinking that went into each piece, each decision. Readers would have to come into my classroom, talk to the students, read with them, and write with them to be able to fill in the stories.

I *can* show the diversity of thinking that goes on in one classroom. Through these portfolios I am attempting to share a picture of these students as writers and readers. (The actual portfolios contain all the finished pieces the students produced throughout the year and the rough drafts that contributed to each piece. They contain all the self-evaluations, the complete reading list, and the entire reader's-writer's log.) *These* portfolios contain only the students' most effective pieces, their best-liked books, a sampling of log entries, and a sampling of evaluative comments.

I want these portfolios to stand as evidence of what thirteen- and fourteen-year-olds know and can do. Each time I share the actual portfolios of my students with teachers in a workshop I ask: "What do these students know as readers and writers? What are they able to do? What's one thing you learned from this portfolio?" I want teachers to ask these same questions of themselves as they read these glimpses into the lives of adolescents. I hope what happens in the workshops happens here. No matter how many teachers read Andy's portfolio, they will all find something different, something valuable that Andy knows and is able to do. What each teacher sees is as diverse as each of these students. I'm seeking that same diversity in the classroom.

I hope that teachers will read these portfolios in conjunction with the rest of this book and no longer wonder, but *know*, how these voices, how these stories, came to be.

Note: The writing of the students stands at whatever stage it reached, first draft thinking or polished final draft. I have not corrected student errors. I think it's important to see that real thinking is not mechanically perfect. What the students think and say *is* important.

Andy

Andy is obsessed with science fiction and fantasy, to the point where prior English teachers and his parents are worried he isn't getting a broad enough awareness of other kinds of literature and command of language. They are also concerned he doesn't know the mechanics of the language.

Andy hates writing response to reading in his log. His entries are not thought provoking or lengthy. However, they are usually honest. He does so much writing that is obviously connected to his reading that the log response doesn't seem that important.

At the beginning of the year I worry that Andy reads and writes too much fantasy/science fiction. I have a hard time responding to it. But I am patient. After several months I ask him to try another genre of reading. If he doesn't like it he can go back to his favorites. I ask him, too, to try a personal narrative.

Although Andy's writing is frequently flavored with the vocabulary of fantasy, by the end of the year, he has such a variety of writing and reading that I'm no longer worried. When I ask Andy what he has learned about himself as a writer this year, he grins: "I didn't think of myself as any kind of a good writer, you know—I didn't really write. I don't know how it happened. I just kind of got interested and now I write a lot, and I'm writing for a magazine . . . "

Most Effective Writing

NIGHT

The sun was low on the horizon.
Shadows crept longer, twisting in and out of
majestic oaks, their trunks tinted a golden hue,
a reflection of the fiery orb that was completing
its daily journey across the heavens.
With one last flicker it sank
beneath the distant tree tops. Instantly darkness
enveloped the wood, swallowing it in utter
blackness . . . the night had claimed the forest.
Sounds echoed all around; the nocturnal
animals were now active in their dark
domain. Scamperings and rustlings crowded
the wood as the creatures awoke and went
off to their nightly work. A soft evening breeze
parted the translucent gray clouds, releasing the
pale moon light from its wispy prison. It shone
upon a silent clearing located off the
beaten paths that wandered throughout
the forest. A small gurgling brook
laughed merrily meandering on its way,
crawling around rocks that were strewn all over
the forest scattered in random
patterns. The brook ended at an out-cropping
of rock, that hung over a mighty river, and
plunged into the chaotic froth joining
the raging waters as they carved at the

                              solid granite banks in the dead of
               night. The stream's source was a spring that
     bubbled up from the depths of the earth
          to surface in this glen. A well, made
     of oblong slabs of granite, hewn from colossal boulders
               that were discarded from the icy glaciers
                    which once covered the wood a mile high,
                         sat a few inches from the spring.
          A giant oak tree stood right next to
          the old dilapidated well, silhouetted
against the giant full moon like a
     gnarled hand clutching at it
               and its silver beams trying to grasp
                    them from the starry sky. Moss hung from
                         its twisted branches and wagged in
                              the hush breezes that filtered
                    through the trees, waving like many
          old men's beards dangling in the
                              obscurity of the evening.
          Bright rays of moonlight struck the
     placid pool of water at the very bottom of the well
turning it into a lake of molten silver, which occasionally
          rippled when the silver drops of cold
                    water plopped from wet
rocks above. This idyllic spot stood untouched by mankind
                    and the problems of the world,
               except for the mysterious constructor
of the well . . .

Empty stairs

Everytime I arrived at Grandma's house I'd leap out of the car in eager anticipation and run around to the back while the others went in the front door.

Always I'd find Grandfather sitting on those stairs leading into the old house. He would always be there, sitting hunched over the evening news, pretending like he was reading it. And then as I would turn the corner he would act surprised and check his watch and exclaim, "My, it's four! How time flies!"

I would run up and jump on his lap and wait for a story. We'd talk and talk while the others wondered where we were. Then as the scent of supper finally filled our nostrils, we'd pack up our thoughts and head in. So much time was spent on those crumbling old stairs.

No tear ever fell on them. No sadness ever filled the air, until the time came when I turned the corner and Grandfather wasn't there.

For the birds

Chapter 1 The Beginning

Dex gazed out of the tiny window at the billowing mist below that hung over the jungle like a thick blanket. A sudden crackle of static broke the intense silence of the cabin and brought Dex back to reality.

"Figen sus cinturones de asientos por favor," said the captain over the loud feedback. "Nosotros estamos empuesamos nuestro decente al aeropuerto de Porto Velho . . . "

Dex, comprehending the thickly-accented Spanish, clicked his seatbelt together and gazed out the window searching for the airstrip. He watched the twin propellers slice their way through the swirling clouds as the plane began its descent through the atmosphere. As the plane emerged beneath the overcast, Dex got his first full view of a tropical rain forest. It was immense. The lush tree tops formed a green carpet that stretched as far as the eye could see in all directions. Finally his keen eyes perceived a mar in the solid green carpet, a small brown strip of packed dirt . . . the airport. He knew this wasn't a first class airline but *this* was ridiculous . . .

[after reaching his hotel room] Dex, sighing, threw himself on the bed along with his suitcase and lay there moaning, staring at the ceiling.

"I'm *never* gonna catch it, Bertha," he mumbled, pulling out his huge knife and addressing it. "Ever since I got fired from that stupid job as zoologist because I couldn't find *any* rare birds! Well *excuse* me, Mr. Jennings, but this time I don't need you!" cried Dex in anger.

"Yes, this time I'm going to prove *you* wrong! *This* time I'm going to *find* a bird!" he smiled to himself. "But not a puny rare bird, but a *unique* bird— the Kiirii. Yes, the one and only *Spectrum Bird* that lives somewhere near here!" Dex plunged Bertha into the geographical map laying on his bed. The knife stabbed, ripping the map, right in the center of Machu Picchu, a mountainous region in neighboring Peru.

Query letter

Roger Moore
Editor, DUNGEON Adventures
TSR, Inc.
P.O. Box 110
Lake Geneva, WI 53147

Dear Mr. Moore:

I would like to propose an idea for a series of three modules to be published in your magazine. My idea takes place in the ADVANCED DUNGEONS & DRAGONS game world. Here is the plot.

There has been a major jail break at Ironrock Prison. Fourteen of the most dangerous criminals have escaped. Among them is the dreaded assassin, Vorpal.

The player characters have been contacted and told to help in the recapture of these escaped convicts.

The adventure starts, as the PC's must track down these villans before they can do any harm. The gang, lead by Vorpal, will lead the characters on a wild goose chase through the nearby town of Twin Rivers. As the PC's move through the dark side of town, trudging through an underground sewer system, looking for information on the wharves, and questioning the local Assassin's Guild (a dangerous job), they'll be the target of ambushes, traps and cold-blooded murder. Finally with a good amount of luck, they'll confront the gang and have a huge fight on their hands.

Now the module can stand alone by having the characters kill (or be killed by) the group, or they could continue by having the gang be driven off into retreat.

If the next module is played, the gang will leave in such a hurry, that in the frenzy, they will be divided. One half of the group, lead by Vorpal, will run out of town and head for the woods. The PC's will continue the chase into the forest and onto the Mirkmist swamp. During the chase they have to pass many obstacles, mainly deadly swamp monsters that roam around looking for just this kind of opportunity to have some player character munchies, to catch up with the convicts. When they do, another fight will arise. Again this module could be played by itself as a wilderness adventure or else be played along with the series.

If used in the series, the PC's would (if they can) force the villans to surrender and make them tell where the other half of the group is headed. With the information given to them the characters will find out the rest of the gang is headed for their mountain hide-out by Hidden Cove, and that Kapaxian (the chaotic Cambion), leader of the second group, plans to use their previously gained treasure and magic (their horde and hide-out were never found) to summon forth a powerful demon lord to destroy the city.

Now Vorpal knows that Kapaxian has no chance of controlling the gigantic demon and willingly tells the PC's everything he knows about the hide-out and draws a crude map to Hidden Cove. With this information the characters head for the mountains to stop Kapaxian. The final confrontation is in the huge underground hide-out, and the player characters must race against time to save the city.

As you can probably tell this module is aimed for any kind of class or race willing to risk their lives in an extremely difficult adventure. The series is for characters of 10th–15th level of ability. It should bring a fast paced adventure full of fun.

I took the opportunity to write a series of modules with one plot that will lead the characters through three very different environments that will pose different problems for the players.

Along with high excitement this module will provide a variety of new monsters and magic items like: "Demon Zombies" with limbs that break off and keep on fighting; blood-sucking Mosquito Men; the deadly grasp of a Tendril; the shocking abilities of an Earthshaker, and the terribly deadly powers of the gigantic demon lord or the magic of *Gauntlets of Thunder & Lightning, Arrows of Explosions* or *Energy Whips*.

I sincerely hope you find this series worthy to be published in dungeon or even as a true module.

Sincerely,

Andy

Reading

Number of books read: 15
Best-liked books:

> Weis and Hickman: *Time of the Twins, War of the Twins, Test of the Twins, Dragons of Autumn Time, Dragons of Winter Nights, Dragons of Spring Dawning*
>
> Piers Anthony: *Yielding a Red Sword*
>
> Jean Auel: *Clan of the Cave Bear*
>
> Douglas Adams: *The Hitchhiker's Guide to the Galaxy, Restaurant at the End of the Universe, So Long and Thanks for All the Fish*

Reader Response:

Test of the Twins: I love the Dragon Lance series. I feel like I'm a part of them. I'm with them on all their adventures. I laugh with them, and I cry with them . . . This book is so *sad*. I mean I cried for a half an hour when Raistlin died. I'm so attached to this series and now I'm going to read the Chronicles.

Dragons of Winter Nights: Stories are neat when you have a twist. This twist should really draw the reader's attention. It might be a mysterious twist, a scary twist, or in this case, a funny twist. The heroes were travelling to a seaport without a sea! Now that's funny. In the story I'm writing these three girls find a lost kid about 4 years old named Norton. Norton has a secret that is so important people kill for it. I think that's a pretty good twist. I hope it will draw the reader in and make him or her wonder what it is. What could a little innocent boy know that would cause men to kill for it? . . . In the story now, a village is being attacked by a flight of dragons and troops. The authors describe the attack beautifully—chaos everywhere, screams, shouts, dust flying, people running around in terror. It makes you seem like you're there in the crowds of people fleeing troops pillaging homes. It's this describing the scene so well that makes me love this book because it's so real it's like you're there watching. Pretty soon you start talking to the characters in the book! . . . Emotion is also a major element needed for the recipe of good writing. You must show the likes and dislikes of the character. In this story Sturm wants to become a knight of Solaminia more than anything and even though he's not right, he acts it. He's the most honorable, honest, just, truthful, caring person in the story, living his life of chivalry. And now he's being tried by another knight who's strict and not likable. During the trial it seems he has no hope becoming a knight . . .

Over the weekend I've become a totally shrewd businessman! Why you might ask? It all happened when I put 1 + 1 + 1 together and got fame, success, and fortune!

The first 1 stands for my need for money! The second 1 stands for my skill at writing! And the third one is my hobby of playing the Advanced Dungeons and Dragons Role Playing Game . . . Over the past 5 days I've been putting my idea into action. I have collected many magazines, books, novels, etc that pertain to AD & D as possible. Now I'm starting to read them. My idea is to combine those three ideas: money, writing and AD & D. First of all, I've played AD & D since I was 8 and I know the rules. I've loved to write ever since 8th grade (long time ago) and I've loved money since I was born! Now TSR, the masters of AD & D and many other role-playing games are making a new magazine called Dungeon . . . In this new magazine they feature modules (scenarios that have a plot, monsters, treasure that you use for playing AD & D) written by other people that send in their work in hopes it will be published. First I bought issue #111 of Dragon, which had the rules, guidelines, etc for sending in modules to Dungeon. After reading these I collected all the resources I didn't have (I have a lot) and read looking for an idea for a plot for my module . . . Now finally I came out with an original idea. A big idea, big enough for three modules—a series! (more money, more writing, more AD & D). My idea was a prison break which is an original idea. Now there are three kinds of AD & D adventures! city, wilderness, dungeon—and the series has them all! first the prisoners, all dangerous characters of evil, go to the dark side of town where the PC's (player characters) must find them and tell them, or take prisoners, using clues, etc. Then you follow one half of them through a forest to a swamp to get them into the sea, mountains, and finally into a dungeon at the climax. I'm starting to write a letter to the editor of Dungeon to see if he likes it!

I like to write because you can put your feelings and thoughts on paper and you can put your dreams into words (The pen is mightier than the sword.) . . . It's hard to write about things you don't know about. And it's hard to write about a subject you don't like. You can't force writing. You must take time and patience.

The hardest part for me is focus. I can't stay on one story for more than a week. I just have too many ideas, it's hard to choose and write it before I come up with a new one.

When I write about a subject like my *For the Birds* story I need to read up on certain subjects like tropical forests, Brazil, Peru, etc so my story will be more realistic . . . It's taking me so long to get started because of all the research! I've read about marijuana, cocaine, farming, the Amazon, South American Indians, Peru, Machu Picchu, parrots, macaw, cockateil, trees and jungles. And I've had to look up Spanish sentences and words and places where his journey takes him. I try to find the most desolate towns. But I found out that giving a character a name is the hardest part . . . it should represent the character, sort of sound like what he likes and how he acts. I chose the name *Dexter Eugene Hazardson* because I feel it's just like what he is. Now Dexter is a name which gives a wimpy feeling, but I have a strong character so I need a good strong name. Dex is his nickname which I think is a change from Dexter. Every character should have a nickname. Eugene shows his civilized side, which is small. It tells you he's an odd ball with a middle name like that . . . somewhere an upper middle class family. The last name is important too. Hazardson, he's a hazard . . .

For the Birds is my writing at its best. I spent a lot of time and research to make it sound realistic . . . All my stories have potential. I think no story is worst. They're all just in different stages of development.

I never knew I could write well until this year. Encouragement helped me the most. I also liked the attention and compliments to stick with a story. Long-piece writing is hard for me.

Reading gives me a ton of ideas for writing. Sometimes I memorize certain lines that are great and place them into my story, using them in a slightly different way. For instance, right here on page 202 (from *Dragons of Winter Nights*) is the classic line which inspired me to write a story which got an A, a poem which got an A, and a score of 8 out of 8 on a national writing test. Yes, the historical famous line . . . are you ready for this sentence? It's a big one. Close your eyes. Brace yourself . . . "And then the clouds parted. Solinari, the silver moon, though only half full, burned in the night sky with a cold brilliance. The water in the pool turned to molten silver."

I don't think reading is hard at all unless you're reading a book that you don't like or a subject you're not interested in. Reading cannot be forced. You read out of free will. I like to read because it puts you into another world . . .

Gillian

The first time I kneel down beside Gillian for a conference, she ends up in the girls' room sobbing. Her first piece of writing is about the day her father left the family. She reads it to me, then breaks out crying. "I can't do this," she says. "Yes you can," I say. "But not this piece, not this time. Someday . . . " She let the piece go, and hasn't looked back since.

I include Gillian in this portfolio section, not just because of what she did in eighth grade, but what she has done since. Several months into her freshman year in high school Gillian walked into my room for a writing conference. She had two folders with her. "Why two writing folders?" I asked. "One's for school, the other's my real writing," she said.

That folder of "real writing" contained initial drafts of her personal narrative "Winged Dinosaurs" and her poem "After His Death I Walk the Beach with Grandpa." In her sophomore year she returned again with her short story "Skeeter," which she had begun in her writers' workshop with teacher Richard Tappan.

Throughout her school career Gillian has found little success in math or science classes. Through writing she has found her voice, and let others learn and feel and think from it.

Most Effective Writing

Opa

The lake is calm. The only movement is a warm summer breeze lightly skimming the water. The high trill of a bird breaks the silence.

Opa and I stand at the water's edge, our pants rolled knee high and our toes slowly sinking into the moist sand. Fishing rods in hand, we slowly make our way towards the familiar dock. Along side it, the old boat, faded with age, bobs peacefully in the dark blue water.

Opa pulls the crawfish in. There are about twenty-five fish, caught in the seaweed-green net. They wriggle around, lashing their stubby tails. Opa reaches out with his large, tanned hand. I place mine, small and pale, in his. There is an opening of water between the dock and the beach. I leap aboard the dock with Opa's help and he hands me the net of crawfish.

Then Opa slowly climbs down the dock and into the boat. I hand him the net and my rod, and he places them in the bottom of the boat. Once again, I reach for his hand, and he pulls me aboard.

The water by the bank is shallow. We use our oars to push off from the sandy bottom. We decide to row to a semi-secluded spot on the boarder of a swampy inlet. Once there, we drop the mini-anchor and grab our poles.

Opa hands me a worm. It squirms and wriggles between his thumb and forefinger, reminding me of last night's spaghetti dinner. I reach out and carefully grasp the worm around its dry, scaly middle. I focus on a big pine tree just across from the boat and quickly loop the hook through the worm's middle. I cast and wait.

What seems like hours pass. The sun sinks on the horizon like a flaming golf ball perched at the end of the lake, ready for tee-off. I, once again, reel in and cast out, this time on the other side of the boat. I glance down at the pile of fish laying in the bottom of the boat. All Opa's. As I do, I feel a sudden tug at the end of my line. The red and white bobber bounces furiously in the water, then disappears below the surface.

When the bobber appears again, I know it is time to reel in. I turn the handle round and round, suddenly realizing how much line I had cast out. It seems like miles of never-ending twine. The pull on the other end gets stronger and harder. So does mine.

Opa sits beside me, with the green net in his hand. It is the same net the crawfish have been in, but Opa has used them all. Unlike me, he isn't afraid of being bitten by his bait.

I can see the tip of the fish's head over the side of the boat. It breaks through the water like a groundhog poking his nose through fresh, spring soil. Opa bends down and with one swift movement, scoops the fish into the net.

My first fish—a little six-inch sunfish.

I use the rest of my strength to paddle home. I need to tell Grandma about the "whale of a fish" I caught, before Opa comes up the path, carrying the truth.

Peacefulness

Peacefulness is a calm lake
in the hands
of the setting sun.
It is a hot fire
with flying sparks
twinkling
in the cold air.
It is the slow
ticking
of a grandfather clock
measuring the silence.
A freshly mowed field,
framed with wild flowers.
Myself,
sitting on a deserted beach,
watching the waves
form on the horizon,
dark curls
laced with white.

Winged dinosaurs

Standing on the cement dock, I shiver with fear. I stare out over the water to the green raft. To me, it bobs up and down like a buoy in a hurricane. In reality, the lake is calm and peaceful.

Jonas, my seven-year-old cousin, has challenged me to a race to the raft. I am twelve. I can't refuse the challenge of a seven-year-old. But I am afraid, afraid of what lies in this lake. I can't see the bottom. I have vacationed here every summer for seven years, yet I have never swum in the lake.

What would happen if I back out now? I wonder. No, I can't, I say to myself. Jonas asked me in front of the entire family. I might be laughed at. I'm the oldest. I don't want to be laughed at. I can't be afraid.

When I was younger, I confessed to my aunt that I was scared of the lake. She promised me there was nothing to be afraid of. She assured me that anything I was worried about was more afraid of me. She told me that monsters slept during the day in dark places and only came out at night when everyone was sleeping.

My aunt didn't know that I saw the lake as a dark place as black as night, with plenty of room to hide the sleeping monsters. She didn't know about the winged dinosaurs with fangs, the furry bats or the scaly alligators that lurked in these murky waters. I knew. In my child's mind I had seen their shadows slithering beneath the surface. I had felt their quick nibbles on my ankles when I once braved a ride in an inner tube. From then on, I had kept the secrets of the lake to myself and built sandcastles on the safety of the beach.

The cottage door slams shut. My relatives weave their way down the wooded, rocky path to the contest. I feel like a piece of meat being thrown to the lions. To the spectators I'm the fearless gladiator. Jonas is the underdog. The cheering is for him.

I take a deep breath and clench my fists at my side. My stomach tightens and I have to tell myself to breathe.

"Ready, Gillian?" Jonas asks.

"Sure," I say.

I stare at the black water. I know the dinosaurs, and bats, and alligators lie in wait. Why am I doing this? I ask myself for the hundredth time.

I unwrap my towel from around my waist and drop it to the dock. Lowering myself down over the cement steps I dip one foot in the water. Like a jester preparing the audience for the real contest, Jonas performs aquatic dives, flips, and somersaults. I'm still trying not to cry as I lower my second foot into the lake.

Jonas, seeing that I am finally in up to my knees, literally dampens my thoughts by throwing his body continuously against the water. He looks like a beached flounder flailing helplessly on the land. His purpose is to soak me. Yet, I'm only afraid he's waking the sleeping creatures. I shiver with fear.

"Come on, Gillian. It's not that cold," my aunt yells. Little does she know.

"I'll count to three and then we dive," Jonas says, as he lines himself up next to me.

"One," he yells. I stare ahead. The raft seems years away. Plenty of time for the dinosaurs to drag me under.

"Two," he shouts. Hungry black shadows writhe and twist under the surface.

"Three!" Jonas dives like a dolphin, his moves free and natural. I plod in and sink like an old sheep dog on a hot day. Petrified of staying below the surface, I frantically pull myself up, clutching at the water like rings on a swing set. I don't care about winning. Now that I can't turn back, I just want to reach the safety of the raft.

My left hand reaches over my head and dips into the water. My right hand follows. Much to my astonishment, I'm doing the crawl. I skim the surface like a waterbug, convinced I'm being chased by slithering snakes. I don't look down for fear of being snatched. I don't look ahead for fear of losing.

From the shore I hear shouts of encouragement. "Come on, Gillian, touch the raft! It's right in front of you."

I turn, looking for Jonas. But I don't see him. He must be on top of the raft. At least I tried, I think.

I look ahead to the green boards. Jonas isn't there either. Confused, I turn around and see him struggling to catch up. I've not only survived the monsters, I'm actually winning the race.

With all my strength, I plunge my hands deeper into the water. Just as I'm about to touch the raft, I pause, and wait for Jonas's outstretched hand. As we touch the raft together, I realize what my seven-year-old cousin has helped me to do. I've not only made it to the raft, but the monsters have been put back to sleep.

After his death
I walk the beach with grandpa

My pale feet
graze the hard sand.
Thoughts of grandpa
follow me like the inescapable patterns
left by my footprints.

Ahead, the happy cries of small children
echo across the beach
mixing with the cries
of swooping gulls.
The sun is disappearing beneath
a hazy line
and blanketing the beach with darkness.
A cold wind catches me from behind
and I shiver in its arms.

A little girl and an old man
walk by me, hand in hand.
I close my eyes
in remembrance.
The little girl is me, eight years ago.
And the strange man? Grandpa.

Though he's been gone only two days
the time has dragged endlessly.
I try to run away
from the burden of his death
but it grows heavier and clings
like my shadow on a long summer day.

Now, for the first time, I cry
for both of us.
My walk along the beach is for him.
I want him to know he lives on
if only in my memories.

Our relationship, filled
with friendship and love,
was taken for granted
until it was almost too late.
He lay in the hospital bed,
a silent, plastic figure,
his only movement
the raising of his hand
to acknowledge I was there.
Unable to sit up,
he stared at the ceiling.
"I love you," he whispered,
"and I will . . . " he paused,
trying to keep us together, "forever."

I promised him he wouldn't die.

Now, as I walk the beach
we once walked together,
I think of my broken promise,
and wonder if I'm forgiven.

"Hey, little white girl in the red dress, where are you?"

The boy's voice echoed through the hall and down the stairs to where I sat, wiping away my tears with red-ribboned pigtails. I heard whoever it was call me again. "I said, where are you? Girl, I know you can hear me!"

A hall door opened and a teacher's voice joined his. "Sidney Harrison, I should have known. Now pipe down and go back to your classroom."

My palms turned cold and clammy between the step and my body. I had heard of Sidney Harrison; he was the one my new neighbor, Kate Greenwood, had told me about. He'd been in the third grade for three years. He was the toughest kid in school. He made everyone call him Skeeter. He beat kids up, Kate said, for no reason, especially the ones who didn't call him Skeeter. Even the sixth graders were afraid of him.

There I sat, head bent, tears streaming down my face, both in fear of Sidney, and in humiliation because I couldn't find the office. The attendance sheet that I was supposed to deliver sat next to me on the step. What did that awful teacher expect of me? I was new. It was my first day. And to send Sidney to find me . . . What was she thinking?

"Girl, why the hell didn't you answer me?"

I said nothing. I stared in shock. Before me stood a tall black boy with nappy hair, small squinting eyes and a permanent scowl on his dirty face. He wore jeans torn at the left knee, orange high-top sneakers, and a white tee-shirt carrying a faded Mobil emblem.

"Girl, can you hear? Do you know who I am? Tell me your name before I smack you!"

"Gillian," I breathed.

"What's that? They call you Gilligan?" he snorted with laughter. Then he spit. A few drops of saliva landed on one of my new black patent leather shoes. I wanted to lean over and wipe it off, but I didn't, in fear of the threatened smack.

"Didn't you hear my question? They call you Gilligan?"

"No," I whispered. "Gillian."

"That's what I thought, Gilligan!" He laughed and yelled, "Gilligan!"

He spit again, this time aiming for my right Mary Jane. He hit his target squarely. I looked up at him in horror.

"Don't you know, little white Gilligan, this is the way it goes. I shine your shoe and you give me a quarter."

Oh, is that all? I thought. He just wants to shine my shoe. He must want to be my friend. Won't Kate be surprised . . .

He whipped a soiled handkerchief out of his pocket and snapped it at me. "Pay up! Now, whitey!" he taunted.

"I'm sorry, Sidney. I don't have a quarter," I answered confidently.

"Sure you do! What'd your mama give you for lunch money?"

"One dollar. But it's a bill. I don't have any change." Now that we were friends I was sure he wouldn't take my dollar, especially since he only wanted a quarter. Then his hand shot out and knocked me across the face from my right ear to my forehead. I felt the sting in my eye and began to cry again.

"Shut up!" he said firmly. He grabbed me by my shoulders and shook me. "How'm I gonna shine your shoes if you don't pay up?"

"I don't need a shoe shine," I whimpered. "My shoes are new."

"Don't tell me what you need!" His loud, angry voice echoed around us and I could see his veins pulsating under his skin. "I know what you want and don't want, so shutup and gimme your dollar!"

I tried to get up but he had a firm grip on my shoulder. He threw me back down, slamming my back against the stair. I looked at him, tears streaming down my face. Sidney had to be the cruelest boy in the world. I could already feel a bruise forming on my back and my eye hurt so much from when he'd hit me that I could barely open it.

"Now!" he yelled. "I want my money now!" I reached into my pocket for my small, red leather change purse. I wondered where all the teachers were? I wondered why my mother would send me to a school as horrible as Woodbury Park? My hands shook as I clicked open the gold clasp and pulled out my neatly folded dollar bill. Sidney snatched it greedily from my hand and looked at me knowingly.

"I knew you'd come around," he grunted. He whacked me again. My head slammed into the hard iron bannister as he ran up the stairs. I hurt more than I ever knew I could, but for the first time that morning, I was mad. I stood up, holding on to the railing for balance.

"Sidney Harrison, you listen up!" Knowing it was already too late to back out, I gathered my courage and continued. "Just because you're in third grade for the third time doesn't mean you're better, just dumber!"

I heard his footsteps halt in the hall above me. A split second later he flew down the stairs. I ran up to meet him on the landing. He was taller than me, but by rising to the tips of my toes, we were face to face.

He raised his fists. "You wanna rethink what you just said, girl? I hope you weren't talkin' to me!"

He reminded me of a steel pole, slim, but powerful. "Yes, I was talking to you." Afraid that I'd made a big mistake, I hoped my face didn't reflect my fear.

"I can't believe you, girl. I thought you knew your place. What do I have to do to you?" He knocked a fist into my throat and held it there. "You better take that back, now!"

I pulled away, ready to follow his advice, but then changed my mind. "What can you do to me?" I said. "If you beat me up you'll get suspended and you'll miss school and you won't learn anything so you'll be stuck in the third grade forever! I'll have graduated from college by the time you make it through sixth grade!"

He dropped his fist. He stepped back, looking humiliated. I should have realized I'd said enough, but I kept going, partly in fear of what he might do if I stopped. "You must be so stupid! All you have to know to pass is how to read and multiply a little. I bet you can't even spell your name—Skeeter!"

He leaned against the wall and slid down it as slowly as the tears slid down his face. I stood there for what seemed an eternity watching his tears mix with dirt and streak down his hollow cheeks.

I pulled my white handkerchief out of my pocket. It was folded into a perfect triangle with my initials embroidered in red on one side. I bent down and pressed it into Sidney's hand. He unfolded it and brought it to his face. I cringed as he ran it across his nose and then handed it back to me.

We stood together. He stared at me, with a mixture of hatred and embarrassment in his eyes. "Let's go," he whispered. "The office is proble' waitin' for the attendance sheet." I didn't move. I was frozen to the spot. I had no idea what to make of his behavior.

"Come on," he said gruffly. He looked me straight in the eyes, "Gillian."

Reading

Number of books read: 49
Best-liked books:
    Torey Hayden: *One Child, Somebody Else's Kids*
    Anonymous: *Go Ask Alice*

Susan Wexler: *The Story of Sandy*
Mary Higgins Clark: *Where Are the Children?*
Hans Pieter Richter: *Friedrich*
Rose Zar: *In the Mouth of the Wolf*
Mary MacCracken: *A Circle of Children*
S. E. Hinton: *That Was Then, This Is Now*
Lois Duncan: *Daughters of Eve*

Reading Response:

Five Were Missing (Duncan): I didn't expect it to be so suspensful or fullfilling. It was sad and I often surprised myself by realizing I was holding my breath.

A Circle of Children: It seems to me that both Mary MacCracken and Torey Hayden are everything I've always wanted to be . . . I don't know what it is but something somehow draws me toward children, especially children with emotional problems. Can you imagine the sensation of being the first person that a child ever said anything to, much more, being loved by a child in the condition some are in? That's my goal, I want to be a teacher like Hayden and MacCracken.

Hayden's book *One Child* really turned my whole perspective around. What I mean is I have always liked children and being around them, and in the past year I have become really interested in disabled children and have been reading up on them. But at the end of this book I wasn't sure I still wanted to be a "special teacher." How could I live through that? Watching a child go through both physical and sexual abuse would hurt me more than ever imaginable. By the end of the book, when I read the poem to Torey—I decided that I did want to teach children like Torey did.

Good literature usually has a point that the "quickie love stories" don't usually make. Romance books are usually "boy meets girl" and they kiss, break up, get back together and live happily ever after, but the books I've been reading lately are with more than one plot and they make a point that's not easy to forget. I do like lovey-dovey books though, it lets me forget about thinking for awhile.

The Story of Sandy: (In her log, Gillian puts all her quotes in red ink and her reaction in blue ink.) This book is described as "The unforgettable book by a foster parent who refused to believe her little boy was hopelessly retarded."

At first the story was very confusing to me. The two parents (Tom and Mary) could not deal with a retarded son, so he went to stay with "Sukey" (the author) and Joe (her husband) . . . When he came to live with his foster parents Sandy could not do much, he acted like a new baby. Here is something I really liked—it got to me and made me sad.

Undressing Sandy was more like unwrapping a package than handling a three-year-old child.

When the tight little snowcap came off, his hair hardly showed the glints of red which had won him his nickname. It was damp with perspiration and matted against his scalp. I ran my hands through it in an effort to fluff up some curls around his pale face . . . Under his snowsuit he wore a pair of tight knitted trousers which were pretty well soaked with urine in spite of the plastic pants and diapers underneath.

I thought it must feel good to get out of such a heavy, clammy outfit and to stretch and feel air on your body. It flashed through my mind that nearly a year ago, during one of her "motherly" interludes, I had learned in a letter from Mary that Sandy was beginning to urinate into a little jar. I called Joe to look in my jelly cupboard and bring me a big, empty glass. He and Tommy soon came in together with a large instant coffee jar, shining and clean and capacious enough to meet a little boy's needs.* When Joe saw what I had in mind he lifted Sandy gently to the floor and steadied him on his uncertain legs. Joe let him lean against his shoulder and cheek, so that Sandy felt Joe's skin warm against his thigh.* Holding the jar below his plump little genitals, he said, 'There we are, Sandy, all set.' Almost immediately a pale-yellow stream trickled into the jar. The small boy in Joe's arms stiffened and straightened, as if the accomplishment gave him pride and power.

My favorite part of that excerpt is where the stars are. I think Joe, for the first time, felt like Sandy was real, a person that he could relate to, and become friends with. Someone just reading that part may think I've gone crazy by reading all that into someone holding a little boy while he goes to the bathroom in a jar, but if you read the beginning of the book you really get a sense of the charecter's feelings and reactions. Maybe I didn't think of all that when I first read it, but that was one area that I reread about 3 times . . . I first wondered if the situation was his parents. Now, I don't know. His parents didn't do anything wrong. I think he really *did* feel love for his father who very much loved him. It was obvious to the reader, that it was very hard for Tommy to give his son up, even to people he had cared for and trusted for years. Mary (the mother), on the other hand, I am still not sure about. My impression was that she had a semi nervous break-down and she didn't know how to deal with her son.

Mary comes home from a holiday in New York with a book (*The Little Auto* by Lois Lenski) tucked under her arm and "As soon as Sandy saw me he patted the floor next to him and said, 'Down.' My heart gave a little leap, for we were immediately back in the groove of our hard-won rela-tionship and had lost nothing by the four-day separation. I sat right down as commanded and Sandy reached up and pulled at my hat, a sign that he wanted me to stay. I opened the book at the first page and read it slowly . . . Into his mouth went his comforting thumb, and he leaned confidingly and heavily against me. The sound of my voice, the sound of the pages being turned must have added to the spell, but there was no doubt that the story had meaning too for Sandy. I was able to read through to the last page. When I closed the book, Sandy reached to open it again and burrowed more closely into me." I especially liked that part because I can easily relate to it. When I babysit I bring books to read to the kids. I try to bring books they don't have or haven't heard of. When they *really* like the book they nuzzle into you and then make you read the book over and over . . . When she talks about how he snuggles up to her and sucks his thumb, it made me remember when my mother used to read me the *Babar* series— that's *exactly* what I would do, and reading that clicked something in my mind, bringing back the scene very clearly.

This next passage is taken from a chapter called "What of Sandy Now." " . . . when put to bed he was allowed to have the light, a big draw-ing pad and crayons at his command for another hour. From the kitchen where Pa-Joe and I were finishing our supper we heard singing—a choir of joy to ears that had listened so long for it. I waited well past the bedtime hour I had set to interrupt it. Just when I was going in he called 'Sukey, I drove you a pictcher' and there it was spread out on his knees . . . I asked what it was called, and he said at once 'Summer comes fast very quickly.' The page was bright with color—green grass, oddly shaped trees, other odd shapes that were birds in the branches, cloud puffs in the sky and a huge sun, the color of his hair."

The reason I like the description of the picture is, not only because it really makes you see the picture, but because it reminds me of the picture Dibs painted in *Dibs in Search of Self*. He says " these are the streaks and stripes of my thoughts." In many ways Sandy reminds me of Dibs . . . in their personality. They were both considered dumb, stupid, retarded. They really weren't, they just needed a little time and encouragement, but eventually they both got there. "There" being able to talk, walk, and do activities. They both turned into young men with big goals which they fullfilled.

Give Me One Good Reason (Norma Klein): I think that this whole section was really one of the strongest, if not *the* strongest point in the book. ("You mean, we should get married to give my child a name?" I repeat, trying to make sure I've got it right.) I really agree with her. I think it's pretty disgusting when people think of women as no-name, no-brain imbeciles who don't know anything but housecleaning and changing diapers . . . It reminds me of when my parents got divorced and the judge said to my mother, "What will she use for a name?" My mom said, "Mine, Nye." The judge then said, "Well, I just give her my blessing . . . with her mother's name and in this society . . . " I remember driving home from court and she was REALLY mad! She said that in 30 years when I was the head of the most famous psychology clinic, she would call him up and tell him where he could put *his* name! I remember being so proud of her, even though I was only 4½ I knew what she meant . . . She told me, "When you're older, and even now, never forget who you are. You are yourself and no matter what your name is, or what you look like, even what kind of society you grow up in, you are always going to be your own person and whatever you do I'll be proud of you, always!" . . . I've finished this book. My mom and Gab should meet. I felt like I knew her—like I know my mother now. They seem to have the same principles, and the same personalities in a way. Gab really stood up for what she thought was right. If she didn't like something, she didn't write it on the bathroom wall—she stood up and said how she felt . . . I would reccomend this book to Kari, because she has problems right now and I think this book might make her think. I want her to do that, about her possibilities, about her chances.

Homecoming (Voigt): I said in earlier entries that this book is scary. It is! I can't believe people are really wandering around homeless and their lives are really like that! Why am I complaining all the time? I have it *easy!* I think it's bad when I can't stay for a basketball game 'cause my mom needs my help! They don't even have a mom! When I read it, my first reactions started out as—how could this girl take care of three kids while they traveled? Now I think, I would like to have the stamina to be the person she is/was. I feel funny. As if I'm making her out as my idol. She isn't really, just a person I admire. I mean, she's only a charecter in a book.

I feel close to her though, like I know her. When I'm reading I sometimes turn around quickly, like someone is watching me! I feel like that now. Why?

On the Beach: There is a poem in the beginning of the book . . . This is the way the world ends, Not with a bang but with a whimper . . . (T.S. Eliot) . . . Wierd huh? It makes me wonder, do poems at the beginning of books make any difference to people without reading logs? I mean, I never read the poems until now—when I look for them. The poems help me figure out and understand the book easier/better. I would never even notice them unless I looked for the author on the first page like I do now . . . I'm a little worried about this book. I mean, because of what it's dealing with—it's so scary! "A book to read. A book to read again. A terrifying prophecy. A reason to act now." *What is prophecy?* A prophecy is something someone describes that may happen in the future. As an example, the author is describing a nuclear war and maybe there'll be one in the future . . .

Peer response to Gillian from Mandy: Gillian, you really describe your feelings really well throughout these entries. The way you described Mrs. Andrews as an orange (" . . . the grandmother in *Homecoming* reminded me of Mrs. Andrews—like an orange—hard to peel but with a soft inside . . . "). Most of all what really surprised me was that you depended on your log to write down your feelings about the TV show "Our House." It shows me that you kind of look at your log as a friend that you can confide in. It really makes me think on how I don't use my log as a friend but just an assignment. Maybe it would help if I looked at it the way you do. By reading your log it helped me in that way. I wish I was as good a writer as you are. I love your stories!!! Love, Mandy

Dicey's Song: Kristin told me that this book had *no* plot *at all.* She said it was really boring and she said it was even worse then *Homecoming.* How could she say that? Punch her out for me, will you? I loved *Homecoming* and I think *Dicey's Song* is really good too. There is a plot! It's about Dicey finding friends, building a relationship with her grandmother, and going through puberty. James is turning into a *really* smart kid who is going to the "top of his class" in every subject. Maybeth is overcoming her shyness and learning to do math and read more. And Sammy is keeping out of trouble for the first time in his life.

Friedrich: I finished this book and I'm just about as disgusted and disappointed as I've ever been. I'm disgusted in the way that people acted and thoroughly disappointed in our country for just watching it happen. Just because our country prides itself on freedom doesn't mean it can't help free others. It was so gross when the nazi boy helped wreck the jew house and when Friedrich died—I wanted to scream and throw things. And the worst part is, the way Friedrich died is one of the better ways most jews died!

This morning on the bus I had just finished reading the book and a kid on my bus stood up and said, "Take us home, we don't want to go to school!" My bus driver said no and the same kid said, "Take us home you jew!" I wanted to stand up and slap him.

In the Mouth of the Wolf: "If you are ever on the run and have to hide the best place is right in the mouth of the wolf." Herman Guterman's advice to his daughter Ruszka . . . I wrote that 'cause I was wondering what the mouth of the wolf is. Does it mean like if you're running or hiding from someone you should hide in the most obvious place since it's so obvious they wouldn't think of it? . . .

Roll of Thunder, Hear My Cry: Unlike the last book (Zibby Oneal's *The Language of Goldfish*), my mother strongly approves of this one. Maybe that's why I'm hesitant to continue. I hate involinarily rebeling but it seems to be what I'm doing. As I read, my eyes do not stay on the page and I think of things that I should be doing . . .

Self-Evaluation

Writing is not something you really do, it is mostly something you are. I write for myself, to feel proud I've accomplished something.

I really like to write and I feel that it has gone very well for me this year. I realize that it takes longer to write than I thought it did. At the beginning of the year I thought I could just churn out pieces on command and they would be perfect. Now, through *Opa* I realize it doesn't work that way, and it takes drafts and drafts of non-perfect writing to come up with something even close to perfect.

You can't come to a blank piece of paper after not writing for a week and turn out a final copy of a book. It takes a long time. I think one has to be patient most of all.

I like to hear what you [Mrs. Rief] say. There are so many things that you pull out of us in our writing that I'm glad you share your writing with us. I would feel stupid sharing my writing with you if you never had anything to share with me. I'm glad you write. That encourages me . . . I think that in 6th grade with Mrs. Bechtell you could tell she cared enough to share her writing and, like you do, she wrote with us, and when we read, she read too . . . I love English when I have a teacher who doesn't try to act like she knows more about everything and when she is wrong, she admits it.

When I see a book I want to read, I don't think about anything but that book. When I was reading *Where Are the Children?* I only thought of that book all day. When I was in science I was wondering what was going on in it . . . When I read, the characters come alive, I can see the children's hair and hear the voices.

When I am reading a book I think about the stories I'm writing and try to pick up different plot ideas, charecter descriptions, and sometimes, like in *Dicey's Song* I wrote a grave inscription on the top of the page so I'd remember it. I may start a story with it but I always write quotes and sometimes descriptions so that I remember them when I'm writing. I think reading also helps my writing in the terms of spelling and also in the way of sentence construction.

Writing helps my reading alot but in a somewhat negative way. When I've written alot and then I sit down to read I feel more critical and the book has to be pretty much perfect for me to want to keep reading. I think that comes from all the conferencing I've done and it's made me kind of a literature snob. The positive side is that I don't spend as much time on books I'm not really enjoying because I'm hoping something better will happen on the next page. Now I know that there's a better plot on the shelf than on the next page.

Janet

Janet is socially versatile, well liked by her peers and adults, and a strong athlete. She is sensitive and compassionate to all those around her. She comes from a family that places high value on education.

Although Janet is an academically above-average student, she needs motivational nudging as a reader and writer. "I need deadlines," she says. I give her deadlines.

Even with deadlines Janet sometimes gets overwhelmed by extracurricular activities and lets the homework slide. Push as I may, she writes little in response to reading for almost a month. Angry with herself, she goes back and rereads two books when she realizes she values and misses that response.

All Janet's pieces start out as narratives. She often gets stuck in the middle, unable to find a direction or an ending. After working with Elizabeth Kirschner, our poet-in-residence for two weeks, Janet discovers poetry as a way out of the narrative dilemma.

Janet teaches me to risk poetry. I teach Janet *intent*, when she worries about the mechanics in a poem. "You can break the rules, if you know the rules," I say. She chooses no punctuation in "Black River."

"I want the reader to slide through at her own pace," she tells me.

This year Janet is just discovering herself as a decision-maker in her own learning. She has trouble articulating what she knows about her own writing, but she is beginning to make discoveries through books and her peers' writing.

Most Effective Writing

BLACK RIVER

the air bites my lungs in iciness
as I exhale a soft veil of swirling mist drifts into the air
and is lost in the deep density of the night
my heart beats in excitement
the bright crimson berries on the weeping crab tree
encrusted in a crystal cocoon
contrast softly with the pure white snow
icy crust
crackling and crunching under the rubber soles of my boots
brings something from deep inside me into the open
I laugh
nothing is funny
my sled is poised
set to go
glancing down the hill I see a sleek flat slope
the light from the house slowly diminishing
into an endless black hole
I take a running start
I race down the hill against wind and time
once gusty and sharp
the wind is now a smooth black river
it hits my face
drying my eyes

filling my ears with rushing sound
it slides over my body and is left on the snow
I slide for an eternity
forever I am going into the deepest blackest river
cold
forbidding
but feeling warm and secure
in my down parka
the sled slows to a stop
I lie there in a deep sleep
every muscle in my body
gives a last grind and drops
like a wet rag
snow
nipping at my waist
cutting at my wrist
the tears
once warm and streaming
now frozen on my face
I stand up and look around
in one direction
that black forbidding river lays before me
clean and cold
I can imagine the crisp air
flying past white diamonds
in the sky
Where do I go?
I smile
shrug
and head uphill for another run

FOG

5am
the earliness of the morning weighed me down
like water in a cup
the harbor was still
so still the silence hummed
the fog rested at the rocky entrance to the harbor
billowing veils
paused
restrained
at the red buoy that sat angled in the water
then,
like a bag of sugar bursting
it was released
floating
drifting
unrolling before me over the stillness
it surrounded me
i could only see the water
a cobalt gray
a red flag
sagging at the top of a sailboat
i breathed in the fog
it was cool and refreshing like silver water

i bathed in the stillness
and then
my heart dropped
i remembered the sun would come up today

A DAY IN JULY

Do you remember? I do.
A day in July
We were on the beach
our pants rolled to our calves
the sand and salt made our ankles itch
like lemon juice on a rash
and the sun made my cheeks burn
Our conversation was not heavy
We did not talk in circles to pull us down
Our words were light and I can't remember them
They were lost like the bubbles on a crashing wave
As the shadows lengthened
so did the focus of my eyes
I knew the day would end
My heart was in my throat
like sea-weed choking a tidal pool
I wanted to stay on that beach with you
until the sand covered our ankles
and the moon held the tide from us

Reading

Number of books read: 32
Best-liked books:
    James Herriot: *All Creatures Great and Small*
    Torey Hayden: *One Child, Somebody Else's Kids*
    Jean Auel: *Clan of the Cave Bear*
    Irene Hunt: *Up a Road Slowly*
    John Steinbeck: *Of Mice and Men*

Reader Response:
    One Child: This was a story of victory. Everything happened right and it was true. I am also reading *Eric.* In *One Child* the woman who is telling it, Torey Hayden, tells it with such innocence . . . she is not calling that much attention to herself. In *Eric* I get the impression that the mother Doris Lund, who wrote it, wanted her son Eric, and herself to be stars and famous. To have people think that they should pity them and realize that what these people did was so wonderful and courageous . . . she is humble but in a way that she wants everyone to notice how humble she is . . . in *One Child* that's avoided . . .

    Clan of the Cave Bear: One of the things that interested me the most was the part about memory . . . how the clan members, especially the medicine woman and religious leader have acquired all this memory and how it is something they were born with . . . the part where she wanders into the cave where the mogars were having their ceremony I didn't understand that well, like p. 427. "Ayla tasted the primordial forest again, then felt it turn to warm salt . . . this feeling of being and remembering the dawn of life. Her innermost and earliest levels matched . . . she felt the individuality of her cells and knew where they split and differentiated in the warm, nurturing waters still carried within her . . . "

I finished 495 pages . . . that's the longest book I've ever read. And I was totally hooked . . . the writer kept me wondering why she had come. And at the end you finally know why.

Call of the Wild: I swear I'm the most violent critic but one thing I always notice when Im reading books is the words the author uses. I noticed on page 173 and 167. He used ominisly (*ominously*). I know this is a long story and he uses a lot of other words twice somehow this one stuck out . . . the first time I wasn't really sure what it meant. so I stopped and pondered . . . when I read it the second time I remembered it more . . . I was thinking that Jack London should of used a different word but than I thought I should have known that word and *then* I thought he should write it for people who wouldnt know that word as well as people who do and t h e n I thought it just depends on what kind of reader he wants to write this for . . .

This is driving me CRAZY! How could you go out and buy so many good books at one time? There are so many books I want to read I'm getting paranoid I'll forget what books I want to read or else I won't be able to find them after schools out, and the scariest thing I'm worried I won't have enough time. I have only 76 years if I live till I'm 84. So now I'm going to abandon *Red Badge of Courage* because for two reasons. One is I'm going to be around all summer and it's my copy and two, right now I'm not really into it and I'm wasting so much time. some books I want to read (these are only a few) *The Catcher in the Rye* by J. D. Salinger, *Amish Adventure* by Barbara Smucker looks neat I love the amish. I loved the movie *Witness. The Pigman* Paul Zindel, *Nobody Else's Kids* by Torey Haydn . . . and *The Little Prince.*

Of Mice and Men: . . . so sad. I had no idea what I was getting into when I picked up the book but it looked short so I tried it. I love Steinbeck's style of writing. My mom said some people don't like it because it's short and not elaborate but I loved it. I could find enough poetry in there to make me see a place or feel a character . . . I think this books ending is the main part. I also think what mainly carried me through the begining was the character Lennie. I don't know how Steinbeck did this but I felt like I knew exactly how George felt about him. I loved him the same way . . . One significant thing I picked out was the relationship to an incident in the begining of the book and then to the end. A man in the bunkhouse had an old dog that had arthridis and another guy took him out and shot him because "he had to do it, it was the best thing" and then at the end George had to shoot Lennie . . . I wanted so bad for it to end happy but then after I thought about it the book wouldn't be a classic if it hadn't ended the way it had.

Self-Evaluation

"Black River" is my best piece because it's what I really feel. It's not something that I made up. It's honest. For me it's honest. I found the words that fit. There's nothing out of place.

Good writing is something that makes you feel something and that you remember. It makes you feel like you are there or that it happened to you. It's not like you have to be in it, but you can watch it happening.

It's difficult for me to evaluate my own writing. I've read it over so many times I don't know what it sounds like to the reader when they first read it. I don't know if it has the same meaning to a reader 'cause this is one thing I'll always remember, sledding at my house at night. And I don't think the reader will feel the same thing I do. If I say bad things about it, I feel like a martyr. If I say good things, I feel like I'm bragging.

There's a difference between enjoying and appreciating writing and reading. I *enjoy* reading Becky's writing. It's witty, sharp. I *appreciate* Karen's and Amanda's. Their writing is developed—sometimes even off-the-wall, like totally plucked from nowhere—so unpredictable . . . If you want to pick a piece of writing just to read, if you want to laugh or feel something, I think that's enjoying. But if you're really in a writing sense, if you really want to find a good piece of writing and feel like you've accomplished something by understanding it, then I think that's appreciating a piece of writing.

Jay

Jay is a model independent learner, who deals with all the idiosyncracies of adolescence with a rare sense of humor. She loves to learn and loves to live. She takes learning seriously, but not too seriously. She makes everyone around her find the joy in their lives, no matter what the circumstances.

This is her first experience in a process-oriented, student-centered classroom. Her discoveries about different classroom approaches are evident in her first piece of writing and her evaluations about what works for her in this classroom. She has a message for teachers.

Her letter to Josh is in response to his mom revealing his problem with drugs and ultimate transfer to a rehabilitation center. Jay's writing is for authentic reasons.

Socially, she gets along well with all peer groups, yet is her own person, often choosing the most extraordinary clothes or hairstyles. She is a fine musician and extremely interested in pursuing the visual arts. Her parents are educators.

Most Effective Writing

Leaf-rot, Biomes, and the Elastic Clause

Third period

"Joy, I am so sick of these stupid vocab sheets!" I whined, as Joy and Connie dragged their chair/desks over to mine in the back left corner of Mrs. Gifford's science room.

Nearly every other day, Mrs. Gifford gave us vocabulary sheets. We had to give the definitions of 25 unrelated words from the glossary of the seventh grade science book. The normal procedure was to find the word in the science book glossary and copy it, word for word, on the sheet. That had gotten Connie and Joy and me A+'s all year. But the last month of school had just started, and I was sick to death of the damn things.

Connie dropped a science book on my desk, and dropped into her seat. "Well, JJ, can ya' tell me what leaf-rot is?"

Joy interjected, "For 1,000 dollars, can you, JJ, tell me what leaf-rot is?"

"Yes, Chuck!" I said enthusiastically. "Leaf-rot is a small purple fuzzy creature with big yellow feet!"

"Yayy!" Joy shouted in a whisper, clapping. "No, seriously though, you could write it."

"Yeah, I bet when she grades she doesn't even read the answers! She just looks at the paper and if all the spaces are filled you get an A," Connie added.

"I don't know, on the first one? She might notice on the first one. But I'll put it in the middle, for biome."

"I know what!" Joy exclaimed. "Let's get everybody to do it!"

"Our buddies Mahesh and Lyle won't do it," Connie said, pointing to them.

"I don't know, Mahesh can't stand Mrs. Giff. She's the only teacher he won't brown-nose," I added.

We split up and circled around the room. Mrs. Gifford had left, who knows where. She never told us anything. I went and told Wade, Sonya, Tori, and Mahesh, although I never did find out if he really did it.

We went on with the dumb worksheet. Mrs. Gifford returned, not noticing the six staples Wade had put in her coffee mug. He's such a serious student. He probably didn't fill in any other definitions on the whole page.

Sixth period

"Here are some worksheets for you to do," Mr. Krueger, the French teacher turned world and history teacher, said, as he passed sheets to the first person in every row.

"Holy schmoke! Deja-vu!" I said, as I passed the sheets to Ryan McCaskill behind me.

"Connie," I said, poking her in the back with my pencil, "more vocab sheets!"

"I know, Mrs. Giff and Mr. Krueger are in cahoots! They're out to get us with stupid and useless vocabulary words!"

"I didn't think Mr. K. preferred women, and besides, she's married," Ryan quipped from behind me.

Connie giggled. Our whole World and History class had our private jokes on the gender preference of Mr. Krueger. It was seventh grade. What do you expect?

Connie and I pulled our seats around to Ryan, and we began to do the stupid sheets together. Ryan would look up vertical climate, Connie would look up Amendment 13, and I'd look up Guam. Then we'd copy from each other. And so we proceeded from 1:46 to 2:16 p.m.

It was last period on a Friday in the last month of school, so we were hyper, and they were still feeding us vocabulary sheets. So when I got to the Elastic Clause, I had no desire whatsoever to know what it was.

My mind began to wander back to Wednesday when I was talking with Jenny Smith, an old and worldly eighth grade friend. "So J-squared, just about time to be done with seventh grade, huh?"

"Yeah, I made it!" I cried, clenching my fists to the heavens. "God, I am so sick of Mr. K. and World and History class! How can the French teacher teach us anything about the Constitution and Micronesia?"

"He was probably at the Constitutional Convention, he's so decrepit."

"Maybe, but he definitely got his clothes, at least his rather noticeable ties, in the late 60's."

"I remember last year, I had Mrs. Two-minutes-people." She was referring to Mrs. Ostgaard, who, every morning, walked around two minutes before the homeroom bell, snarling, "Two minutes, people!" at everyone she passed.

" . . . and we're having our Constitution test. Of course I hadn't studied, so when I got to the Elastic Clause, I had no idea so I put my own definition."

Connie's voice brought me back to World and History, Friday afternoon. "JJ, did you find the Elastic Clause? Class is almost over, and we're almost done."

"Oh, yeah. The elastic clause is a clause dealing mainly with elastic, that was written by elastic farmers in northwest Oregon."

On Monday I received my papers back in science and World and History. I got an A+ on both. Not a thing was marked wrong.

DEAR JOSH,

I've had trouble starting this letter. I just don't know what to say. Maybe because I feel a bit betrayed. I don't really know what's going on. But this is your deal, and if you understand, that's what's important.

Nahanni was really upset. It took many hugs of reassurance and general love until she was somewhat back to normal. Then she left for Georgia for

April vacation, and I won't be able to talk to her for awhile. But I think she feels somewhat better.

Me? I don't know what I'm feeling. I'm sorry I never really tried to understand you. All I could do is what you asked me to, and correct your spelling. I took it for granted that you'd always be at school, in my life. If not for me, or Nahanni, for you. However, it's your turn now, to shape your own life. Take these broken wings, and learn to fly, I know you can.

You can make a new beginning for yourself. In your new world, you can't be categorized as one of "them." Take this opportunity to make a new Josh. But whatever you do, don't lose sight of who Josh is. The Josh I know, respect, and love. (You know the kind of love I mean, the trusting kind.) Don't make yourself to be something you're not. Know your boundaries. But maintain your goals. Do your best. (I sound like a gym teacher, but it has some real meaning.)

Take care, and thank you.

Love,
Jay

P.S. I've found a book for you. *The Runner* by Cynthia Voigt. It may not be *The Outsiders*, but it'll make you think. (God forbid a book should make you think.) Anyway, I think you would like it.

School daze

help
i am going to be
refracted and i can't
count a present progressive
is staring me in the face with verbs crying
to be conjugated algebraic
sentences are chasing
me as i run into a mercantilist
policy a phosphorescent light appears
in front of me trying
to help me in this mass
confusion i stare at it hoping
for enlightenment
but nothing
jumps out at me so i
run head on
into a quadratic
monomial factor a voice
out of nowhere echoes around me

"You're eighth graders now!"

then why do i feel as befuddled as ever
?

Reading

Number of books read: 58
Best-liked books:

> Harry Mazer: *Hey Kid! Does She Love Me?*
>
> Phyllis Reynolds Naylor: *The Year of the Gopher*
>
> John Rowe Townsend: *Noah's Castle*

John Bellairs: *The Spell of the Sorcerer's Skull, The Revenge of the Wizard's Ghost*

Nat Henthoff: *The Day They Came to Arrest the Book*

William Sleator: *Interstellar Pig*

Madeleine L'Engle: *A Ring of Endless Light, A House Like a Lotus*

Torey Hayden: *Murphy's Boy, The Sunflower Forest*

Anonymous: *Go Ask Alice*

Charlotte Perkins Gilman: *Herland*

William Goldman: *The Princess Bride*

Robert Cormier: *The Chocolate War*

Ellen Raskin: *The Westing Game*

Elie Wiesel: *Night*

Brett Lott: *The Man Who Owned Vermont*

Pamela Sargent: *Earthseed*

Roberto Innocenti: *Rose Blanche*

Forrest Carter: *The Education of Little Tree*

Reading Response:

Spell of the Sorcerer's Skull: I love John Bellairs. I've read the *Spell of the Sorcerers Skull* at least five times, and each time I love it more . . . I really like Johnny. I can really relate to him. Being shy and wanting to be spontaneous and loud, but too afraid of what everyone else will think. And he and I get along with almost any body. He with Fergie, me with anyone. . . . I also love Professor Childermass and all his crankiness. I wish John Bellairs would have put his fuss closet in this and other books. I remember being in absolute hysterics when I read about the Fuss Closet in *The Curse of the Blue Figurine.* It had real potential . . . I hope I get as scared reading it this time as I did the first time . . .

I really like writing in my reading log. Just being free to write what I think and feel about what I've read helps me to understand things I didn't when I was reading."

Dirk Gently's Holistic Detective Agency: I love Douglas Adam's style of writing! His comparisons are great! Like when Richard broke into Susan's flat and "Lots of little synapses deep inside his cerebral cortex all joined hands and started dancing around and singing nursery rhymes." And when Richard thought he saw a man carrying a dead body, but it was "probably just a sack of something nutritious" . . . I absolutely *love* the vocabulary Douglas Adams uses! Rancid, galumphing, bramble hacking expeditions, its great! I really laughed when Dirk was saying that impossible was not in his dictionary. In fact everything from herring to marmalade was not in his dictionary because in a fit of rage, the secretary had torn several of the pages out . . .

Why Me?: I wonder where this is set, what country? I know it is some place with English influence because of some of the vocabulary. They ordered jelly and sweetbiscuits, Sarah and Jane call their mother mum, and a lot of the dialogue "I don't want to get cross . . . I shall feel different . . . Don't talk such rubbish!" . . . I would like to hear more about Jane. How she feels about having a diabetic sister. Does she think Sarah is a whiner? Does she feel Sarah's gotten spoiled? Is she afraid of getting diabetes? Does she care at all? . . . I know what Jane thinks of Sarah. She doesn't like her. She resents the fact that Sarah gets what she wants because she's diabetic. They often fight and get violent, throwing each other around, literally. I don't like Sarah again. She has no self-control, and is hurting herself. Eating sweets all the time, then she gets tired of people telling her not to do it . . . She will not let herself be nice to Jane. If Jane says something nice, Sarah turns it into a fight, pointing out Jane's faults. Like the title suggests, I think Sarah wallows in self pity . . . I ended up liking Sarah after all. After she got her dog, Charlotte, she matured a lot. She didn't pick fights with Jane anymore. She was able to give herself injections and began to accept diabetes . . . this book really captured what it's like between 12 and 13. Sometimes you're moody and childish, sometimes you're happy and mature.

Let the Circle Be Unbroken: Between social studies, learning about South Africa and this book, I'm learning about segregation, the hard way. White people hating the blacks, for no good reason . . . Now I am feeling really sorry for TJ. In *Roll of Thunder, Hear My Cry* I couldn't feel too sorry for him because of the way he started being friends with the whites. But now, as he is practically on death row, I feel very sorry for the dumb jerk . . . GRRR!!! That makes me SOO MAD! TJ got sentenced to the death penalty. He's black. The jury, judge, lawyers and all witnesses except TJ were white . . . It is so unfair. I know that TJ did not kill that shopkeeper. Mr. Jamison, his lawyer, was right in saying that whites had trained blacks to do what they said and not to question it. I think RW and Melvin did that to TJ. He thought that since they were white whatever they did was ok. And that he should feel honored to be a part of their plan, legal or not. I think what RW and Melvin did to TJ was worse than awful, terrible, despicable or disgusting. To set up a boy for murder (a black boy killing a white man, no less) just because he is black. I can't think of a strong enough word. I ABSOLUTELY DESPISE RW AND MELVIN!

Herland: . . . a nice idea, a country made only of women. And no matter what Jeff, Terry or Van (3 main characters) think, I think the women are beautiful. Not stunningly gorgeous, some not even pretty, but beautiful people. They have made such a beautiful country without men! From the descriptions, I have a fabulous idea of what this country looks like. Neat and tidy, but welcoming, unless you are a man. If there were such a place as this, I'd love to go! . . . What an interesting history of a country! Van found out from some of the women and some books that most of the men were killed in a war, 2000 years before. Then an earth quake killed the rest. After the earthquake, the country was walled in. No one could get in or out. After several years one woman had 5 daughters. A "gift from the gods." Then her 5 daughters had 5 daughters each. Eventually only these type of women were left, no "normal" people and that's how it's been for 2000 years . . . It's weird how the 3 main characters are 3 opposite people. If all their characteristics were pooled, you would have a "regular person." But as it is, they are all rather one-sided . . . It's not that the women distrust the men, they've just forgotten what men are like. Since they have no need for men, they simply don't think about them. If a woman has a "true desire" for motherhood, she will be granted one daughter. If a woman really wants a child and solemnly vows to raise her child according to the country's methods, she is pregnated. If the "wisewomen" don't see the mother fit to raise the child alone they send other overmothers to help raise the daughter . . . Although I really liked it, it ended too soon. It just said, "We finally left Herland." I want to know what happened! All three men had married women. But Terry's way of treating his wife was found unfit by the "wise women" so he was asked to leave. Jeff decided to stay with his wife. But Ellador persuaded Van to leave with Jerry and to take her with him. Terry was unofficially divorced. But I want to know how Ellador did in the USA! Did they ever go back? Did they ever see Jeff again? Did Ellador love the rest of the world as much as she thought she would? I can't write to the author because the book was written just before World War I. What happened? . . . The author was a real feminist and wanted to show her beliefs on how much better a female dominated society would be. They don't distrust men, they just have no use for them."

The Chocolate War: The characters are beginning to come together now. So is the story. The first few chapters of this book were very hard for me. There was no relation between the characters at all. and I could not figure out what Archie and the Vigils were about. I know now that they give out these assignments to guys at school. I don't know why. Maybe to see if the kid will go against teachers? Are there consequences if a kid refuses? Why aren't they kicked out of school for what they do. *Everyone,* including Brother Leon knows what the Vigils do, so why don't they get punished? . . . Cormier's style of writing is very complex. He never says anything, he shows it all. He never outright said that Jerry's refusal to sell chocolates was an assignment from the Vigils. He just has Jerry thinking "Why did Archie pick me?" . . . Oh my god! How cruel! I didn't think that Archie would be that unforgiving! I know the Vigils didn't let anyone in the whole school with anything against them but simply refusing to sell the candy, that's not so terrible! Even after all the other guys in the school began selling again, except for Jerry, Archie still got his revenge. He is so inhuman. Kicking Jerry when he was down, in all aspects of Jerry's life, football, school, at home. But I still cannot hate Archie. I have to admire his talent for psychological attacks. He has an amazing talent for concocting such schemes. A good author, like Cormier, doesn't let you hate the antagonist. I did hate Janza though. He simply does not give up, thinks of no one but himself . . . all brute force. If someone had him at a psychological disadvantage, he would just hit them. With deadly force . . . That's sort of a weakness. A person can more easily bounce back from physical beating than from psychological . . . But if you really are a mastermind like Archie, it is harder to find someone who you can't beat.

Dicey's Song: I love Dicey's name. Dicey Tillerman, it has hard consonants, but mostly it flows. I wonder where Cynthia Voigt came up with Dicey. It's a beautiful name. I think I have tried to read this before, but couldn't stick with it. The same for *A Solitary Blue.* I couldnt read it, it was so slow. Nothing happened. That was about 5th grade, 3 years ago, ages. Now I can appreciate the gentle flow of a story that takes the time to smell the roses. Something doesn't have to be happening ALL THE TIME.

Many Waters: This is about the 40th time I've read this book. I remember we bought it last winter. We (my dad and I) waited in line for 30 minutes for Madeline L'Engle to sign it. And we were about 5th in line. We waited so long because everytime someone new came to Ms. L'Engle, she'd chat for awhile, sign the book, chat for awhile longer. And no one got upset about waiting so long. It was snowing outside and we were packed all snug inside Cover to Cover. (A bookstore about 5 blocks from my house.) And there is nothing cozier than a bookstore during a near blizzard at night. She was a pleasant woman. I remember gray hair falling from a loose knot at the nape of her neck. All black clothes and silver Hopi jewelry . . . You know, I don't think sending the flood and having Noah, his family and the animals repopulate the world worked, d'you? Crime rates, Aids, terrorists, drugs, rude cab drivers, pollution, the ozone hole, wars, etc. etc. If God's reasoning was that peoples' hearts were turning bitter, why hasn't he done anything recently? The anti-Christ? Or did God give up on humans? Or is he too busy trying to help individuals, he can't see the forest through the trees? I'm disappointed. Maybe he's leaving it up to man to do something. Great world I'm going to inherit.

Self-Evaluation:

My best piece was my latest, my worst was my first. I had the most experience possible in a school year behind the best piece . . . I've learned what I write best. Humor. I usually did it but I wasn't aware of it. I've learned that the best way for me to write is to just do it. Don't brainstorm, it doesn't help me organize my thoughts. Just write, then try to put them in order. If I put my mind to something, it gets done. If I want to write, I do.

I think you have to write about something that is important to you in order to write well. If you don't care about it, it won't be good. I learned that writing about things I miss, makes me miss them less. I learned also that small things, one thought, can grow into a big piece.

As I looked at my writing, I found that I liked my pieces where I didn't brainstorm. When I did brainstorm, the piece came out sounding like a list to me . . . You helped me by not being a teacher. Not telling me what I did wrong or write (Ha Ha. A joke) You told me what you *liked.*

For me a good piece of writing, like a good book, grabs your interest and doesn't let it go until it's over. And even then you're still thinking about it.

I know you shouldn't choose a book by its cover, but I do. If the blurb on the jacket interests me, I'll start reading it. The title also tells me a lot. I will not read *Toasters I Have Known,* or something dumb like that. If the author didn't take the time or the effort to find a good title, forget it.

The best books I read were best because I didn't read them, I felt them. I could be in the book.

I've found that juvenile books bore me. They're trite and stupid. I *can* read adult books. They don't completely baffle me. You can tell from my log entries, when I was reading juvenile humor, I was bored, the entries are boring. But entries about other books, other genres, I got more stimulated. My mind was working and my entries aren't boring.

When I read really good personal narrative, I tend to write personal narrative. When I try reading in new genres, sometimes I try writing in new genres. When I read less, I write less.

You know you have good writing if it reads well. If I read something I really like, a good word, good description, a good topic, I try to work it into my own writing. If you aren't a good reader, how will you know if a piece reads well? If you are not a good reader, you will probably not be a good writer.

Mrs. Rief, you helped me by not telling me what to do or how to do it. You helped by suggesting and recommending books, new ideas, and the questions you asked in my log. You were more a reliable source than a teacher.

John

John is a precocious young man, who arrives in class in September with a seventy-two-page book he has written entitled *The Punmakers*. He has twenty-eight pages of illustrations that stretch down two walls of my room. He advises me he is a "science fiction nut." I'm worried. I admit to John I don't do science fiction well. I don't write it. I don't read it. I don't enjoy it. And, worst of all, I don't understand it. John doesn't seem worried, even though he will have to teach me about science fiction.

John tries to teach me. He often spends recess in my room discussing the Ottoman Empire, Douglas Adams's humor, and his own book, *The Punmakers*. I would describe our one-sided conversations (John talks, I nod) as nothing less than "cerebral chats."

Despite his obsession with science fiction, John tries nonfiction at my suggestion. Ironically, he says his best piece of writing all year is a nonfiction piece because it's "honest."

Most Effective Writing

Mr. King

During my life in Maryland, when I was six or seven, there was a small cottage across the street. The cottage's boundary was marked off by a wooden fence that was entirely made of logs that were tapered on each end. A gravel driveway led to an unused garage.

In this cottage lived an old man with a sparse forest of silvery-grey hair on his head. The man's name was Mr. King. I never knew what his first name was, but at the time it didn't seem especially important.

Leading up to his door from the driveway, there was a path of cement stepping stones. What was unique and special was that when the "stones" had just been poured, Mr. King seems to have dumped on a multitude of colorful marbles, like irregularly planted seeds.

Almost every day, my brother and I would go to visit him. We would go toddling up the marbled walk, and we would visit with him for awhile. Somewhere, deep in the bowels of his six room house, he had cached a huge supply of Fruit Stripe chewing gum, the gum with the multi-colored bucktooth zebra with a disproportionate snout on the package. Mr. King would then, smiling, dole it out to us as a reward for visiting him, a pack at a time. Assuming there were five sticks to a package, my brother and I probably consumed over twelve hundred sticks of Fruit Stripe chewing gum while Mr. King was there.

Then one day, I don't remember exactly when, our supreme benefactor was gone. There was none of the blue smoke that was supposed to accompany the disappearance of a great being. Soon afterward, Bucky the Zebra was gone from the candy racks that lurk beside checkout counters in grocery stores. It seems that when Fruit Stripe chewing gum became obsolete, so did Mr. King.

Mr. King died of diabetes. Now it all fit into place. Mr. King gave us the sweets which he could not have himself. They tore up his marbles, and a new family moved in. The only trace they couldn't erase was my memory of Mr. King and his obsolete chewing gum.

AURORA

I waited patiently
watching the silent black Northern sky
dotted here and there with the pinpricks
that were the stars.
Then, all of a sudden
it happened.
The scintillating curtain of color
swept through the sky silently.
It offered thoughts of spring
far away on the Alaskan tundra.
The rain of beauty
then stopped,
fading off into the distance
behind the looming piney treeline.
I trudged again homeward, but
I no longer felt cold.
I felt greatly warmed
from deep within myself.
It was as if the skies had opened
to reveal Heaven.

Reading

Number of books read: 25+ ("LOOOOONG BOOKS")
Best-liked books:

James Thurber: *The Thurber Carnival*

Robert Asprin: The Myth Series

Douglas Adams: The Hitchhiker Series

Tom Clancy: *The Hunt for Red October*

Reader Response:

The Thurber Carnival: Thurber's "Snapshot of a Dog" is a bouillon cube version of Farley Mowat's *The Dog Who Wouldn't Be*. What I mean is, it's a lot smaller, it doesn't taste the same, and it is a lot stronger.

Notes from week-long camping trip: Now I truly comprehend the meanings of the expressions "soaking wet" and "freezing cold." At Lost River we went into the caves only because we expected them to be warm and dry. We discovered instead that water flows downhill and stays in a place with no downhill exits. I discovered that my jacket isn't waterproof, and that my opinion-forming brain cells are exceedingly warped by certain conditions: WET! COLD! DIRTY!

Field and Forest wasn't as bad as I expected it to be. I think that everyone despises me, however, because I answered so many of the questions. I learned however: how to use a sling cycrometer, how to find soil pH, how to tell soil depth and moisture . . .

My sleeping bag has drowned itself. As I bobbed myself to sleep, I deeply regretted not getting my sleeping bag dried. I dreamt of swimming the 40000 meter and just recently woke up feeling like a fungus of some obscure family and species found only in the lost world of Atlantis.

The piece that's my best is a nonfiction piece. It's my best because it's the most believable to me. It's real. It happened. It was important to me and I liked it the most.

My others are fiction. They're not as believable. They don't hold any feelings or memories for me, or for anyone else. If you let someone read it, and they don't like it, like five people or more, you know it doesn't work. It's not worth it to write a piece that only one person in the whole school likes.

Karen

Karen is a prolific writer and avid reader. By the end of eighth grade she has written more than twenty-two pieces of writing and read more than fifty books.

She reads as a writer—questioning, analyzing, predicting, wondering, experiencing, and enjoying. She has filled two notebooks in response to reading, wondering about the authors' intentions and reasons for writing what they write and how they write. She, too, works out her own experiences, feelings, and questions in her writing in an attempt to figure out where she belongs in this world.

Karen's poetry and self-evaluations are particularly astute. She knows herself well and is able and willing to articulate those beliefs through her writing.

Most Effective Writing

The nursing home (see chapter 1)

All the openings are round
A circle
    Laughs because
  A gold-flecked eagle
  Has just hatched.
The triangle
    Frowns
  Like a sharp-tongued governess
  Yelling at a child
    who won't eat spinach.
Then a hexagon
    Cheers for
  A young girl who
    has picked her first daisy.
A square
    Asks to belong
  In a world
    where all the openings are round.
The pentagon
    Screams
  When two thunderclouds
  Crash together
    with a crooked bolt of lightning.
An oval
    Dreams
  Of two pure white swans
    floating on a clear glass pond.
The last octagon
    Dives into
  The ocean's depths
    to find a small green spider.

Stars are born

The first star glimmered.
A mouse
  warmed his feet
 in front of a crackling fire.
A teddy bear floated on a breathless pond,
  sending out ripples of air.
Scarlet wheat waved goodby
  to a magenta goose.
A frog
  buried itself
 in the clear ice of a white immobile lake.
Clouds became mountains
  of mashed potatoes and the butter
 slipped slowly off the edge,
  falling on a silent green airplane.
A nymph,
  wrapped in layers
  of thin white gauze,
 changed to a sleek silver owl.
Cold air
  whistled
 as it flew through the child's frail frame.
The dream of a forest fire
  whipped through the dark green woods,
 turning the sky a brilliant orange-red.
  And the sky filled
 With the cold light of a million stars.

If

If boughs were leaves,
  and leaves were boughs,
  Would we wear hardhats in fall?
And if we called paper crayons,
  and crayons paper,
  Would we draw white sheets
    on waxy backgrounds?

If flowers were weeds,
  and weeds were flowers,
  Would our gardens be filled with thistles?
If yes was no,
  and no was yes,
  Would we be allowed to stay up late?
And if young was old,
  and old was young,
  Would babies have wrinkles
    and grandparents none?

(Untitled)

"Out!"
"B . . . b . . but."

"If you're not interested, you can leave!"

The door opened and I found myself suddenly sitting in the hall outside the science room. Embarrassment hit me like an iron fist. A gaggle of sixth graders passed by staring curiously. I felt myself turning fire-engine red, setting off my green sweater like a Christmas stocking.

Uneasiness grew as one of my friends placed herself in front of me like a 50 pound weight and asked what crime had been commited. Grinning sheepishly I admitted I had been caught doodling. The girl walked on, a slight, smug smile on her face. I considered how unjust my punishment was. After all, I was only doodling to relieve my anxiety. And believe me, I had plenty to worry about with Mrs. Anderson bawling out our science class.

"Out!"

I heard the familiar word bellowed even through the heavy door, and soon I was receiving company. As I curled up like an ostrich hiding its head in the sand, Zan came barreling out of the room. While Zan paced the hall, talking to anyone and everyone, I began to wish very hard for the earth to open and swallow me up.

Crossing my legs, I started to feel sick to my stomach as an 8th grade class came pouring out of a nearby room, talking and laughing. Lady Luck was with me for an instant. I had no good friends in the departing group.

Despondently I studied the little sketch that had been my undoing. Abruptly the door to the forbidden room opened once more. It was time to go! Shuffling into the classroom I gathered my belongings and scurried out, avoiding the fuming teacher completely. As I wandered up to English, the waves of hurt slowly began to recede.

Reading

Number of books read: 51
Best-liked books:

> Anne McCaffrey: *Dragonsong*
>
> Robert O'Brien: *Z for Zachariah*
>
> Alice Walker: *The Color Purple*
>
> Henry Simon: *The Victor Book of Opera*
>
> Barbara Cooney: *Miss Rumphius*
>
> M. M. Kaye: *The Ordinary Princess*

Reader Response:

Eight Cousins: Louisa May Alcott's books are fun to read because they combine an almost too sweet exterior with a slightly mischevious interior. Under all the morals that are obvious and not so obvious, lies the idea that one can have fun, simply. Rose makes a nice change from modern hero-ines. She shows that a person can be all girl and still have ideas of her own.

The Picture of Dorian Gray: I had one main question about this book. It was answered at the end. All through the story I wondered why Dorian hid the painting instead of destroying it. At the end of the book Gray, too, hits upon the idea, while studying the painting and contemplating about what it has done to his life. Dorian takes the knife he used to kill Basil and stabs at the painting. The servents then heard a shriek. Here comes the wonderful twist ending. When Dorian is found, he is an ugly old man while the painting is once again a portrait of Dorian in all his wonderful youth. I thought this was one of the best endings I have ever read.

Early Poems of Robert Frost: I'm reading slowly because I want to get the full impact of his poems . . . I think one or two of my poems are as good as one or two of his beginning poems. Many of his early poems are amaturish, and very different from his later poems, such as "Death of a Hired Man," which I find extremely good. I like the symbolism in that poem. You know something ominous is going to happen.

The Talisman: Stephen King is a master of horror . . . The images King writes of live in my head. I wonder what his motive is for writing this type of book. Is King mentally disturbed? Does he take sadistic pleasure in turning blood to ice? Or does he sense that people want to be scared and play up to our unlikely need? I feel that this is most likely. I personally enjoy being scared. King delves into the deep recesses of a person's mind and brings to life all the hidden terrors and horrifying memories. I think that one of King's books is a great cure for any case of guilt because one feels thay can't be as awful as the characters in the book.

The Color Purple: Celie begins to love Shug. As the relationship develops I wonder if Celie will find out who she is. I think she might be lesbian, but even so it might be good, because if Shug responds it means Celie will finally get some love in her life . . . I'm glad Celie was happy at the end of the book, I mean I wanted her to be. The only problem is that she seems almost too happy. My idea is that the author exaggerrated Celie's happiness at the end of the book so that it would be in sharper contrast to her misery at the beginning of it . . . I was right. Celie is a lesbian. If anything this adds to her personality . . . it allows the author to explore the topic. Instead of seeming revolting, it seems to be a beautiful part of Celie's life, mainly because it brings love into her life for the first time.

Clan of the Cave Bear: You know, it's too bad. Jean Auel lost her plot in that last book (*Valley of the Horses*) and had to resort to sex to keep the reader involved.

Parents and teachers aren't too smart. Instead of banning books, which makes all kids want to read them, put them on a required list. Kids wouldn't touch 'em.

Self-Evaluation:

To write well you have to be able to relate to the writing. You have to have an idea of what you're talking about because if you're trying to write about something you know absolutely nothing about, it's going to sound like you know absolutely nothing about what you're talking about. The main thing is you've got to be able to feel what you're writing.

My poetry surprised me. I didn't think I was able to express myself and show me so well. "All the Openings Are Round" is one of my best pieces because it shows a lot of me in it. That's a reason other people may not think it's so good, but I understand it all. People who know me are going to like it better because they know it shows more of me.

My worst pieces of writing I wrote when I was not in a writing mood. I don't even want anybody to hear this last piece it is *so bad.* I was trying to get something out and it didn't work. It didn't show any of me. It really stunk. It sounded as if I was being really self-pitying.

[I've noticed] my endings have changed the most. In elementary school I'd write reports and the teachers would say "Karen, change your ending." All my endings sounded the same, like "in conclusion," your basic ending. This year though I'm starting to get my endings better because I'm trying new things, new kinds of writing, and it's changing the endings.

Matt

Matt is well liked by the majority of his peers, despite the fact he is an integral member of a group of students who take little interest in academics, but find "antagonizing the system" their main agenda. He "hangs out" more with older students from the high school than with his peers. In a school setting, Matt is unmotivated and often disinterested.

However, Matt is an articulate, caring, sensitive young man, who knows a lot from life experiences. He is one of the best responders to others' writing, is seldom disruptive in class, and contributes well orally.

He has an older brother he emulates and lives with his mom and dad.

Most Effective Writing

(Untitled)

After you hear what I have to share, I hope you don't think you read a book about a bunch of losers or vandals. It's more like Adam taking the fruit from the tree when the serpent tempted him. That's what happened to me, the serpent being temptation, exposure and curiosity.

The serpent promised Adam power. I was promised the Garden of Eden, which symbolized popularity, recognition and a little authority. Adam's plan basically backfired. He lost purity and gained shame. My path wasn't as easy as picking a fruit off a branch like Adam did. I had a long and rough road with many obstacles along the way. Me? I'm not always sure what I lost or gained . . .

"Hold this, Matt. I'm going for the jump."

"You're too stupid to jump," replied Dan. "I'd call it more like falling."

Dave threw me a round object. It had a picture of a bear on it. It said in white letters KODIAK SMOKELESS TOBACCO. It also said, WARNING: THIS PRODUCT MAY CAUSE CANCER AND TOOTH LOSS. But I blocked that out, as I was dying to try it.

I put the tin of Kodiak in my cutoff jeans. I could only put it in one pocket. All the others had rips in them. Suddenly something hit me. I was always the leader of the playground. I did what I wanted and not what other people wanted me to do. Now I was going to try dipping because Dave was. I was wearing cutoff jeans because Dan was. Maybe it was cause I wanted to. At least that's what I wanted to think. Or maybe it was 'cause things at the time weren't going too good. Maybe it was 'cause my former girlfriend was going out with my brother. That really sucked.

Oh well, I had to think and there was no better place than in the water. But first I would ask for some chew. There was another reason I wanted to try it. My brother Jake smoked and I figured if I chewed he would stop smoking to try to turn me away from chewing.

Dave was swimming toward shore. He yelled, "That was a trip. Now I need some dip." That must be another word for chew, I thought.

Dave strutted over, all six foot two of him. He looked kind of goofy walking. "Break out the dip," Dave shouted.

I opened the chew. It looked like the peat moss fertilizer my mom used in the garden every year. Dave put three fingers in the container and took some and loaded it in the front of his lip. It stuck out like someone just socked him and he had a fat lip.

I said, "How do you do this shit, anyhow?"

"Little Matt wants to try," Dave said, acting like he had something over my head.

"Yeah, I do," I said, sticking my fingers into the container. I wasn't about to wait for instructions and let Dave get higher on himself. I stuffed the dip in the front of my lip. I observed Dave. He spit a brownish liquid caused by the chew. I spit. It was brownish. I must be doing something right, I thought. It felt like my lips were on fire. I had heard the first few times you did it, it felt like it was a clear day with no clouds in the sky. But after two minutes, it felt like my head was a cloud. I was very lightheaded, to the point I had to sit down on the railrod tracks.

Dan and my brother were still in the water swimming. Dave sat down near me. "Boy, am I dizzy," he said. This was a load off my chest because I didn't want to think I was the only one who was dizzy. You know what I'm saying?

"Time to go," yelled Jake. We walked towards the jeep. Dave picked up an empty can. "Come on, Matt. They're leaving." We ran for the jeep. I took out my dip. Dave kept his and spit in the can. I threw mine on the ground as we left.

Mr. and Mrs. MacBeth—the perfect marriage?

I'm fed up with the whole thing. How long are we going to drag this out? You know we all think that MacBeth is pond scum and Mrs. MacBeth is a wench. They should have gone to a marriage shrink long ago. It's apparent after reading the book, watching the movies, and the play, we don't think they're the world's best couple.

Reading

Number of books read: Doesn't list any
Best liked books: Doesn't list any

Reader Response:

March 29: I read a book called *Snakes of the World*. It tells about different snakes and how they adapt to different environments and in captivity. I learned that a ball python is probably the best pet snake to own. This is because it is gentle-natured and stays (grows) to 6 feet or under. Another book I have been reading is *Owning a Python*. It tells how to make a comfortable situation for the snake. It gave me the idea of changing my closet into a home for my snake. The whole thing will cost about $200. It's worth it. I love snakes. Another book I've been reading is *Snakes of North America*, which tells about the snake in our area and others. I've also been reading some Longfellow. He writes about Indian legends. It's pretty neat . . . I haven't got deep into the legends. It's still telling about the Song of Hiawatha, or something like that. It's an old book.

April 2: I've been reading another book on snakes, on pet snakes and on some of the diseases like scalemites. I learned that there needs to be a proper ventilation system in a snake's cage. The cage must also have a large dish of bathing water and another for drinking. These must be changed daily. The snake must also have a scenic view so it won't go blind . . . I learned stuff about poisonous snakes also. Some collectors keep them as pets. STUPID! They say they make bad pets. If you want to hold them you have to drain their poison . . . They say ball pythons make good pets and so do reticulating pythons. But if you hold a reticulating python wrong, it is likely to hurt you or kill, seeing they're up to 32 feet of muscle overall. I would say the ball python is the best. They only get to 5 feet. They are thick and it is all muscle. The reticulating python got its name from a Mexican form of sewing . . .

"House upon the hill, Moon is lying still." (The Doors)

April 3: I read about owning a python. The kid was 13 and had a normal bank account a 13-year-old has. (Not much.) Well, his friends thought the snake was neat so they got road kills for the snake. It died of Rotten Mouth. It told of other dangers too. It also clearly said to keep the snake away from babies (no kidding!). It told about building a suitable place for the snake. The one I plan to build is just fine.

"The outlaws live on the side of the lake. the minister's daughter's in love with the snake." (The Doors)

I've also read some more of Longfellow. He told about a man and a bear wrestling. I finished the part about the Song of Hiawatha . . .

Self-Evaluation

I write about stuff that I can remember real clear and that sticks in my mind . . . true things. The writing periods we had in class helped the most, where you can write about anything. You don't say it has to be fiction or nonfiction. Some people write plays, some write fiction. I write nonfiction.

[My writing] isn't easy to finish because there's so much more that goes on with each story I told. So much telling why something happened or didn't happen. There's always something that needs to be in the front or the back of it.

Reading helps my writing . . . I started this book *Single Season* . . . at the very beginning there was a quote from Led Zeppelin and I got a good idea out of it . . . I get my ideas out of anything. Sometimes you get good ideas from books . . . I don't write with much punctuation and that's probably because I don't read enough.

Comments

For the entire year, Matt handed in only what's here—with the exception of three additional responses to reading. For purposes of clarity, I edited his spelling and punctuation. I never changed a word or the order of words. The introduction to his story originally looked like this:

after you here wate I have to share I hope you don't think you read a book about a bunch of loosers or vandles its More like Adam taking the frut of the tree when the serpent tempted him Thats wate happend to me the serpent being temptation exposior and cureosity The serpent promesed adam power I was promised the garden of Eden that simballised pupularity recanishon and a little athority Adames plan basicelly backfired . . .

During the year I offered Matt all kinds of options so he could succeed. I based my expectations on what he told me he could handle. He opted to write half as much. He chose to write a letter to me once a week in response to reading. He said he could only read at home, but it was too much trouble to write it down on a list. He still seldom handed anything in. He stayed in at recess. We signed weekly contracts with each other. He wrote nothing. I told him he needed to write only in class. The minute I left his side, he stopped.

I praised what he did. Perhaps too much. I helped him edit his words, telling him what he had to say was as well written as S. E. Hinton's writing. Even, or especially, the lavish praise didn't help. Nothing worked.

In mid-April I wrote a narrative evaluation to Matt's parents:

> In September of this year, after hearing a few pages of Matt's writing, I fully expected by this time of the year he would have completed a book—his first novel. I mean that in all seriousness. Matt's writing is lively, realistic, filled with a natural voice that invites readers to listen. He has a story to tell and a unique way of saying it. His writing is filled with natural metaphors about real people in real situations. He knows how to tell a story. With Matt it is a natural talent.
>
> But Matt chose not to use that talent. I would have been excited to work with him in developing his writing for publication. I think Matt wasted an opportunity. Other students accepted the invitation. Yet none has the writing ability that comes naturally to Matt. My offer stands with him. Whenever he decides to write and get some help, I would love to work with him. But, the initiative and the commitment have to come from Matt.
>
> What disappoints me the most is that Matt didn't find the class valuable enough to participate in. He didn't recognize the resources in other students. There is so much he could have learned this year about development of writing and the mechanics of the craft. He already knows the hard part—getting meaningful words down on paper. I know he reads at home about subjects he cares about. I wish he could have cared enough to find a way to talk to me through writing about that reading. I wish I could have found one book that really engaged him.
>
> In class Matt was a good responder to the play *MacBeth*. He understood the concepts, the characters, and their evil motives well. He asked and answered good questions. He is an honest person, who, for the most part, is sensitive to his peers and caring about the feelings of others. I enjoyed having him in class. I just feel really badly that I couldn't reach him as a reader or writer . . . Perhaps the best thing he could do to help other students like him, is to tell me what I could have done differently.

I never heard from Matt's parents. Perhaps they never received the letter. During the fall, after eighth grade, I wrote Matt a letter asking how things were going at the high school, did he need any help with any writing, and asking again why things didn't work for him in eighth grade. I never heard from him. He sends "Tell Mrs. Rief 'hi!'" with other students. I send the greeting back. I wish I knew how to reach students like Matt. I wish he'd tell me.

Sandy

The first week of school Sandy chooses to read Maslow's *Toward a Psychology of Being*. She is on intellectual overflow and I have no idea how to respond. In her log I write mundane comments like: "Nicely said!" "Good thinking!" or ask questions like: "What makes certain parts stick with you?" I am desperate. After two weeks I am more honest: "Extraordinary writing, Sandy! I hope I can keep up with you. You are teaching me . . . I'd love to see your poem. How can you say you're not a writer? Abstract thoughts are always more difficult than the concrete. Have you ever read any of Paulo Freire's work? He has a lot to say about educational psychology."

When Sandy is home from school, she likes to spend her time alone, usually reading (her favorite author is Erich Fromm), writing, listening to music, drawing, or riding her bike. When she can, she likes to go see films and attend lectures discussing current world issues.

At school, Sandy is more comfortable with teachers than with her peers. Her choices of books and topics for writing are so *heavy* that I work hard at getting her to take life less seriously. I move her seat to a table where students are serious about learning, but can laugh with each other too. I want Sandy to laugh more often. In May she does, over her own unique interpretation and response to a social studies project: *Dick and Jane Visit the Massachusetts Bay Colony* (Figure A.1).

Most Effective Writing

A QUESTION OF TIME
(response to *Romeo and Juliet*)

There's no need,
no time for reconciliation
between hatreds—
the frost that kills the flower.
A plague on both houses
is fatal justice.
Running through a labyrinth of
misunderstanding
out of breath
out of patience
out of time.
They stumble that run fast.
There is no escape
from rips in the fabric of fate
as it unravels, it becomes
deaf to peace.
It's a question of time
it's running out,
it won't be long.
It's easy to make
the mistake
 of letting go,
but passion lends power.
Like the lightning

that illuminates the sky for just a moment,
love cut short is a flicker;
and when night's candles are burnt out,
blazing love is extinguished by eternal dark.
It's a question of time.
It's always too late.

(Conclusion of untitled essay written after visiting a local nursing home)

I think I have gained some compassion for the elderly. I don't want to end up in a nursing home where everyone *exists* but doesn't *live*: where there are no goals to strive for, nothing to live for. In our youth-oriented culture, people are afraid of getting old. This fear leads to an intolerance of the elderly and their needs at a time when being with family is most important. I think our society needs to gain a better understanding of what the elderly can offer and implement ways in which the elderly can become valued participants in our culture.

Afternoon garden

Sitting in my private garden
around the kitchen table
not noticing the sharp movements of
preparation
or the smell of wool and
the cooking turkey mixed with anticipation,
absorbed by the no-wax linoleum,
the hard wooden chairs have become
a part of me and
if I close my eyes, up and down are
erased like a mistake.
But the sound of hollow happiness
and greeting
remind me that nothing is real.

(Untitled)

A is for apartheid
B is for bomb
C is for Cambodia and Vietnam
D is for destruction
E El Salvador
and
FBI fascists fighting for more
G is for Grenada
H Hiroshima
and
I have come to bring your subpoena
J is for the Jews
K the KGB
L is for the lives that were lost to greed
M was a mistake, and so was
Nagasaki
blown into
Oblivion with a walkie-talkie
P is for Pakistan
Q is for the Queen

R is for religion and all that deceives
S is for secrets
T is for TV
and we're
Under attack and Under siege
V is Vietnam (I know I said it before)
and the last one I can think of is
W for war
now
X
Y and
Z are our last hope
let's get it together before we go up in smoke

Figure A.1 Sandy's unique interpretation of a social studies project

continued

THIS IS JANE

DICK AND JANE LEARN ABOUT THE MASSACHUSETTS BAY COLONY FROM TV.

DICK AND JANE THINK IT WOULD BE FUN TO SEE IT FOR THEMSELVES

DICK AND JANE BUILD A TIME TRAV'LER WITH A SPATULA, MACARONI, PAPER MACHÉ ... AND THE RED WAGON

Figure A.1 Continued Sandy's unique interpretation of a social studies project

THE TIME TRAV'LER
GOES FAST-
ZOOM! ZOOM!
really FAST

THE TIME TRAV'LER
STOPS IN 1636, IN
CAMBRIDGE, MA. OUTSIDE
A COURTHOUSE.

GO INSIDE!
GO INSIDE!

DICK AND JANE GO
INSIDE.

DICK AND JANE
SEE A LADY. SHE
IS CRYING.
DICK AND JANE
REMEMBER HER
FROM TV. HER NAME
IS ANNE HUTCHINSON.

ANNE TELLS HER FRIENDS
THAT THEY DO NOT NEED
RULES BECAUSE THEY KNOW
GOOD FROM BAD. THIS
MAKES THE MEN WITH
POWER VERY ANGRY. THIS
MAKES THE MEN WITH POWER
THINK OF ROGER WILLIAMS. THE
MEN WITH POWER SEND
ROGER AWAY FOR SAYING
THINGS THEY DO NOT LIKE.
ROGER GOES TO RHODE
ISLAND TO GET AWAY.
GET AWAY, ROGER!
GET AWAY!

continued

THE MEN WITH POWER ARE SAYING BAD WORDS TO ANNE. THE MEN WITH POWER ARE MAKING ANNE SAY BAD THINGS SHE HAS DONE. THE MEN WITH POWER SAY ANNE CANNOT LIVE WITH THEM ANYMORE. THE MEN WITH POWER SAY ANNE MUST GO.

RUN, ANNE, RUN! RUN TO RHODE ISLAND!

ANNE RUNS TO RHODE ISLAND WITH HER FRIENDS. ANNE IS SAFE.

DICK AND JANE ARE HAPPY THAT ANNE IS SAFE.

DICK AND JANE GO HOME IN THEIR TIME TRAV'LER AND ARE GLAD THAT THEY CAN SAY WHAT THEY WANT AND NOT BE PUNISHED.

THE END

Figure A.1 Continued: Sandy's unique interpretation of a social studies project

Reading

Number of books read: 19
Best-liked books:

Jacker: *Black Flag of Anarchy*

Unamuno: *Three Exemplary Novels*

Hesse: *Siddhartha*

Moore: *Emperor of Ice Cream*

Uris: *Trinity*

Koestler: *Darkness at Noon*

Reader Response:

Toward a Psychology of Being (Maslow): In fourth grade I read a book called the Vandal which I belive can be found in the Middle School Library. I don't know if I conciously understood it, but I understood it enough to have to return it before finishing it because I was plain scared. I wanted to be safe in my ignorance, like I would hang myself with knowledge if I read it . . .

" . . . the ability to abstract without giving up concreteness and the ability to be concrete without giving up abstractness . . . the strong tendency is abstracting to relate the aspects of the object to our linguistic systems. This makes special troubles because language is a secondary rather than a primary process . . . because it deals with external reality rather than psychic reality, with the conscious rather than the unconscious." I have always felt this way and that's why I'm never happy with the things I write because I can't convey how I feel. In only one instance have I accomplished that, in a poem I wrote called An Afternoon in Utopia. I have always been fascinated by linguistics. since 5th grade I've always wondered how language affects the minds ability to organize thoughts, memories, feelings and how it affects the culture in which it is spoken. I always have trouble expressing myself, like I can't talk fast enough, and sometimes I talk backwards, or I'll use words to mean what I want them to mean and not how their supposed to be used.

The Struggle for Black Equality: I got this book from Mr. Allen after "mock trials" last thursday. I think we were talking about the stuff that somehow gets left out of the American history books, the kind of stuff that people don't think really happens in America (the Sacco-Vanzetti case, or the Haymarket Riot), and even after you learn about it, you still find it hard to believe. So I was saying that my dad saw Malcolm X at the first talk he gave for white people to attend at BU, and that he thought Malcolm X would have been a good leader for our country if he had lived, 'cause as he got older he mellowed out a bit. I don't remember exactly how this fit into the conversation, but a lot of the people that we associate with the Civil Rights movement as being big advocates of it, were really interested in helping blacks get their rights, but more interested in their career in politics and keeping everyone happy. A lot of southern towns completely disregarded Federal law and fought peaceful demonstrations with fire-hoses, police dogs and water hoses that pumped water at 100 pounds of pressure. Martin Luther King, Jr. accomplished a lot with the peaceful demonstrations, but the inner city blacks up north were a bit more agressive like in Watts. I read about the riots in Watts last year in *Rivers of Blood, Years of Darkness* by Robert Corot. One thing it mentioned in this book that I don't remember in Corot's was although nothing was planned like the marches and sit-ins, the blacks for the most part, only destroyed stores and buildings and facilities that were run or for whites and rascists; they left librarys and things left open for them, respected property of institutions that had helped them and apartments where the inhabitants had written "soul brothers" on the windows. It's hard to imagine all of this really happening since everyone in this town is pretty comfortable. My mom, when she was in college, belonged to a sorority for awhile so she could meet people. She had two friends who wanted to be in the sorority, but never were accepted into it since one was black and one was a jew. My mom was really put off by this so she tried to quit the sorority . . . I can't relate to that kind of discrimination. It's never happened to me that I can remember, and I don't see it around me, so I can't imagine what it's like. I can feel for blacks and other minoritys who are discriminated against, but I can't put myself in their shoes.

Darkness at Noon: The most striking sentence so far, to me, has been on page 23. It's not a sentence I see now, but a train of thought of the protagonist, Rubashov. Rubashov is being held in prison as a political captive. "He tried to hold on to the the hatred he had for a few minutes felt for the officer with the scar, he thought it might stiffen him for the coming struggle. Instead, he fell once more under the familiar and fatal constraint to put himself in the position of his opponent, and to see the scene through the others' eyes . . . The old disease . . . Revolutionaries should not think through other peoples' minds. Or, perhaps they should? Or even ought to? How can one change the world if one identifies oneself with everybody? How else can one change it? He who understands and forgives—where would he find a motive to act? Where would he not?" . . . This reminds me of myself. If I think I have an answer to something, I often try to argue the other side, which usually leads me into more confusion.

The Inner Nazi (a critical analysis by Hans Staudinger): The first belief I had that was disspelled by this book was that Hitler's expansionist ideas were motivated soley by his lust for power. This book pointed out that if power was his only motive, he would not have invaded every country he had the opportunity to, but instead would've held on to what area he could securely dictate. (Most people would agree that Germany alone is a pretty good catch.) Hitler knew he could easily control Germany without worry, but he decided to go on, to attempt to conquer other countries, explained in this book by his dedication to his Nazi ideology. He really believed in what he was doing . . . I wonder why I am so drawn to Hitler's insanities, and not the Jews' suffering? It's not that I sympathize with the Nazis, but I think it has something to do with "evil" being more interesting than good. I think I've gotten a bit apathetic towards melodramatic stories and this doesn't exclude WWII stories. I know it was real and all that, but the stories all sound the same after a while. After WWII, everyone said, "Oh, that won't happen again, we won't let it" but it is happening, and it's happening all over the world (S. Africa). It really makes me wonder, when Americans come together, we write letters, make phone calls, demonstrate, petition and protest when Coke changes their formula, but we can't end apartheid in South Africa.

Darkness at Noon: Rubashov's talks with No. 402 are interesting. One especially meaningful one to me is on page 126. It starts with Rubashov and alternates between him and No. 402.

R: I am capitulating.
402: I'd rather hang.
R: Each according to his own kind.
402: I was inclined to consider you an exception. Have you no spark of honour left?
R: Our ideas of honour differ.
402: Honour is to live and die for one's belief.
R: Honour is to be useful without vanity.
402: Honour is decency—not usefulness.
R: What is decency?
402: Something your kind will never understand.
R: We have replaced decency by reason.

You see, when I was younger—say ten—I definately thought that it was much better to live and be useful than to die and cease to be useful. It seemed pretty reasonable to think that, after all, what good are you to your cause once you are dead? Unless you are a very prominent figure in society, no one is really going to take note of your death. Well, that's what I thought. Now I don't know. Whether anyone takes notice of your death I think is irrelevant, it's what *you* believe. If you sacrifice what you believe to live, than can you say you really believe in it? If you can't, than what you "believe" is merely a matter of convenience, isn't it? What separates man from the animals is his conscience. If a man doesn't follow his conscience, is he really a man? It all becomes a matter of survival, like the animals . . .

Self-Evaluation

I chose my best pieces of writing based on the amount of positive feedback I received from teachers and peers. I devoted a lot of time to working out the problems that inevitably came up while writing the pieces, and so I feel good about them. 'Afternoon Garden' was chosen out of pride. It was the first real poem I wrote this year. I received positive feedback from Elizabeth and others, which made me more confident about the worthiness of it.

Good writing should be thoughtful and the idea behind it should be direct enough to get it. I enjoy writing that raises questions in my mind, but good writing does not necessarily have to do that.

What helped me the most to become a better writer was a demand for a quantity of writing each trimester, which forced me to keep writing.

When I "finish" a piece of writing, there's always a funny feeling about it. As soon as I do it, it's no longer the best I can do. I can't think of what to do to add to it to improve it, but I'm never quite satisfied with it. It seems to evaporate on contact, making way for the next piece of writing. I don't save much of my writing. I only have one piece at the most that I keep from each year.

I save my journals . . . I have my journal from last year, and it serves as a source of ideas for writing, and insight into how much I've changed. I found in it dreams that had been forgotten and lies I thought I would forget. The interesting thing is, it was the truth I forgot; the lies I remember still. Not real earth-shattering lies, just an unsucessful attempt at self-deception. I guess that's why I write, if that's a reason . . .

Writing is also a release for me, a release of old ideas. Sometimes I get really dogmatic about something, then I write about it. When I see it all in perspective then I can see how dogmatic I was being. Or if I'm clinging to ideas because I don't have a better one to replace it with. I don't want to be without a protective coat of morals, so I write and can let go of the ideas. I write to get things out of my system. If I really hate my parents, I'll write to get it out.

I read to learn. I don't read for pleasure in itself, but I get pleasure from learning. The best books I read made me think of myself, and made me think.

I don't know how my writing affects my reading. They seem very separate right now to me. In November I wrote some BS about my reading/writing. (I think one thing that has helped me as a writer is the concious connection between reading and writing. When I read something I like, I now think "What makes this good? How could I bring this into my own writing?" When it loses my interest, I think "What makes this so boring? Do I do this in my own writing? How can I change it?") When I wrote that, I just said that stuff because I thought it sounded good, and I was having fun trying to sound sincere. Maybe I subconciously think that stuff.

Scott

Scott does not take learning seriously. He is willing to settle for mediocrity: so long as he maintains a B and so long as he's done. Scott's writing remains undeveloped from initial draft to finished product. When I ask in conferences, "How can I help?" Scott's reply is often, "You can't." If he's not ready, I can't. So I walk away. When I hear that response too often, I remain at his side. "Would you read what you have to me? . . . I'd like you to try this . . . " Sometimes he tries, often he doesn't.

His response to books tends to be lazy, more of a brief summary than a reaction. Evaluation lacks much self-reflection. "I think that my spelling is really bad and I need to work on it alot. I believe that I also need to write more and make better final drafts." Even here I ask, "What makes a good piece of writing? What can you do to make yours better?"

In Scott's log I try to push his thinking with carefully worded questions that focus on one thing at a time. In response to *One Child* I write: "I like the way you compared Torey Hayden to Ms. Brown. What does Torey do that works with Sheila? What has Sheila been subjected to in her young life that might make her act the way she does?"

My response as he's reading *Roll of Thunder, Hear My Cry:* "It's inconceivable to me that millions of So. Africans *believe* in white supremacy/apartheid! In what ways does this book remind you of the movie *Cry Freedom?*"

Scott often ignores my questions. He has to be given the time to answer them as soon as I hand the log back, or they will never get answered. After reiterating some of my questions ("Why do you think there is still apartheid in So. Africa? What took so long for equality in America? Why do Blacks still have problems here?) and reminding him of my questions for several weeks, Scott finally responds.

I think the reason there is still apartheid is that when they started it the Blacks hadn't had a chance to learn so they weren't that smart. But now the leaders of South Africa are seeing that everywhere in the world that the race of someone has no bearing on there intelligents. I personally think they are scared that the blacks will take over the country. I think that if they just gave them a chance they could prove to be usefull in helping the country grow. But since they won't give the blacks a chance they will never find out. The reason I think it took the U.S. so long for the exact same reason and I think that reason is really dumb.

Despite his lack of drive at this point in his life, he has a wonderful sense of humor; he keeps everyone laughing, as he pokes fun at himself— never at anyone else's expense.

I continuously push Scott to write what he knows about—himself and what he cares about. It takes him all year to accept the advice. What takes so long, I don't know and he can't articulate it.

Once Scott starts writing for himself—things that matter and hold meaning for him—the writing changes. The writing is not provocative but his sense of humor and unique voice is evident in his best writing.

Hockey

The whistle blew echoing through the rink. I glanced through the stands—about ten people. I felt lonely like no one cared. Then a sigh of relief ran through my body. Only ten people to see me if I messed up. Again the whistle blew. I looked at the ref with his black pants and striped polyester shirt standing at center ice. I skated toward him and went into my position.

I hit my man in the leg with my stick. He did the same to retaliate. In the few seconds it took to face off, I already disliked the man I was playing against. The puck flashed to the ice—the action had started! I skated my hardest. I think I always skate my hardest at the beginning of games. I made the first hit on my man. He may have been four inches and thirty pounds more than me but he was not as stable on his skates. I hit him solid and then rose up with my shoulder digging into his gut. I gave a mean short smile as satisfaction filled my body.

My ego trip did not last long, however, for I was soon sandwiched by two guys. As I crumpled to the ground, I could feel the energy leave my body. I struggled to get up, not really sure of what was going on or where I was. This always happens after someone puts a good hit on me. My head was throbbing and I was dizzy. I regrouped and managed to sight the bench. I pushed off with my left leg—the stronger of the two. As I approached the bench, I mumbled and then began to yell my position. The coach looked preoccupied so I began to look for Chris, another kid on the team, who plays the same position. I yelled to him and I knew he heard me because he was scrambling to get his gloves on and get out on the ice. I rammed into the boards at full speed. Somehow I thought I could do a half flip in midair and land on my feet. I squirmed in my short flight to do so but abruptly landed head first into the bench. No one looked—they were all content with watching the game.

I fidgeted on the back bench. I found it hard with all of my equipment on to fit the straw of the water bottle into one of the vents in my helmet. I squeezed the bottle hard because there was little water left in it. The water felt good, clean and refreshing. The water ran down my face and took a jump off my nose. I tried to get some of it in my mouth but I found it hard with my mouth guard in. I didn't really watch the game. I was dazed, catching my breath, waiting for the next shift, the next hit . . .

Eating blues

Characters: Stacy, Debbie, The Voice

Place: Digestive System

Time: Night

Setting: Debbie and Stacy met each other at a restaurant and pigged out. They are now dreaming the same dream with each other in the other's dream. They are getting a tour and lecture on why they should not pig out. They are now traveling through the human body.

Stacy: (sees large intestine) Whoa! Watch out, Debbie, large intestines headed your way.

Debbie: (jumps large intestines) Thanks, Stacy, they could have killed me.

Voice: Hello, I'm your guide.

Debbie: Our guide for what?

Voice: Well, haven't you ever wondered where those three turkey pot pies you eat each night go, huh, Debbie?

Debbie: Come on, hold your voice down! I don't want the whole world to know.

Stacy: (giggling) Three pot pies!

Debbie: Well, *you* must be here for a reason too!

Voice: Yeah, Stacy, what about those seven Sara Lee cheesecakes you crave every day of the week?

Stacy: So, I'm no angel. (Steps in something resembling dog solid waste) Oh, sick! What is that?

Voice: FFC

Debbie: WHAT!?

Voice: Fat, Flab and Cellulite.

(They both start sinking in FFC) Ahhh!

Both: (Wake in a cold sweat) I promise never to eat more than 100 calories a meal again!

(Untitled)

It was a cold and rainy day. Maybe this was the reason that I wanted to stay home. I knew that school would be the usual drag. So I said I was sick. With that simple remark, my mother believed me. (I am a very trustworthy kind of a boy.)

I went back to bed while my mom got ready for school. (She's a teacher.) Before she left my mom called our neighbors, just in case I had to call anyone while she was gone. They told her that someone had attempted to rob their house while they were away the weekend before. When my mom heard this, she decided it would be better for my general welfare, if she stayed home. I only wish she had told me!

I woke up at 11:30. I listened to the radio for awhile and then headed for the bathroom. On my way back from the bathroom I swore that I heard the clanging of dishes. I didn't think much of it until I heard a cabinet door opening and closing. I WAS SCARED!

I flipped up my mattress in case the robber had a gun. Then I dialed 0. "Hello, get me the police."

"Hold on, dear, I'll patch you through," said the operator.

"Please hurry," I replied.

"Hello, State Police Headquarters," a high scratchy voice said. "Can you hold please?"

"NO!" I whispered, so the robber couldn't hear me. Elevator music kicked in. I couldn't believe it. Someone was robbing my house and the #$%^& police had put me on hold. I was real glad my parents had paid their taxes.

"Hello," a deep voice said.

"Someone is robbing my house," I said, with tears in my eyes.

"Son, where do you live?" the voice asked. I gave the man directions.

"Son, well have a squad car over there in a few minutes. 'Till then, just keep talking to me."

Seconds, minutes, they all seemed like hours passing by. What the policeman said next really scared me. "Son, we can't seem to find your house." But that didn't matter.

"Sir, sir, he's coming up the stairs," I mumbled.

"Who is son?" the officer asked.

"The r-r-robber," I managed to squeeze out. Click! the latch of my door lifted. My very short life passed before my eyes. I saw the first fish I caught. The time I got the toilet seat caught on my head. The door OPENED and in walked my MOM.

Number of books read: 14
Best-liked books:

> Anonymous: *Go Ask Alice*
>
> Torey Hayden: *One Child*
>
> Sal Lopes: *The Wall: Images and Offerings from the Vietnam Veterans Memorial*
>
> Douglas Adams: *The Hitchhiker's Guide to the Galaxy*
>
> Mildred Taylor: *Roll of Thunder, Hear My Cry*

Reader Response:

Go Ask Alice: At first this diary was no big deal just like I think of any girl's diary. On the last few pages I read she had LSD slipped in her drink and liked it . . . It is so bad what drugs do to people. I'm surprised that always being high she can write in a diary. If she wants to quit so bad like she says she does why doesn't she just go and see someone. You would think her parents would be able to tell she was on drugs. It is so sad that people feel they have to do that to be cool. I have been worried since 5th grade about high school and the pressures. If the first party I go to has drinking I don't think I'll go to any more. If not drinking and doing drugs means I can't have a social life then I guess I won't have a social life.

Self-Evaluation

One thing I have learned is that it is better to just write down what you think . . . to write about things I like, not things other people like . . . and to put feelings in the writing. I try to do that in my own writing. I don't know if it is in my writing—it takes other people to tell me about it to see it.

In reading I know not to read a book if I don't like it. I can always stop a book in the middle. I think I've read more books for this reason.

Tricia

Writing is Tricia's passion. She is the student the elementary teachers were always talking about. "Wait'll you get Tricia." She is willing to take on difficult topics and is honest in her feelings toward them. She is an independent learner who knows how to fit the task to meet her needs.

Tricia's head and heart often work faster than her pen. She is able to laugh at her own spelling, fully aware that what she has to say is so important that she'll worry about spelling in its rightful place—last.

"Summer Night" is a piece she began in 6th grade with her teacher Dolly Bechtell. In 8th grade she knows she can make it even better. The most important thing is, she still *wants* to work on it.

Tricia's parents are divorced. She lives with her mom and seldom sees her dad. She must pay for most of her clothes and books through babysitting money. She takes life seriously.

Michael Ginsberg, a professor at the University of Kentucky in Louisville, and one of the readers of my initial manuscript, noted: "Imagine what Tricia must be carrying around in her head and her heart. I'd want to know how you work with someone who has so much to say."

I just let her say it!

Most Effective Writing

Picture perfect

Characters: Harry L. Tuckerman—39, proud, tall, stubborn
Lauren B. Tuckerman—37, stubborn, fed up, sad

Place: Dining Room

Time: 7 o'clock in the evening, late autumn

Setting: There's a table in front of a bow window. At both ends of the table there is a chair. Over the window hangs a painting of the scene (what's on the stage) only with a husband and wife at the table drinking coffee and talking. The wife is talking while the husband attentively listens. It looks like a Norman Rockwell painting. The room is decorated in pale orange and yellow; it's dimly lit. A clock with a pendulum hangs to the right of the window. It smells like fresh coffee and pie. The table is set with a tablecloth and dishes. There is a spotlight on the clock.

Action: The clock strikes 7. (spotlight shuts off) Suddenly everything gets bright as if a light switch was turned on. Lauren walks on. She walks as if angry, upset. She carries a tray. On the tray are coffee and two slices of pie. She places the tray on table. Lauren seems to be muttering to herself. She looks up to the painting and shakes her head, then looks away. She stands on one of the chairs and determinedly takes down the painting. When standing back on the floor she puts painting under her arm and pulls the left chair out from the table. She puts the painting on the chair and removes a can of spray paint from the pocket of her flowered apron.

In a flash she shakes the can, removes cap and spray paints black over the picture of the husband. Then in a matter of fact way, hangs it back up, replaces the chair and pours coffee.

Her husband, Harry, walks in. He goes directly to the chair his wife just used, and sits down to drink his coffee.

Lauren: Aren't you going to ask why I've painted you out of the picture? (sits in other chair)

Harry: I wasn't . . . aware that I was. (doesn't look up from his pie and coffee)

Lauren: What do you think of my haircut?

Harry: (glances up) Beautiful, dear. (eats some pie)

Lauren: (stabs pie with fork) I didn't get one. What's wrong with you? You didn't even ask why you're painted out of the picture. (sounds hysterical)

Harry: I think I already told you I wasn't aware that I was.

Lauren: That's just IT. You're never "aware." (gulps coffee) We got that painted 16 years ago. Do you remember?

Harry: Hmmm.

Lauren: (stirring coffee) The day after you asked me to marry you, you said, "Lauren, we need a way to always remember how happy we are now." We sat for the painting right here. Remember the painter?

Harry: Sure. (finishes off pie and opens newspaper)

Lauren: We used to have the greatest conversations. In high school you were captain of the debating team. I think that's what made me marry you instead of Billy Linky. (Looks over at Harry, who's reading. Sighs. Looks at painting, says to herself . . .) The painting's more realistic this way.

Harry: Coffee. (Lauren pours him a cup. Silence.)

Lauren: (yelling—sounds desperate) What happened?

Harry: (looks up confused) At work?

Lauren: No, DAMN IT! Haven't you listened to one word I've said?

Harry: (calm) Can't say that I have. (looks up at the painting) Why did you paint me black?

Lauren: My dear, dear husband (sarcastically) . . . Do your ears work? DO THEY? Do they work? (Takes empty tray, holds it high in the air, then drops it. It makes a loud crashing sound.) Could you hear that? Maybe I should make an appointment to have your ears checked tomorrow.

Harry: What in bloody hell has gotten into you, woman?

Lauren: Eureka! (jumps up) Finally I've found it, after years of searching . . . emotion. Harry Tuckerman isn't made of ice. What a surprise!

Harry: Calm down, Lauren. Sit down and take a couple of deep breaths.

Lauren: Why? It's not as if we have children to hear us yelling!

Harry: Oh, so that's what this is all about. It's not like we haven't discussed this before. So just sit down and relax.

Lauren: What!? And lose your attention? I have it for the first time in months. I'm not about to "calm down." Not until you hear what I have to say.

Harry: (sternly, grabs for her arm) Woman.

Lauren: (reaches for her arm again) Keep your hands off me. I mean it. If you come near me I'll throw this knife at you. (Picks up the pie server. Harry sits down. Lauren sits down.) Harry, I've come to the end of my rope. This is the straw that breaks the camel's back. However you want to put it, it's over.

Harry: But—

Lauren: I've thought about it for a long time.

Harry: Wha . . . (looks confused) Lauren . . . Lauren, I . . . (tears are in Lauren's eyes. Harry looks at painting, then at Lauren.) I'm sorry.

Lauren: You probably don't even know what you're sorry about. (wipes tears from her face) Anyway, you're too late. (Gets up and walks off stage. Lights go out. Stage is black. Harry leaves stage. Spotlight on painting comes on. There's a diagonal black spray-painted line between Lauren and Harry. Both Harry and Lauren are painted black.)

Are we creating our own destruction?

The sky's so blue, the snow white, the trees bare. The world appears clean, clear, and simple. With time everything changes. It always does.

The sky fills with black storm clouds. The snow changes to an industrial gray slush filled with road salt that eats away at our cars and poisons our drinking water.

The trees are silhouettes now; they have a sensible and logical look to them. The placement of each branch makes sense. But with spring, buds form and grow into leaves flopping topsy-turvy on the breeze, changing the sensible to an illogical crazy quilt in motion. After many springs have come and gone the trees will die, some with limbs outstretched reaching for the sky, others in a buzz of flying chips. Men named Skip, dressed in red-checkered flannel shirts, dusty black workboots, and fluorescent orange hard hats operate the means by which the trees fall.

Take a breath of fresh air. How fresh is it really? How much of Antarctica melted away in the time it took you to take that breath? In some far distant land how much of a rain forest is stolen away? With each felled tree and bush the amount of air left to breathe lessens.

Lush forests, thick with the scent of orchids and the amaryllis, the air filled with the sound of animals cawing and screeching in the trees, once flourished on the earth. Now, in their place, are flat, arid, lifeless brown deserts. The only sounds are those of the wind as it lifts layers of sand to scar the landscape.

The Bible tells us that God loves his creatures too much and too dearly to let us kill ourselves off. It's a nice thought, which I wish I could believe without any doubt. But somehow the melancholy, pessimistic voices of the scientists as they preach of skin cancer, radiation, rising temperatures, oil spills, deformed babies and heart attacks pounds on incessantly like an undercurrent through my mind, impossible to escape. Why do we listen to them? Maybe because their voices are more like our own whining tones than the voice of hope.

Out west and all over the globe animals, whose veins pump blood filled with the instinct to survive, are shot down with a small metal bullet by men who don't deserve the title of "man." These hunters' veins pump blood that has a need for money and power but lacks the ability to feel compassion.

Even the food we eat is poisoned by our actions. Magazines and cheery-faced news-anchors repeatedly warn us to watch what we eat. Grapes from Chile are tainted with cyanide. An apple a day no longer keeps the doctor away!

The chickens we eat are raised in boxes. Calves are placed in tiny stalls with no grass or sun to play in, then pumped full of drugs. We eat it, calling it veal, somehow removing the steaming meat before us from the image of it once having slept in its own diarrhea and excrement.

Everywhere you look the gold in life is tarnishing, its sparkle swallowed by shadows.

In the distance animals die, forests are lost, human rights disappear, but none of that ever touches my life. Now that will change. This year the grass will brown, the water and electricity will dwindle away. I'll feel the decline. The carefree spring days that are soon to come will feel an eternity away.

Even those carefree spring days are soiled. The birds sing in the warmth of the sun, their song awakens the flowers below the snow. As it melts away a garden of Doritos bags and Budweiser cans appears before the daffodils and crocuses show signs of life. That colorful metal and plastic garden remains long after the petals have fallen and the stalks have become brown and weathered.

So what do we do as we march on, creating our own destruction? We pen off letters to Senators in Congress. A good idea, one that our country was founded on. But even if enough letters were written that they reached from Madbury, New Hampshire to Washington, DC there's nothing that can stop what we've started unless each of us the world over is willing to change the way we live.

As long as we have hope and are willing to try, we won't be lost. For with both, we may have days where the sky's so blue, the snow so white, and the world is clean, clear, and simple.

Summer night

The breeze was slowly sweeping away the day's humidity. The leaves of the birches rustled in its wake. I fumbled for the latch of the screen door. Half running, half stumbling, I arrived at my hammock. I curled up in it, trying to escape my thoughts. They hurt too much.

The pine trees over our house created a lacework through which the half moon shone. It shimmered on the leaves of the birches, covering me with a milky canopy. Why? I wanted to scream.

Through the open windows of our mobile home, the sounds of the baseball game on TV drifted out into the semi-silence of the night. Looking through the window in our door I saw my mother's garnet ring shining in the light as her pen glided across the page of a letter she was writing. I brushed a tear from my cheek. I hated her. She'd done it. I burned with anger, but my entire body felt weak I was so hurt. If she hadn't dated Ed, maybe Jeff wouldn't have remarried his wife. Maybe my mom would have married him and things would have happened like they did in my daydreams. How could she throw away the only person I wanted her to marry since she and my dad divorced five years ago? I hated her so much I never wanted to speak to her again.

The wind picked up and I shivered, clutching my arms in an attempt to stay warm. I jumped out of the hammock. My bare feet hit the sticks and grass covering the ground. I walked carefully, wincing in pain each time I stepped on a rock as I made my way to the back of the house. I stole past the light from the living-room, past the canoe, and past the dark empty windows of my mom's room. Faintly, I could see the trail between the shed and the woods. I walked in partial darkness with only the filtered light from the moon and my memory to see by.

The ladder loomed ahead of me, leaning against our house. My mom would be mad if I climbed onto the roof at night, but I had an even stronger voice kicking me, telling me I had to go up. Looking both ways I took a step onto the ladder. It was old and gray from the weather. It swayed under my weight. It creaked, as if trying to decide if it should just give in and fall over. It held. I reached the top. Long ago, before I could remember, my dad had stepped on the top step and it had split in two. Precariously, I took the giant step from where the rungs of the ladder ended to the roof. I felt for the supports jutting up from the flat roof, making my way across to the front of the house.

The sky was crystal clear, as if swept by a broom. The stars reminded me of something I'd read in a book by L.M. Montgomery, that the stars were created by cutting holes in the cloth that forms the night sky. That's what the sky looked like right then.

The moonlight shown on the pond where the spring peepers and crickets serenaded the grasses of the field. The anger eased away slowly like a snail. The pain stayed. Instead of feeling like I was being stabbed over and over again, it was like the dull pounding of a drum resonating through my entire body.

"Garnet," I said softly. "He gave it to her. It should have left with him."

Reading

Number of books read: 33
Best-liked books:

Corrie ten Boom: *The Hiding Place*

Elie Wiesel: *Night*

Thyra Bjorn: *Papa's Daughter*

Byrd Baylor: *Moon Song, The Desert Is Theirs, I'm in Charge of Celebrations*

Scott O'Dell: *The Road to Damietta*

Antoine de Saint-Exupery: *The Little Prince*

Reader Response:

The Hiding Place: It's weird, human nature that is. Some people can kill and hurt people because of lies they've been fed. But even in the face of all this others think not of themselves but of persecuted people. People during the Holland ocupation would risk their lives to harbour Jews. Other people risked their lives for the truth and once having it could still find it in their hearts to forgive and even pity the murderers.

Sweet Whispers, Brother Rush: The most important thing I've learned from reading is that every kind of person has a place in this world.
... When we wrote our oppinions on the response groups I said that I thought we should swap papers, and read each others. You mentioned, "But does someone else read it the same way you do?" I think that if other people don't read it the same way it's not effective in its present form. There's no better time to find out then before you've made it final. You won't always be around to read your piece to everyone ...

Flowers in the Attic: All of V. C. Andrews use the same themes over and over. In fifth grade I did, but you don't expect a published writer to do the same things in every book (There's a happy family. Then the parents change and life becomes bad. There are some family secrets brought out. The heroine is the one most affected by it. The relationship between the children is always the same. At one point or another the heroine becomes involved with a rich relative—the heroine wants revenge and to show her evil parents she is worth something, she has some talent.) ... Before I started reading this book I knew there would be incest and cruelty but here I am on page 200 out of 411 pages and there has been cruelty but it's more expected than in the Heaven series. It's written better. It's like when people ask you if there can be swearing in a story and you say "if it's neccesary and appropriate." Just from the type of people the grandparents are you aren't shocked/hurt when they prove they're heartless . . . V. C. Andrews must have written these books a while ago because in all of them a girl gets her period and she's never even heard of it. Nowadays talk-shows, commercials, sitcoms all talk about it.

Peer response (Lizzie) to Tricia: Tricia, you give so much detail in your entries I feel like I am right there in the story, listening! Does Carrie marry Alex? I hope she does too. It is so awful that her grandmother tells her that she is evil or inferior. When I read that part in your entry I thought about apartheid and how the whites teach the blacks that they are inferior when there not. How does Cathy cope with the death of a husband and a baby? I would hate to be Cathy!

. . . Cathy has finaly found hapiness and revenge. But I pity Cathy, all the men she has ever loved have died. I'll draw you a diagram:

Cathy   loved—father (died in auto accident)

        loved—Chris (but it was incest so she couldn't go through with it)

        loved—Paul (was engaged but she called it off)

  married/loved—Julian (auto accident, then committed suicide)

    affair/loved—Bart (her mother's husband, died saving the evil grandmother from a fire)

    affair/loved—Paul (married him but he had a series of heart attacks—died in sleep)

lived with/loved—Chris (act as husband and wife—acts as father to her two children—still alive)

Tricia's response: *One Child* has realy proven to be a good book. I don't think I'll go back to V. C. Andrews books. There are so many that more can be learned from. They're fun to read but after the stories faded in your memories you have a feeling of having wasted your time.

My main goal as a writer is to make people understand how I feal and think.

. . . I honestly fealt I would throw up if I continued reading. I was at the part where Shiela's uncle Jerry rapes her (attempts to) and uses a knife since he can't . . . By the time they were at the hospital I put the book in my bag, closed my eyes, lay my head on the table and tried to get rid of the nausea I fealt . . . real stomach "I'm going to throw up" sickness. Then it spread to my head and it was the same feeling I fealt before I fainted after getting my ears pierced only this was worse.

Please notice how I spelled *felt*. I think this year is the first that I've actually tried to correct my spelling. Most years I've simply accepted that I can't spell. But by having words I use over and over again be listed each time I do it wrong I realised I should at least try. Congradulations! You are the first teacher who's ever gotten me to do that (try to correct my spelling).

Night: You succeded Mrs. Reif. Night was a Five easily. I started the book on Good Friday. I'd just been home from church for a cuple hours.

Every time that he lost faith in humanity it hurt me more than any other book I've ever read about the war camps or the war. I wanted to yell at him "Don't give up . . . don't stop believing, feeling, hoping and loving . . . just please, please, don't give up . . . But when I got to page 94 I cried. I'll copy the words down that inspired me to. "I woke up from my apathy just at the moment when two men came up to my father. I threw myself on top of his body. He was cold. I slapped him. I rubbed his hands crying:

> "Father! Father! Wake up. They're trying to throw you out of the carriage . . ."
> His body remained inert.
> The two gravediggers seized me by the collar.
> "Leave him. You can see perfectly well that he's dead."
> "No," I cried. "He isn't dead! Not yet!"
> I set to work to slap him as hard as I could. After a moment my father's eyelids moved slightly over his glazed eyes. He was breathing weakly.
> "You see," I cried.
> The two men moved away.

When I read that I felt so relieved that he hadn't died. Inside I knew Elie would have given up on life and died right away if the father had died. But even stranger than that feeling I suddenly saw myself in Elie's place trying to wake my father. I could see him. I felt the urgent need to save him. I started to cry. Not sob but tears were coming down my face. It's not been since when I read *Where the Red Fern Grows* or maybe one of the Anne of Green Gables books in 5th or 6th grade that I've cried over a book.

But another part . . . hit me even harder. "When I got down after roll call, I could see his lips trembling as he murmured something. Bending over him, I stayed gazing at him for over an hour, engraving into myself the picture of his blood-stained face, his shattered skull." I gasped and all that I could do was cry. I read that over and over. "Gazing at him for over an hour engraving into myself the picture of his blood-stained face, his shattered skull." I sobbed, my entire body shook. Tears streamed down my neck . . .

My response: Tricia, I cried over this book too—for Wiesel—and for the death of my mother. I've written to Wiesel, but never sent the letter to him. I wrote about my mom. I really miss her. I'm so angry at someone, or something, for letting her die so young—of cancer. She was a beautiful woman . . . I wonder if you'd want to write to Mr. Wiesel. You have some significant things to share with him. Writers like to hear how their words have affected a reader . . . I'm sorry *Night* made you cry, however, it's being touched like that which makes us all better human beings . . . You seemed pretty sad today. Is everything okay?

From Tricia: Don't worry about me. Tuesday was just a bad day. My locker wouldn't open. My mom had strep throat so my older brother had to drive me to school. But worst of all my science major was that day. I was up late the night before working on it. But the major went fine, my locker opens, and my mom's recovering . . . Crying because of a book doesn't bother me. I think of it as my body realising how good a book is . . .

One of the lines that stood out in Sharon Flitterman-King's article "The Role of the Response Journal in Active Reading" was "the response journal is a sourcebook, a repository for wanderings and wonderings, speculations, questions . . . " My own response to the article is the woman analyses her students much too closely. "Glen seems to be groping for something, for language that will capture and express the sensations that Blake's poem triggers in him." I read that and was turned off. I thought it was awsome that he used a word like "unrighteousness." Personaly I like how you use the logs much better than how this woman does. It seems more like the logs are all for her benefit. The things she's stated as pros for the students have just sounded like 'busy work' to me.

An American Childhood: I don't think I'll continue with this book. I just can't get into it. There are so many metephores that the actual occurences are hidden. You have to decifer it to figure out what's happening and how she feals.

Papa's Daughter: It's not difficult to read but it leaves you feeling good inside. Your eyes may have rivers coming out of them, but still your heart is dancing. Mrs. Rief you should really (did I spell it right?) read *Papa's Wife* and *Papa's Daughter*. If you'd like you could borrow our copy since we got ours after searching second hand bookstores . . . I've found another 5! She and Eric (her husband) are perfect for one another . . . Possibly the book struck me so much because today I looked thrugh someof the old picture albums . . . Mrs. Reif (I'm not sure if it's ie or ei—sorry) have you ever noticed that it's easier to get across sadness in a piece of writing than happiness? I'll make a goal of writing something that makes the reader (feel or feal) happy.

Response from Tricia's Mom: I'm very glad that you have enjoyed Papa's Wife and Papa's Daughter so much. I'll have to remember to search for more of the series. Don't worry about writing things that I "shouldn't" know, my mom still doesn't know all I did (at least I *hope* she doesn't). I'm sure your dad would prefer to live close to you and your brother; it's the need to work that has taken him away . . . I've always *felt* I should read Thoreau (the tyranny of the shoulds); I'm impressed that you are tackling him at your age. I'm proud of you. Keep up the good work. Mom.

Self-Evaluation

"Picture Perfect" is my best piece because the characters are the most realistic and it's the easiest to relate to. There's action in the piece, which I usualy leave out. You hopefully can imagine the set in your mind's eye. The language fits the characters and it's able to keep your attention.

I must believe in just write until inspiration hits because I have so many (other) bad pieces.

Having time to write in class helped the most. As soon as I get interested in a piece I realised I'm willing to write at home . . . Conferencing needs to be done with more than just one person. Different people see writing differently.

Also having so few deadlines is a real help. In 6th grade we had a lot of them so I handed in a lot of mediocre work. It's also great how much confidence Mrs. Rief has in us . . .

Learning to spell isn't the worst thing in the world. Spelling is actualy of *some* importance. To make revisions it's easiest to make them (or at least as many as possible) on the first draft. In other words, don't bother recopieing every time you make a change.

Reading helps me see things in a new light, go new places, escape from the usual humdrum life, and become worried about someone else's life. The hardest part is finding a book I like. I need to be forced each night, otherwise without meaning to I forget.

The books I read are usualy written at least 10 years ago. I don't care for the modern realistic fiction. They're all about bratty teens who want popularity or sex. I'd rather read cleaner books that make you think on how you could improve your character.

I get ideas on how to solve problems in my writing by what I read. If I've been reading something that makes me think my writing reflects that. If I'm in the mood to read one genre most likely I'll also be in the mood to write about it.

I [noticed] I read more often now, read harder books, and pay more attention to what I read. The most important things I can do as a writer are show people how I feel inside and say things I wouldn't dare say in person. I can make people see the places or scenes I write about. Hopefully I can entertain and make them think.

Appendix B

Best-liked Books

AUTHOR	TITLE	GENRE
Adams	The Hitchhiker's Guide to the Galaxy	scifi
Adams	Life, the Universe and Everything	scifi
Adams	The Restaurant at the End of the Universe	scifi
Adams	Dirk Gently's Holistic Detective Agency	scifi
Adams	So Long and Thanks for All the Fish	scifi
Angelou	I Know Why the Caged Bird Sings	autobio
Angelou	The Heart of a Woman	autobio
Angelou	Gather Together In My Name	autobio
Angelou	All God's Children Need Traveling Shoes	autobio
Anonymous	Go Ask Alice*	nonfict
Anthony	Incarnations of Immortality series*	fantasy
Anthony	Xanth series*	fantasy
Arnothy	I Am Fifteen and I Don't Want to Die	autobio
Asprin	Myth series	fantasy
Asprin/Abby	Thieves' World Series*	fantasy
Auel	Clan of the Cave Bear	fiction
Axline	Dibs in Search of Self	nonfict
Babbitt	Tuck Everlasting	fiction
Banks	The Indian in the Cupboard	fantasy
Banks	The Return of the Indian	fantasy
Bennett	The Pigeon	YA fict
Bennett	The Executioner	YA fict
Blume	Are You There God? It's Me. Margaret	YA fict
Blume	Tiger Eyes	YA fict
Blume	Just as Long as We're Together	YA fict
Blume	Then Again, Maybe I Won't	YA fict
Bombeck	I Want to Grow Hair	nonfict
Bridgers	All Together Now	YA fict
Bridgers	Home Before Dark	YA fict
Bridgers	Permanent Connections	YA fict
Brooks	The Moves Make the Man*	YA fict
Buchanan	A Shining Season	nonfict
Burnford	The Incredible Journey	juvfict
Callahan	Adrift	autobio
Cin-Forshay	Walk Through Cold Fire	YA fict
Clancy	The Hunt for Red October	fiction
Clark,MH	The Cradle Will Fall*	fiction
Clark,MH	Where Are the Children?*	fiction
Clark,MH	A Cry in the Night*	fiction
Collier	My Brother Sam is Dead	histfict
Conrad	Prairie Songs	YA fict
Cooper, S.	The Dark is Rising	fantasy
Cooper, S.	Silver on the Tree	fantasy
Cooper, S.	The Grey King	fantasy
Cooper, S.	Over Sea, Under Stone	fantasy
Cormier	Beyond the Chocolate War	YA fict
Cormier	The Chocolate War*	YA fict

Craven	I Heard the Owl Call My Name	fiction
Dahl	Boy: Tales of Childhood	autobio
Dahl	Matilda	juvfict
Duncan	The Third Eye*	YA fict
Duncan	Daughters of Eve*	YA fict
Duncan	Stranger With My Face*	YA fict
Duncan	Don't Look Behind You	YA fict
Duncan	I Know What You Did Last Summer	YA fic
Earthworks	50 Simple Things You Can Do to Save the Earth	inform
Fox	The Slave Dancer	histfict
Fox	The One-Eyed Cat	YA fict
Fox	A Place Apart	YA fict
Fox	The Moonlight Man	YA fict
Frank	The Diary of a Young Girl	nonfict
Fulghum	All I Really Need to Know I Learned in Kindergarten	nonfict
Gardiner	Stone Fox*	fiction
Goldman	The Princess Bride*	fantasy
Gordon	Waiting for the Rain	YA fict
Greene	Summer of My German Soldier*	histfic
Gunther	Death Be Not Proud	bio
Hayden	Murphy's Boy*	nonfict
Hayden	One Child*	nonfict
Hayden	Somebody Else's Kids*	nonfict
Hayden	Just Another Kid	nonfict
Hinton	Tex	YA fict
Hinton	The Outsiders*	YA fict
Hinton	That Was Then, This Is Now*	YA fict
Hinton	Rumblefish	YA fict
Hinton	Taming the Star Runner	YA fict
Holm	North to Freedom	advent
Holman	The Wild Children	histfict
Jones	The Acorn People	nonfict
Keyes	Flowers for Algernon	fiction
King	Firestarter*	fiction
King	The Eyes of the Dragon	fiction
King	The Stand*	fiction
King	The Bachman Books	fiction
King	It*	fiction
King	Cujo	fiction
King	Pet Sematary*	fiction
Kinsella, W.P.	Shoeless Joe	fiction
Kinsella, W.P.	The Iowa Baseball Confederacy	fiction
Kjelgaard	Irish Red et al	juvfict
Knowles	A Separate Peace	fiction
Koontz	Lightning	fiction
Koontz	Watchers	fiction
L'Engle	A Ring of Endless Light	fantasy
L'Engle	A Swiftly Tilting Planet	fantasy
L'Engle	A Wind at the Door	fantasy
L'Engle	A Wrinkle in Time	fantasy
Lee	To Kill a Mockingbird *	fiction
Lipsyte	The Contender	YA fict

AUTHOR **TITLE** **GENRE**

Best-liked Books

AUTHOR	TITLE	GENRE
Little	Little by Little	autobio
Little	Hey World, Here I Am!	poetry
Little	Mama's Going to Buy You a Mockingbird	YA fict
London	The Call of the Wild*	fiction
Lowry	Find a Stranger, Say Goodbye	YA fict
Lowry	Anastasia's Chosen Career	YA fict
Lowry	Anastasia Krupnik	YA fict
Lowry	A Summer to Die	YA fict
Lowry	Number the Stars	realfict
MacCracken	A Circle of Children*	nonfict
Marsden	So Much To Tell You	YA fict
Mazer	After the Rain; Heartbeat	YA fict
McCullough	The Thornbirds	fiction
McManus	They Shoot Canoes, Don't They?*	humor
McManus	Never Sniff a Gift Fish*	humor
Mowat	A Whale for the Killing	nonfict
Mowat	The Dog Who Wouldn't Be	nonfict
Mowat	Lost in the Barrens	nonfict
Mowat	Never Cry Wolf*	nonfict
Myers	The Outside Shot	YA fict
Myers	Fallen Angels*	realfict
O'Brien	Z for Zachariah	scifi
Paterson	The Great Gilly Hopkins	YA fict
Paterson	Bridge to Terabithia*	YA fict
Paterson	Jacob Have I Loved	histfic
Paulsen	Hatchet*	adven
Paulsen	Dogsong	YA fict
Paulsen	The Island	YA fict
Paulsen	Sentries	YA fict
Paulsen	The Crossing	YA fict
Pike	Fall Into Darkness	YA fict
Pike	Slumber Party	YA fict
Pike	Spellbound*	YA fict
Pinkwater	The Snarkout Boys and the Avocado of Death	humor
Pinkwater	The Snarkout Boys and the Baconburg Horror	humor
Rawls	Summer of the Monkeys	YA fict
Rawls	Where the Red Fern Grows*	YA fict
Saint-Exupery	The Little Prince*	fantasy
Salinger	The Catcher in the Rye	fiction
Smith	A Tree Grows in Brooklyn	fiction
Steinbeck	Of Mice and Men*	fiction
Strasser	The Wave*	YA fict
ten Boom	The Hiding Place*	nonfict
Tolkien	The Fellowship of the Ring	fantasy
Tolkien	The Silmarillion	fantasy
Tolkien	The Hobbit	fantasy
Tolkien	The Return of the King	fantasy
Tolkien	The Two Towers	fantasy
Voigt	Dicey's Song*	YA fict
Voigt	Homecoming	YA fict
Voigt	Izzy, Willy-Nilly*	YA fict
Voigt	The Runner*	YA fict

AUTHOR	TITLE	GENRE
Walker	The Color Purple	fiction
Weis/Hickman	The Dragonlance series*	fantasy
White	The Once and Future King	fantasy
Wiesel	Night*	autobio
Yolen	Sister Light, Sister Dark	fantasy
Yolen	White Jenna	fantasy

* the most popular books

(posted by the book shelves)

1. You may borrow no more than two books at a time.
2. Sign the date, your name, and section to the card in each borrowed book.
3. File the card alphabetically by author in your class section file box.
4. If you damage or lose the book, it will cost you the full price plus .50.
5. You may keep books for a reasonable length of time, usually two to three weeks.
6. When you return the book, pull the card from the file, cross out your name, put the card back in the book, and put the book on the shelf in the correct alphabetical order by author.
7. If the book was especially good, you may want to write a few sentences on the Recommended List posted on the wall by the book shelves. Sign your name, so other students will know who recommended the book.
8. Thank you.

Late- or Missing-Book Notice

_____ has checked the following book out of my classroom library and either lost or damaged the book.

Title _____

Author _____

You owe _____ for the replacement of this book. You will not be allowed to borrow any more books or participate in any upcoming field trips until the book is returned or the financial obligation is met.

Thank you. _____ for Linda Rief

(adapted from
Mary Ellen Giacobbe)

1. Fold the paper you choose for your book in half to form rectangles. Fold each page separately, putting them together, one page inside another. Up to six pages works well. Fold a piece of heavier construction paper into a rectangle that is about ¼″ larger all around than your pages. Put this evenly around your folded pages. Put this booklet aside until you get to step 5.

2. Cut two pieces of cardboard, of equal size. Make sure the size you choose is about ¼″ all around larger than the construction paper covering the pages of your book. (You can choose any size, although 6″ × 9″ covers standard paper folded in half.) Tape cardboard together leaving ⅛″ space in center.

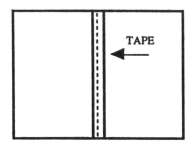

TAPE

3. You will need *one* piece of paper, wallpaper, cloth, etc. for the cover. The piece should be cut approximately 1″ larger all around than the size of your two pieces of cardboard taped together.

4. Open the cardboard, taped side down. Glue cardboard to cover paper (wallpaper, cloth, etc). Glue stick or rubber cement works best. Leave equal overhang on all sides. Fold the overhang and glue to cardboard, folding the corners first, then the sides. Put the cover on a hard surface. Top with several heavy books until glue dries.

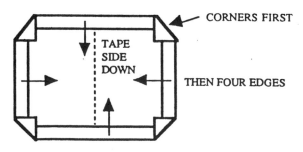

CORNERS FIRST

TAPE
SIDE
DOWN

THEN FOUR EDGES

5. Open your booklet (from step 1) flat. Mark the midpoint on the fold. Make two marks one inch from the ends of the inner paper. Make a mark halfway between each end mark and the midpoint. Poke holes through all thicknesses at marks.

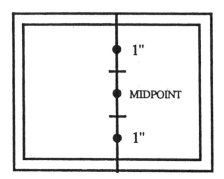

6. Thread a needle with dental floss or heavy-duty carpet/button thread, about 20″ long. Beginning at the back of the construction paper at the center hole, follow the pattern below, leaving 2″ to tie.

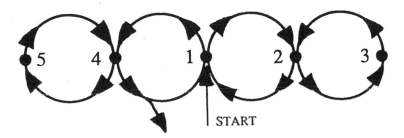

7. Open cover. Glue the construction paper to the covers. Put your newly glued book on a hard surface with three or four heavy books on top while it dries to keep edges from coming apart.

Japanese binding
or blanket stitch

Students using this binding must leave a ¹/₂″ margin down the left side of every page of their final draft. If they want to cover pages with clear contact paper, that must be done before they bind the books.

1. Pages of book can be any size, single or folded in half. If they are folded in half they are laid one on top of another, not inserted into each other.

2. Covers can be as simple as heavy oaktag: simply cut ¹/₄″ larger all around than the pages of book (and go on to step 6), or heavy duty with cloth or wallpaper covers. If you want a heavy-duty cover, cut two pieces of cardboard ¹/₄″ larger than your pages all around. Cut two pieces of wallpaper or cloth 1″ larger all around than each piece of cardboard. Measure and cut a ¹/₂″ strip from each piece of cardboard, on the edge to be bound.

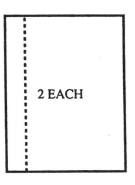

2 EACH

3. Glue one large piece of cardboard and the ¹/₂″ strip side by side with ¹/₁₆″ between them to one piece of cover material. If the wallpaper or cloth has a pattern to it, make sure the back cover is mounted right side up. Glue the second piece of cardboard and strip in the same manner to the second piece of cover material.

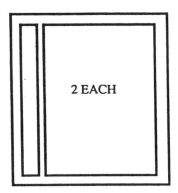

2 EACH

4. Fold the cloth over the cardboard, gluing the corners down first and then the four edges.

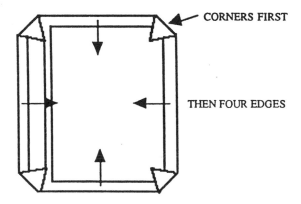

CORNERS FIRST

THEN FOUR EDGES

5. Cut two pieces of heavy construction paper or poster board, $1/8''$ all around *smaller* than the larger pieces of cardboard. Glue this poster board over the inside covers, making sure the $1/2''$ strip is not covered.

POSTER
BOARD
INSIDE
COVER
2 EACH

6. Using the top cover only, measure $1/2''$ in from the edge of the side to be bound. Draw a light pencil mark all along the edge. Place a ruler along that line and place a dot every half inch all along the line.

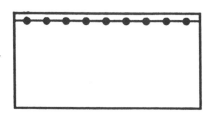

7. Check to make sure all pages of your book are in the proper order. Lay your covers on the pages, front and back. On the heavy duty cover, make sure the ½″ strip edge is along the edge to be bound—front and back. Making sure all edges are even, clamp your book tightly together with clamps provided.

8. Place book on hard surface on top of wooden boards provided. With thin nail, gently pound a hole through each pencil mark along the edge of your book. As soon as nail breaks through the surface of the back cover stop nailing, and remove the nail. The neater the nail holes, the neater the binding.

9. Measure a piece of thread—dental floss or carpet thread—about five times the length of your binding edge. Starting on the inside of the back cover, push your needle through to the front cover and around the edge and out the same hole. Sew to the next hole by going across the front. Loop all the way around that hole. Sew to the next hole across the back.

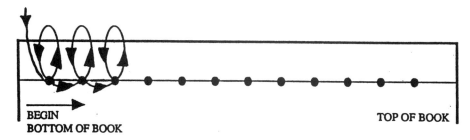

BEGIN
BOTTOM OF BOOK   **TOP OF BOOK**

10. Repeat until all holes have been sewn through. At the last hole the needle should be in a position to weave just back and forth through the empty spaces as shown for the entire length of the book. Stop at the inside back cover where you began and tie off the threads in a triple knot.

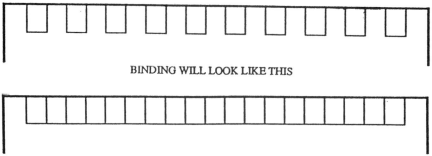

BINDING WILL LOOK LIKE THIS

WEAVE THREAD BACK AND FORTH THROUGH EMPTY SPACES TO LOOK LIKE THIS

(to be submitted with each manuscript)

This sheet, along with a set of instructions copied from the back of *Merlyn's Pen*, is available at the "Publication Center" in the classroom.

Author's full name: _____

Author's grade: _____

Author's home address: _____

_____

Author's phone: _____

Name of English teacher: _____ Linda Rief _____

Name of Principal: _____ Donald Wilson _____

School's full name: _____ Oyster River Middle School _____

School's full address: _____ Garrison Ave. _____

_____ Durham, NH 03824 _____

School's Phone: _____ (603)868-2820 _____

Appendix H

Letter to Parents

Dear Parents,

I want to welcome your son or daughter to my language arts classroom.

Nancie Atwell, a former eighth-grade teacher in Boothbay Harbor, Maine, says in her book *In the Middle:* "I learned that freedom to choose and time to read in school are not luxuries . . . They are the wellspring of student literacy and literary appreciation."

I would add that freedom to choose what is written and time to write are also essential ingredients in a literate environment. Like Nancie, I "know now that writers write reading, and that readers read writing."

Writing and reading *are* the basics. Language skills, such as usage, spelling, punctuation, vocabulary and sentence structure, will be taught through the individualized reading and writing of your son or daughter. I try to create an atmosphere in my classroom where natural and purposeful reading, writing, speaking, and listening take place—a place where your son or daughter becomes a more articulate user of language, both written and spoken. Writing is thinking. I want my students doing a lot of thinking as they learn to manipulate language.

Most of the writing will be done in class, where I can best help the students. At home I expect all students to read for a half hour each night and respond in their reader's-writer's log. For the first few weeks the students will have a lot of choice about what they write and what they read. I want to find out what they are able to do and what they need to know more about, so that I can help them grow as learners.

With this goal in mind, I have several requests. I believe it would benefit your son or daughter to have the following materials available to them. (I cannot require the students to purchase any of these materials. And because these are consumable goods, the school district cannot pay for them. I have researched these materials well, however, and believe they will be useful for years to come.)

Reader's-Writer's Log   price:
(a blank notebook for responding to reading and writing)

Punctuation Pocket or Topics Folder (for works in progress)

                                                                              price:

Merlyn's Pen (One of the finest national publications of student writing— four issues per year—MP has been written up in *Time*, the *Christian Science Monitor* and is a parents' choice publication. I have several students a year selected for publication in this magazine.)

                                                                              price:

Writing Magazine (filled with author interviews, genres of writing, and skills basic to the writing process: leads, voice, point of view, sentence structure, usage, etc.—nine issues per year)

price:

Writers Inc (a comprehensive handbook—concise, up-to-date, and accessible—filled with writing strategies, reading, thinking and study skills, the mechanics of written and spoken language, and an appendix with Hammond maps, charts, lists, and documents)

price:

Even though I have a core of literature anthologies and at least five hundred paperbacks available to your student in my room, I believe the above-described materials would greatly enhance each student's individual needs. If the costs are prohibitive, please let me know and I will make sure your son or daughter finds a way to receive the materials.

If you have any paperbacks you no longer need, I am always looking for contributions to our classroom library. A suggestion you might consider for adding to our classroom book collection and showing the students how much you value reading would be to donate your child's favorite book on his or her birthday—children's literature, adolescent fiction, classic, etc.

Thank you for your continued support. I'm looking forward to a very successful year for all of us.

_____ Student's Name

_____ Check enclosed (made out to ORMS) for the above described materials.

_____ I am available to volunteer. (I frequently need parents to drive on field trips, and/or to help me cover and catalog paperback books.) Name and phone number:

_____

Is there anything I should know about your son or daughter to help make this a most successful year? Any comments or suggestions for the language arts program?

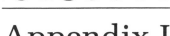

Appendix I   **Reading-Writing Survey**

(adapted from
Nancie Atwell's book
In the Middle)

Name: Section:

Date:

Please answer the following questions as thoroughly as possible. Refer to specific pieces of writing you have done or specific books you have read as you answer each question.

What does one have to do in order to be a good writer?

What is the easiest part of writing for you? What do you do well?

What is the hardest part of writing for you? What do you need to work on?

How do you come up with ideas for writing?

What are the qualities of good writing?

What is the best piece of writing you've ever done? What makes it so good?

What helps you the most to make your writing better?

What kind of response helps you the most as a writer? Who gives you that response?

Why is it important to be able to write well?

What do you like about writing?

What happens to your finished pieces of writing?

How did you learn to write?

What kind of writing do you do *just for you*?

What does one have to do in order to be a good reader?

What makes reading easy for you?

What is the hardest part of reading for you?

How do you go about choosing books to read?

What are the qualities you look for in a good book?

What's the best book you've ever read? What made it so good?

How did you learn to read?

What kind of reading do you do *just for you?*

What do you think the connections are between reading and writing? How does the ability to do one, help you do the other?

Name
Address
Date

Mrs. Rief
English Teacher
Oyster River Middle School
Garrison Ave.
Durham, NH 03824

Dear Mrs. Rief,

I would like to attend one of the teacher workshops with you this year and teach other people about what I know.

I have lived in Canada for most of my life and so I have more experience in French than in English. When I moved here, I was able to compare the Canadian ways of teaching and the American ways.

I am a pretty good student in all subjects and a pretty good learner. I think that if one understands what they are studying, he/she will remember it better. The way I learn best is to read over the material first and then ask myself questions about it to understand it. Of course, it is different with each student.

Reading makes writing easier and vice versa. If one reads a lot, he/she will become better spellers and if one writes often they will have better reading skills. But one should practice both and practice them everyday.

I think that teachers should know a lot of different ways of teaching because each student has his/her own way of learning.

I hope you will read the following résumé and consider my application.

Sincerely,

(Signature)

Marie (plus last name)

Name
Address
Phone Number

Skills: I like realistic writing and personal narratives. I am a pretty good drawer and I have had a lot of experience talking in front of other people. I read and write every day.

Educational Background:

1986–present: 8th grade, ORMS, Durham, NH
1985–1986:   7th grade, ORMS, Durham, NH
1984–1985:   6th grade, Guillaume Mathiew, Charlesbourg, Québec
                 provence, Canada
1980–1984:   1st, 2nd, 3rd, 4th, 5th grade, Felix-Antoine Savard,
                 Chicoutimi, Canada

Related Experiences:
7th grade: I wrote three articles in the school newspaper *The Middle of the River.*
            I wrote a play called "Ann" for a social studies medieval festival with a partner and acted a major part in it. I also made the scenery for other plays in the festival.
4th grade: I wrote a realistic-fiction story which was put into booklet form (French).
2nd grade: I wrote a small science-fiction piece which was put into book form as one of the best in second grade (French).
Kindergarten: I played a small role in a play (French).

Work Experiences: I started to babysit two years ago and I now babysit for two families. I receive an allowance each week for the housework I do. My brother and I are fully responsible for our dog, Kelly.

Biographical Information:
Summer of 1985: Moved to Lee, New Hampshire. I learned most of my English during seventh-grade year (85–86).
Summer of 1984: Moved to Charlesbourg, a French town near Québec City, Canada
1972–1984:   Lived in Chicoutimi, a French city in Québec provence, Canada

Hobbies: I have been playing the piano for eight years now and I also enjoy downhill skiing which I have been doing for three years.

Appendix K

Author/Book Share

Name _____ Sect._____

AUTHOR: _____ GENRE _____

DATE OF PRESENTATION: _____

Excerpts/Pieces to be read (include title of book and page numbers, and WHY you chose each excerpt—what do you want to show about this author's writing?)

MONDAY: _____

_____

_____

_____

TUESDAY: _____

_____

_____

_____

WEDNESDAY: _____

_____

_____

_____

THURSDAY: _____

_____

_____

_____

(If possible, clip copies of excerpts/poems/pieces to be read to the back of this sheet when you hand it in.)

FRIDAY: What are the most significant things you found out about your author/genre that you plan to share? How did you go about gathering that information? List all the references you used, including the titles of all the books you read by your author. Clip drafts of any letter you wrote to your author and/or replies you received to the back of this.

SELF-EVALUATION:

What did you have to do to gather information for this presentation? What kind of *process grade* would you give yourself for effort and "good faith participation" in gathering information and for originality of presentation (range of information, variety of formats, appealing delivery)?

PROCESS GRADE _____ (student) _____ (teacher)

To what extent do you believe you became an expert on your author and taught others? What kind of *content grade* would you give yourself for range and depth of information?

CONTENT GRADE _____ (student) _____ (teacher)

STUDENT COMMENTS:

TEACHER COMMENTS:

(I often put a cartoon that has to do with reading or writing at the top of this sheet.)

Reader's-writer's log

Books: You choose your own books. You are to have a book in your possession at all times. If you don't like the book, abandon it and choose another.

Reading: I believe we learn to read by reading. Therefore, I expect you to read for a minimum of a half hour each night, five nights per week. We will also be reading in class at least one period per week plus the schoolwide Sustained Silent Reading period.

Log entries: I expect a minimum of five entries per week. All entries should include the following information:

> Date
> Title of Book
> Time read for
> From page _____ to page _____

Your written response in the log does not *have* to be in response to the book you are reading. What you write in this log should be what you want to preserve/remember as a reader and writer. Written entries are your thoughts, reactions, interpretations, questions to what you are reading, what you are writing, and what you are observing in the world around you. Your comments may also be in response to the author's process as a writer, and your process as a reader, writer, and learner.

If you are stuck, think about the following:

Quote or point out: Quote a part of the book, your own writing, or something you heard or read, that you think is an example of good writing. What did you like about the quote? What makes you feel this is good writing? Why do you want to save it?

Experiences or memories: How does this book make you think or feel? Does the book remind you of anything? What comes to mind? What kinds of ideas does this book give you for writing?

Reactions: Do you love/hate/can't stop reading this book? What makes you feel that way? What reactions do you have to your own writing, the writing of your peers, the world around you?

Questions: What confuses you? What don't you understand? Why did the author do something a particular way? What would you have done if you were the writer? What questions do you have about your own writing? about observations of the world around you?

277

**Reader's-Writer's
Log Instructions**

Evaluation: How does this book compare to others you have read? What makes it an effective or ineffective piece of writing? How is your writing going?

Response to your log entries: During the year you will be analyzing your response and observations. You will be sharing your logs with me, peers of your choice, and sometimes parents. The response to you is meant to "affirm what you know, challenge your thinking, and extend your learning." (Atwell 1987)

Reading List

(I usually put a cartoon that has to do with reading here.)

Reading List

BOOK TITLE	# OF PAGES	AUTHOR	DATE BEGUN	DATE FINISHED	DEGREE OF DIFFICULTY EASY AVERAGE HARD	RATING 1 2 3 4 5 BEST ONE WORD DESCRIPTION

Conference Sheet

DATE	HOW CAN I HELP YOU? (WRITER)	LIKE/HEAR/STICKS WITH ME (LISTENER → . . .	QUESTIONS	ONE SUGGESTION	WHAT WILL YOU WORK ON NEXT?

- lead
  - is specific and grabs attention
  - gives a direction to the writing
- appeals to reader
  - reader can relate or identify
  - holds reader's attention throughout
- appeals to writer
  - writing is honest, real
  - writer knows what he or she is talking about
- clear focus
  - lots of detail about the central idea
  - "Write about a man, not man" (E. B. White).
- middle
  - logical and well organized
- ending
  - sense of closure, but want to read it again
- word choice
  - simple, clear, direct
  - creates picture in the reader's mind
- style
  - surprising viewpoint for common things
- purpose
  - makes you think or feel
- mechanics
  - legible writing with correct spelling, usage, punctuation, and paragraphing

PERSONAL SPELLING LIST

Responding to Writing   Appendix P

Use this sheet to respond to the enclosed eight pieces of writing. Do not discuss your responses with anyone until we are all done.

Point out what you like, what you hear, or what sticks with you:

A

B

C

D

E

F

G

H

What questions would you ask the writer?

A

B

C

D

E

F

G

H

What's one suggestion you would give the writer?

A

B

C

D

E

F

G

H

Appendix Q

Process Paper

(Case History of the Piece) and Grades

Name _____ Sect _____

Title: _____

Tell me everything you can about how this piece of writing came to be. What do you want me to know about the writing of this that I wouldn't know from just reading it? What are your reasons for giving it a particular grade? (You may continue on the back.)

student grade	teacher grade	Teacher's Comments
PROCESS		
CONTENT		
MECHANICS		

Dear Parents,

I am attempting to find a way to show you what your son or daughter is capable of doing as a reader and writer, beyond the usual cursory statement, checklist, and/or letter grade. I believe evaluation should nurture growth, help students become better users of language (as readers, writers, speakers, listeners, and thinkers), and foster more independence as learners.

Each trimester I will ask the students to bring their portfolios home. The portfolio is a place where the students store and evaluate their best work. It is a place to show who they are as readers and writers. In addition to the portfolio of finished writing, students maintain working folders of drafts in progress, and a reader's-writer's log, which they respond in nightly.

Please find enclosed in your son's or daughter's portfolio the following:

1–3 best pieces of writing, with all rough drafts
(a comparison of earlier drafts to finished pieces is one way of seeing growth in the writer)

self-evaluation for the trimester

reader's-writer's log with list of books read
(For students who strongly object to sharing the log with parents, I ask them to take a response they wrote to one of the best books read, and write an expanded discussion to include in the portfolio)

Mom/Dad,
This is what I've done well as a writer, reader, listener, and speaker (written by student):

Please note that this is what I'll be attempting to do better (written by student):

Please read through the portfolio's contents.
What do you notice that your son/daughter is able to do well?

What could you tell me about your son/daughter as a reader and writer at home that this portfolio doesn't tell me?

This is what I notice that your son/daughter does well:

Please sign this letter and return it with the portfolio.

_____   _____
Parent Signature   Date

Sincerely,
Linda Rief

Self-Evaluation/Year-End Appendix S

Name _____ Sect. _____ Date _____

Self-Evaluation: Year-End   Using your working folder and your portfolio, please answer the following questions as thoroughly and specifically as possible. Please write in ink on white-lined composition paper.

1. Number of pages of writing (rough draft and finished pieces):

2. Number of finished pieces of writing:

3. List all the kinds of writing you tried (essay, poetry, narrative, etc.) that worked:

4. List all the kinds of writing you tried that didn't work:

5. List everything you are able to do (have learned) with respect to

   • your writing process
   • the content of your writing
   • the mechanics of your writing

6. What are the three most important things you are able to do well as a writer?

7. What are three things you wish you could do better as a writer?

8. What's one kind of writing you never attempted that you'd like to try? Why didn't you try it?

9. Arrange all your finished pieces of writing in your portfolio from most effective to least effective. Make sure all the rough drafts that contributed to each piece are stapled to the back of the respective final draft. Talk about the three pieces you ranked *#1*, *#2*, and *#3* in the following way:

   Title?

   What made you rank this piece in this order? Be specific with at least three qualities.

   In what ways is this piece different from the others?

   As you wrote, what were some of the things you changed, or decisions you made, from one draft to the next?

   What made you make those changes?

   Looking at your first draft and final draft of this piece, is there anything you notice that shows how you changed as a writer? Or that you are able to do that you never knew you could do?

10. Looking at your top three pieces and the piece you ranked least effective, what do you notice? Be specific with at least three differences.

11. In the past year, how have you changed or grown as a writer? What have you discovered about yourself as a writer?

285

What have you discovered about your writing? (Cite examples from your writing to show what you mean.)

12. What has helped you the most with your writing?

13. What were your writing goals at the beginning of the year?
In what ways did those goals change as the year progressed?
How well did you accomplish your goals?
What are your continuing goals as a writer?

14. What do you do at home as a writer that I would never know about in school?

15. List pieces you've written in other classes that you would choose to put in your portfolio. How do these pieces show who you are as a writer in ways other than what we do in English?

Using your reader's-writer's log, please answer the following:

16. How many books did you start this year?

17. How many books did you finish?

18. Number of entries you have in your log?

19. Number of pages of writing in your log?

20. List all the different kinds (genres) of books you read (biography, science fiction, fantasy, poetry, realistic fiction, etc.)

21. List all the books you ranked as a 5 on your reading list, by author and title.
Next to each of those titles, talk about the most significant things you remember from each of those books.

22. List everything you are able to do (have learned) with respect to

   • your reading process
   • the content of books

23. What are three of the most important things you are able to do as a reader?

24. What are three things you wish you could do better as a reader?

25. What's one kind of book you've never attempted that you'd like to try? Why didn't you attempt it?

26. Look through the log entries you have on either *one of the best books* you read or on *one of your most significant observations as a writer or reader*. In what way has that book, or what you noticed, kept you thinking, or changed your thinking, about any issue important to you?

27. In the past year, how have you changed or grown as a reader?
What have you discovered about yourself as a reader?
What have you discovered about your reading?
Cite specific examples from your reading list and/or log entries.

28. What has helped you the most with your reading?

29. What were your reading goals at the beginning of the year? How well did you accomplish those goals? In what ways did those goals change throughout the year? What are your continuing goals as a reader?

30. What kinds of reading do you do in other classes or at home that is significant to you as a reader, but I would only know about if you told me?

31. In what ways have you noticed your reading affecting your writing? and/or your writing affecting your reading?

32. In what ways have you improved as a speaker this year? How did that happen? What would you like to be able to do better as a speaker?

33. In what ways have you improved as a listener this year? How did that happen? What would you like to be able to do better as a listener?

34. If you had to write your own year-end report, what grade would you give yourself in the following categories?

WRITING: _____

READING: _____

SPEAKING: _____

LISTENING: _____

ATTITUDE: _____

EFFORT: _____

35. What's your greatest strength as a learner?

36. What's your greatest weakness as a learner?

37. What could you do to overcome each weakness?

38. If you had to name the one thing that I did that helped you *the most* as a writer, reader, speaker, and/or listener in this classroom, what would that be?

39. What could I have done differently that might have helped you more? Be specific and honest. Give me reasons for the suggestions.

Thank you all for teaching me so much this year.

Index